LABORATORY EXPERIMENTS IN COLLEGE PHYSICS

SIXTH EDITION

CICERO H. BERNARD

Emeritus—Midwestern State University

CHIROLD D. EPP

Johnson Space Center—Houston
Formerly—Midwestern State University

JOHN WILEY & SONS
New York • Chichester • Brisbane • Toronto • Singapore

Copyright © 1964, 1972, 1980, and 1987 by John Wiley & Sons, Inc.
Copyright © 1949 and 1957 by Ginn and Company (Xerox Corporation).

All rights reserved. Published simultaneously in Canada.

Reproduction or translation of any part of
this work beyond that permitted by Sections
107 and 108 of the 1976 United States Copyright
Act without the permission of the copyright
owner is unlawful. Requests for permission
or further information should be addressed to
the Permissions Department, John Wiley & Sons.

ISBN 0-471-80578-5

Printed in the United States of America

10 9 8 7 6 5 4 3 2 1

PREFACE TO FIRST EDITION

This laboratory manual is designed for students who are taking their first course in physics, and the order of the subject matter of the experiments is approximately as found in most of the standard textbooks for first-year college physics. The experiments are independent of each other, however, and may be used in any order desired.

In practically all the experiments the apparatus is of simple design and of a type found in most physics laboratories. The author believes that the fundamental principles of physics can best be learned through the use of simple apparatus. If high precision is required in a first course in physics, much of the understanding of fundamental principles is sacrificed to acquiring skill in operating complex equipment. At the beginning of each experiment there are two lists of apparatus. One list includes the items the student will check out of the supply room; the other list includes the apparatus the laboratory instructor will place in the laboratory to be used as needed. These two lists include *every item* of equipment needed, thereby simplifying the procedure of determining all the necessary pieces of apparatus to operate the laboratory.

In addition to the list of apparatus, the instructions for each experiment include a statement of the purpose of the experiment, an introduction summarizing the physical principles involved, and directions for the experimental procedure. A description of the operation and use of the apparatus is included in some of the experiments where such an explanation seems necessary. It is assumed that in most cases the experimental work will accompany or follow the corresponding subject matter in the textbook, and consequently only a brief summary of the theory is given in the introduction to each experiment.

The questions that follow each experiment are designed to aid the student in making more careful observations and to train him or her to analyze these observations and interpret the results. Many of them are questions the student cannot answer unless he or she has been a careful observer. The author believes that the answers to these questions give a very clear indication of the student's grasp of the experiment, and are a very important part of the report handed in to the instructor.

Forms on which the data and results may be recorded are included with most of the experiments on perforated sheets so they may be detached and handed in as part of the report. They are omitted in two or three experiments because the nature of the data prevents their arrangement in tabulated form.

The experiments are designed for a three-hour laboratory period, if all the calculations are made and all the questions are answered during the laboratory period. If only a two-hour period is available and the instructor desires to have the computations completed within that time, the experiments are so arranged that certain parts of each can be omitted. The use of modern calculating devices will also shorten the time needed for completing the experiments.

C. H. BERNARD

PREFACE TO SIXTH EDITION

The general nature of the makeup of this laboratory manual and other useful information about its philosophical approach to experimental work are outlined in the preceding preface to the first edition. The general format of the sixth edition is similar to that of the fifth edition; however, the diversity of the experiments has been increased to include some new equipment including air tracks, air tables, lasers and oscilloscopes.

In the preparation of this edition only one experiment has been deleted in its entirety (cloud chamber). Four new experiments dealing with impulse and momentum, the oscilloscope, the laser, and polarized light have been added. Also, procedures have been added to some of the experiments to include the use of a wider variety of laboratory apparatus. This large number of changes has been accomplished without adding pages to the manual. In order to do this, a number of experiments have been combined and some parts of various experiments deleted.

The general level and approach of the manual remain unchanged. Graphical analysis remains a major emphasis of the manual. Alternate procedures and methods of data analysis have been maintained, allowing flexibility in the choice of experimental approach and equipment. The manual still provides sufficient experiments for a full two-semester program for each of the two levels of beginning physics taught by most institutions.

The authors believe that the questions at the end of each experiment can be effectively used to enhance the learning obtained from each experiment. These questions are designed to accomplish the following goals:

1. To stimulate careful observation of physical phenomena.
2. To develop an inquisitive attitude toward what is observed.
3. To promote precise measurement practice and improve technique.
4. To serve as a guide in analyzing experimental data.
5. To add fullness to the written report relative to the significance of the findings. If the student writes the answers in essay form, the answers supply a smooth analysis and summary of the experimental results. In order to promote good continuity and smoothness to reports, the instructor may suggest a particular sequence for the answers.

Respondents to a questionnaire furnished the suggestions for much of the revision in this edition. All ideas submitted were considered, but it was neither feasible nor practical to use every suggestion received. The indication of experiments desired and experiments presently being used served as a guide in determining which experiments and procedures to add and which to delete or combine with other experiments.

The response to the question about creating two different manuals for the two levels of physics was mixed, with strong feelings on both sides. This unclear choice of the current users of the manual and the extra work and expense involved made it an unwise approach at this time. Also, about fifty percent of the questionnaire respondents indicated an interest in a computer supplement to the manual with a widely varying choice of computers. Without stronger indications of marketing potential this was considered an unwise venture at this time.

Our gratitude goes to Bernard O. Beck and Company, Central Scientific Company, Daedalon Corporation, Metrologic Instruments Inc., Leeds and Northrup, The Nucleus Inc., PASCO Scientific Company, and Sargent–Welch Scientific Company for permission to use their photographs of apparatus. We also express our appreciation to John Wiley & Sons and their physics editor, Mr. Robert A. McConnin, for the excellent cooperation during the preparation and publication of this new edition.

We extend our thanks to all who submitted suggestions for this revision, with special recognition to Dr. John Hubsiz, College of the Mainland, who made many suggestions, proofread some new materials, provided a great deal of information, and permitted one author to use his laboratory facilities to test experimental procedures. We extend special thanks to Dr. Donald E. Simanek, Lock Haven University, for his many suggestions. Dr. Simanek made major contributions to the treatment of errors throughout the manual and to the Introduction to Experiment 46–Polarized Light. We are also indebted to Mr. Raymond Sims, Midwestern State University, for some technical information and Mrs. Glenda Epp for assisting in typing and assembling the manuscript.

CICERO H. BERNARD
Rush Springs, Oklahoma

CHIROLD D. EPP
Houston, Texas

CONTENTS

v

PART FIVE—LIGHT

PART SIX—NUCLEAR PHYSICS

APPENDIXES

INTRODUCTION

I.   General Laboratory Instructions

The student should read the entire Introduction very carefully since references to it will be made in many sections of this book.

LABORATORY OBJECTIVES.   The laboratory is a workshop for students, the place where they get firsthand knowledge of physical principles and experimental methods through the handling of apparatus designed to demonstrate the meaning and application of these principles. Some of the more specific objectives are (a) to acquire training in scientific methods of observation and recording of data; (b) to acquire techniques in the handling and adjustment of equipment; (c) to gain an understanding of the limitations and strengths of experimentation; (d) to obtain experience in the use of graphical representation; and (e) to collect data and to develop confidence in one's ability to compute reliable answers or to determine valid relationships. When one develops the skill of computing answers from experimental data which check with known values of the desired quantities, he or she acquires the confidence needed to perform an experiment and determine some quantity or relationship which was previously not known to anyone.

DEVELOPMENT OF CHARACTER AND SENSE OF RESPONSIBILITY.   Prospective employers and placement offices frequently send questionnaires to physics instructors requesting information concerning the character, attitudes, honesty, and dependability of students. The instructor makes an evaluation of these traits from observations of the student's performance in class and in the laboratory. The laboratory is a place for serious thought and investigation, and the following suggestions should help you to develop the above-mentioned traits.

a.   Be prompt in arriving at your work station and be well prepared concerning the principles of the experiment. If, for some good reason, you are late or absent, report the matter to the instructor.

b.   Work quietly and attempt to make the most careful observations possible by adjusting the equipment so that it will give its best possible performance.

c.   Be honest in making and recording observations. Record data as indicated by your equipment and not as you thought they were supposed to be, if they differ.

Copy no data, conclusions, or computations from any source. If your results seem to be outside the limits predicted by the experimental uncertainties, recheck your measurements and computations. If this does not give the answer, make the best possible explanation for the discrepancy.

d.   Have the entire procedure well in mind and perform the various steps in the order that will make the best use of your time. Cooperate with your partner so that each of you gets experience in manipulating the equipment. Then, each of you should compute your results independently to check on the accuracy of your work.

e.   Always remain at your assigned station and do not disturb other people in the class concerning any part of the experiment. Do not disturb other equipment that may be in the room but is not a part of your present experiment.

f.   Always abide by any precautions that your instructor may have given you regarding the proper handling of the equipment. Delicate equipment may be easily damaged.

PREPARATION FOR THE ACTUAL LABORATORY WORK.   The efficiency of performance in the laboratory depends largely on the preparation made before the experimental work begins. The entire experiment should be read before any measurements are made. It is also advisable for the student to review sections in the class textbook that deal with the principles under investigation.

Laboratory experiments are usually intended as a discovery time for the students so they should not be concerned if they do not know the expected results before beginning. At the start of the laboratory period the instructor will discuss any required special instructions needed for the apparatus being used, including precautions and perhaps some special techniques which should be used to get the best results. The instructor may also choose to discuss the underlying theory at the start of the laboratory period and/or discuss the results at the end of the laboratory period. The details of how to perform the experiment can be found in the "Procedure" section of each experiment.

CHECKING OUT APPARATUS.   A list of apparatus is given with each experiment, and the items listed as special apparatus will usually be checked out of the storeroom by the student. Perhaps only one student will sign for the equipment issued, but all students working as a partnership will be held equally responsible for its care. Check each item of the equipment received, and make sure that you have all articles required and that all are in good condition. Also check apparatus already on the table and compare it with the items listed under general apparatus. Report any irregularities to the instructor or assistant at once.*

MATERIALS WHICH THE STUDENT WILL SUPPLY.

Equipment which is not considered as general laboratory apparatus will be needed at various times. These items consist of graph paper, straight edge, protractor, hand calculator, and watch with sweep second hand. You should always have your textbook available for reference.

PERFORMANCE OF THE EXPERIMENT.

Before beginning the experimental work always read the entire procedure to get a general idea of what is to be done. You should always arrange and adjust the apparatus to give the best performance possible and then make and record readings as precisely as the apparatus will permit. Always estimate one significant figure beyond the smallest graduation on the instrument being read.

Data *should never* be recorded on *scrap paper* and then transferred to your record form. If, after you have recorded a reading, you decide that it is in error and should be discarded, mark through it and record the corrected reading below it. Always record the proper unit beside the number or at the heading of a column when a whole column of readings use the same unit.

Do not hesitate to discuss any details of the experiment with the laboratory instructor during the laboratory period. You may want to question certain procedures or suggest improvements in the method. A good question may be more important than a good answer.

REPORT OF EXPERIMENTAL WORK.

The form of the report required will be designated by the instructor in the course. In any case, the original data should be presented in neat form, such as that suggested at the end of each experiment in this manual. The data should be followed by sample calculations showing the method of obtaining the results. If the experiment requires several computations of the same type, only one of each type need be shown in the report.

Each experiment has a stated ''purpose.'' Use this as a guide to your investigations. Your report should include a separate section in which you clearly and concisely state your results and conclusions. This section should respond to the question: ''To what extent was the experiment's purpose accomplished?'' This discussion should stick to the facts, and all conclusions should be supported by reference to your data and observations. Avoid idle, unsupported speculation.

Many of the questions at the end of each experiment are intended to stimulate thought and to guide the student in drawing conclusions concerning the results. These questions are to be answered in discussion style, and the answers so worded that the reader can ascertain the question from the answer. The sheet containing the questions may be removed from the manual if the instructor prefers that both questions and answers be a part of the report.

PROFICIENCY IN THE LABORATORY.

The factors that will be used to measure your proficiency in the laboratory fall into two general categories as defined below.

1. General laboratory performance including:
   a. Conduct in the laboratory.
   b. Care and technique in operating the equipment.
   c. Ability to grasp the fundamental principles demonstrated.
2. Presentation of experimental results including:
   a. Neatness and orderly arrangement of recorded data and computations.
   b. Interpretation of experimental data and conclusions drawn from it.
   c. Answers to the assigned questions.
   d. Answers to quiz questions if given.

*Near the close of the laboratory period, all items of equipment checked out should be neatly arranged in the checkout box and returned to the storeroom.

II. Errors and Significant Figures

NOTE TO INSTRUCTOR:

For students doing experimental work in physics for the first time, Sections II and III of the Introduction can provide a worthwhile introductory laboratory (or class) exercise. A discussion by the instructor, coupled with student participation, will result in more careful and meaningful observations and better reports on all experiments assigned in the course.

A. ERRORS IN MEASUREMENT

No measurement is absolutely accurate or exact. Human and instrumental limitations cause unavoidable deviations from the "true" values of the quantities we are measuring. The deviation of the value of a particular measurement from its "true" value is called the *error* in that measurement. While we cannot know the error in individual measurements, we can, by taking enough repeated measurements, obtain enough information to make good estimates of the average size of these errors in the data. We can also use this information to predict the expected size of errors in results.

A measurement or experimental result is of little value if nothing is known about the probable size of its error. We know nothing about the reliability of a result unless we know something about the probable errors which were used to obtain that result.

There are many types of errors which affect measured quantities, and there are several ways to classify them. The most general classification is *determinate* (or *systematic*) errors and *indeterminate* (also called *chance* or *random*) errors.

Indeterminate errors are present in all experimental measurements. The name "indeterminate" indicates that we have no way to determine the size or sign of the error in any individual measurement. Indeterminate errors show up as variability in the size of a particular measurement when that measurement is repeated many times (assuming all other conditions are held constant to the best of the experimenter's ability). These types of errors may have many causes, including operator errors or biases, fluctuating experimental conditions, varying environmental conditions, and inherent variability of measuring instruments.

The effect of indeterminate errors can be minimized by taking a number of repeated measurements and then taking their average. The average is generally "better" than any single measurement. This is because errors of positive and negative sign tend to compensate each other.

Determinate errors can be more serious because their effects cannot be reduced by averaging repeated measurements. This is because the determinate error has the same size and sign in repeated measurements, so there is no opportunity for positive and negative errors to compensate each other. Causes of determinate errors include defective or miscalibrated apparatus, a constant

bias in observation or procedure, or even blunders such as failure to include a correction term. Every effort should be made to minimize these errors, including careful calibration of the apparatus and use of the best possible measurement techniques.

One additional note. The term uncertainty is frequently used when talking about measurement errors. As used in this manual the terms uncertainty and error are synonymous.

The words precision and accuracy are frequently used when discussing measurement errors and it is important that the student understand what they mean. A measurement with relatively small indeterminate error is said to have high *precision*. A measurement which has small determinate error is said to have high *accuracy*. A measurement which has both high precision and high accuracy is sometimes called highly *reliable*. These words can be tricky. A precise measurement may be inaccurate if it has a determinate error. An accurate measurement may be imprecise if its random error is large.

It is most important for students to learn how to estimate the experimental errors and to see how these errors affect the reliability of the final result. Entire books have been written on this subject. The following discussion is designed to make the student aware of some common errors and some simple ways to quantify them.

ERRORS IN THE CALIBRATION OF THE INSTRUMENTS. These errors may result from an instrument being used under conditions different from those for which the calibration was made. If a measuring tape is calibrated to be used at 20°C, indicated measurements made at 30°C will not be the correct values. Some very delicate instruments must have the calibration checked at periodic intervals. Instruments may also be worn by use to such an extent that precise settings cannot be made. One must also choose an instrument which is calibrated to give the precision required in the measurement. For example, an ordinary meter stick would not be appropriate for measuring the diameter of a small wire, which may be no larger than the smallest division on the stick.

ERRORS INHERENT IN READING THE SCALE. A student's personal bias is often responsible

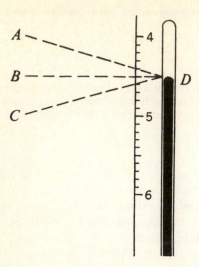

Figure A Errors in Scale Readings Resulting from Parallax.

for inaccurate results. When a series of trials are to be made for a certain measurement, students very often assume the first trial to be about correct and attempt to make all the others agree with it, thus giving more significance to the first reading than to any of the succeeding ones. Other personal errors are introduced because of insufficient care being used in adjusting instruments, inaccurate estimations of fractional divisions, and parallax.

The apparent distance between two objects will depend on the position of the eye. Two objects may appear to be in line when viewed with one eye but out of line when viewed with the other eye or when the head is moved to one side. This apparent change in position due to sidewise motion of the eye is called *parallax*.

If one is attempting to read the position of the mercury level in a tube near a scale (see Figure A) the line of sight must always be perpendicular to the scale. If one should sight along the line *AD*, one would read 4.4; if along line *CD*, one would read 4.6; the correct reading is 4.5, as read along the line *BD*.

The chance of error due to parallax between scale divisions and the object being measured may be reduced to a minimum by placing the measuring scale as near as possible to the object being measured. A meter stick should always be placed with markings against the object being measured to reduce such errors.

Other problems associated with reading instruments might come under the heading of *random fluctuations*. As one attempts to read a voltmeter connected across some circuit element in the ac power line, the needle may fluctuate back and forth while one attempts to get a reading. The same situation exists in attempting to read the scale on a count-rate meter connected to a Geiger tube. Methods of handling statistical fluctuations will be discussed in connection with the appropriate experiment in this book.

ERRORS INHERENT IN THE INSENSITIVITY OF THE INDICATOR TO CHANGES. In some experimental setups, one indicating portion of the equipment may not show sufficient response to changes in other indicating parts. When a certain amount of weight is added in one place, friction in the connecting links may prevent a scale indicator from showing the proper response. It may be that some instrument is slow in responding to a change in temperature, and readings must not be made too quickly. The usual laboratory thermometer, calibrated in 1° divisions, could not be expected to show sufficient response to a temperature change of 0.01°.

ERRORS RESULTING FROM FLUCTUATIONS IN THE ENVIRONMENT. If one is attempting to read an instrument out in the open where the adjustment is affected by gusts of wind, an accurate reading would be difficult to obtain. These types of errors, resulting from changes in the environmental conditions, can only be reduced through proper control of such conditions as temperature, humidity, noise background, vibration, stray electric fields, wind, and so forth. Sometimes these are beyond the control of the experimenter.

ABSOLUTE ERROR. When the size of the estimated error in a quantity is expressed in the same units as that quantity it is called an *absolute error*. The quantity and its absolute error are conventionally expressed in the standard form of this example:

$$\text{Radius} = 3.86 \pm 0.02 \text{ cm}$$

The value 0.02 after the $\pm$ sign in this example is the estimated absolute error in the value 3.86.

PERCENT ERROR. The error in a measurement can also be expressed as a percent. For the previous example, the percent error in the radius would be

$$\text{Percent error in radius} = \frac{0.02}{3.86} \times 100 = 0.52\%$$

In general,

$$\text{Percent error} = \frac{\text{absolute error in a measurement}}{\text{size of the measurement}} \times 100$$

DEVIATION (OR DIFFERENCE). There are cases when we want to compare the results of two measurements assumed equally reliable, that is, to find the absolute or percent difference between the two. For example, you might want to compare two independent determinations you made of a quantity, or to compare your experimental result with that obtained independently by someone else in the class. Suppose two measurements of a length give 4.0

and 4.2 cm, respectively, the exact value not being known. The percent difference is found by comparing the deviation (or difference) with the average of the two. Hence, we have

$$\text{Percent difference} = \frac{4.2-4.0}{4.1} \times 100 = 0.049 \times 100 = 4.9\%$$

Deviations can be expressed as absolute amounts or as percents.

EXPERIMENTAL DISCREPANCY. When an experimental result is compared with another which is assumed more reliable, it is customary to call the difference between the two the experimental discrepancy. These may be expressed as absolutes or as percents.

An obvious question is when should errors be expressed as absolute errors and when should they be expressed as percent errors? Common sense and good judgment must be used in choosing which form to use to represent an error. Consider a temperature measurement with a thermometer known to be reliable to $+0.5°C$. This causes a 0.5% error in the measurement of the boiling point of water (100° C) but a 10% error in the measurement of cold water at a temperature of 5°C. Clearly in this case the use of absolute errors is more meaningful.

There are situations in which errors expressed as percents are more meaningful. For example, suppose one measures the distance between two streets to be 390 m, while a professional surveyor's record shows the distance as 400 m. In another case, a person estimates the width of a table as 1.8 m when it should be 2.0 m. The absolute error in the first case is 10 m, and in the latter case, 0.2 m. The one who measured the table made an error of 0.1 m in each meter measured. On this basis they would have made an error of 40 m in the street measurement. The percent error in the first case is 2.5% whereas the percent error in the second case is 10%.

ESTIMATED ERRORS. The reliability of a given measurement is increased by obtaining the average of a number of *independent* readings. This average is likely to be a more reliable value for the measurement than one single reading. These fluctuations (or deviations) in the individual readings indicate that errors do exist in experimental measurements. The average deviation may be obtained by finding the absolute value of the difference between the mean and the individual values, and then averaging these deviations. If m is the mean value and d is the average of the deviations from the mean, then the measured value of the quantity Q should be recorded as

$$Q = m \pm d$$

If only one reading is made, one may estimate the errors by examining the scale being read and deciding by what fraction of scale division a reading could be in error. This may vary from 0.1 to 0.5 of the smallest scale division.

The percent error, in terms of the above symbols, is expressed by the relation

$$\text{Percent error} = (d/m) \times 100$$

It is important to understand what the average deviation really means. If a measurement is repeated 100 times, only 50 of them would lie within the average deviation from the mean. Thus, we expect only 50 percent of experimental measurements to be within one average deviation from the mean, or if a single measurement is made it has only a 50 percent probability of being within one average deviation from the mean. Suppose the question were asked: ''How much error would have to be assumed to insure that there is a 95 percent probability that a single measurement would lie within the error?'' An answer requires a lengthy discussion of experimental statistics which will not be done here. For this course students should constantly be aware of experimental errors and quantify them when possible, including finding the average deviations as required. Students should also be aware of the fact that they have not been given the full story on errors and they can expect further learning in this area in advanced laboratory work.

B. SIGNIFICANT FIGURES

The digits required to express a number to the same accuracy as the measurement it represents are known as *significant figures*. If the length of a cylinder is measured as 20.64 cm, this quantity is said to be measured to four significant figures. If written as 0.0002064 km, it still has only four significant figures. The zeros preceding the 2 are used only to indicate the position of the decimal point. The zero between the 2 and 6 is a significant figure, but the other zeros are not. If the above measurement is made with a meter stick, the last digit recorded is an estimated figure representing a fractional part of a millimeter division. *All recorded data should include the last estimated figure in the result, even though it may be zero.* If this measurement had appeared to be exactly 20 cm, it should have been recorded as 20.00 cm, since lengths can be estimated by means of this instrument to about 0.01 cm. When the measurement is written as 20 cm it indicates that the value is known to be somewhere between 19 and 21 cm, whereas the value is actually known to be between 19.99 and 20.01 cm.

Referring again to the 20.64 cm measurement, the possible error in this measurement is ± 0.01 cm and was recorded as being nearer to 20.64 than to 20.63 or 20.65. Hence, the uncertainty is less than one part in two thousand. It should be understood that the measurement error is dependent on the measuring instrument and the technique of the experimenter. For example, it is possible that the 20.64 cm measurement is known to lie between 20.635 and 20.645 cm so that the error is ± 0.005 cm. Thus, it is important to specify the measurement error.

C. COMPUTATIONS WITH EXPERIMENTAL DATA

It is usually a simple matter to determine the number of significant figures that should be recorded from a measurement, but difficulties can arise when these numbers are used in calculations. Calculations can produce a large number of figures which look significant but really are not. In this age of electronic calculators people have a tendency to enter all figures and go through with their computations, recording their final results with all the figures they see on their calculator display. Doing the calculation this way will not cause an error; however, recording all the figures in the final result gives a physically incorrect answer. *No mathematical computation can give a result whose accuracy is greater than that of the quantities used.*

As a general rule it is better to carry too many figures than not enough. As an example, suppose we desire to find the sum of the three numbers shown in the first column in the example

81.572	82
1024.7	1025
710	710
1816.272	1817

Since 710 is only known to the nearest units one might argue that all quantities could be written to the nearest units place before adding as shown in the second column at the right. This says that the answer is between 1816 and 1818, while in fact the answer is 1816, which is between 1815 and 1817. This round-off technique failed to take into account the fact that the sum of the first two numbers is known to be closer to 1106 than to 1107.

In addition or subtraction of experimental data it is a simple matter to determine what figures to carry and how many figures must be retained in the final result. The final result should contain the same accuracy as the least accurate quantity used in the calculation. Numbers used in the calculation need only retain, when possible, one more figure of accuracy than the least accurate number.

In multiplication and division of experimental data, things are not so obvious. These computations may involve unrelated quantities (length and time or mass and volume, etc.). It is still true that the accuracy of the final result cannot be more accurate than the least accurate measurement; however, it is not so clear what the accuracy of the final result means in this case, because the units may be different from those entering into the computation. To fully explain how to make correct computations with experimental data requires a discussion on propagation of errors, which will not be done here.

In beginning laboratory work the student usually is not asked to do detailed error analysis. In this case the student should always record the same decimal accuracy as the least accurate number entering into a calculation. In arriving at this result the student should always carry one extra decimal place for intermediate calculations.

III. Graphical Representation of Experimental Data

From an examination of the tabulated values of a number of measurements of related quantities, it is often difficult to grasp the relationship existing between the numbers. A method widely used to discover such relationships is the graphical method, which gives a pictorial view of the results and makes it possible to interpret the data by a quick glance.

INDEPENDENT AND DEPENDENT VARIABLES. In any experimental study of cause and effect the aim is to vary one condition at a time (the cause) and to observe the corresponding values of another quantity (the effect) which is suspected of being related to the first. This existing relationship is most easily interpreted from the graph if the first of these quantities, the independent variable, is plotted on the abscissa scale (X axis) and the dependent variable is plotted on the ordinate scale (Y axis). Very often the values to be plotted are all positive and only the first quadrant of a rectangular coordinate system will be needed. In such cases, the origin should be shifted to the lower left-hand corner of the sheet of graph paper. When possible, draw the axes inside the margins of the graph paper along the first or second large square. This provides more space in which to write the scale and also furnishes guidelines for lettering the names of the variables being plotted (Figure B). Graph paper with 20 squares to the inch is recommended for the curves to be plotted in this course.

CHOICE OF SCALE. Choose a size of graph that bears some relation to the accuracy with which the plotted data

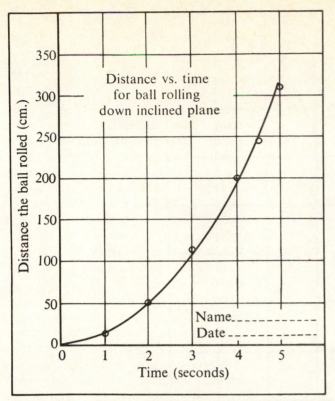

Figure B Sample Graph Showing the General Form that Should be Used in Curve Plotting.

graph. The same procedure should be used for the ordinate scale, but the divisions on the ordinate and abscissa scales need not be alike. In many cases it is not necessary that the intersection of the two axes represent the zero values of both variables. If the values to be plotted are exceptionally large or small, use some multiplying factor that permits using a maximum of two or three digits to indicate the value of the main division. A multiplying factor such as $\times\ 10^{-2}$ or $\times\ 10^{-6}$ placed at the right of the largest value on the scale may be used.

LABELING. After you have decided which variable is to be plotted on which axis, neatly letter the name of the quantity being plotted together with the proper unit (see Figure B). Abbreviate all units in standard form. Then write the numbers along main divisions on the graph paper, using an appropriate scale as explained in the preceding paragraph. The title should be neatly lettered on the body of the graph paper, but it is usually best to do this after the points have been plotted so the title will not interfere with the curve. Explanatory legends should also be shown.

PLOTTING AND DRAWING THE CURVE. Use a sharp pencil and make small dots to locate the points. *DO NOT* write the coordinates of the points on the graph paper. Your table of data shows these. Carefully encircle each point with a circle about 2 or 3 mm in diameter. In drawing the graph it is not always possible to make all of the points lie on a smooth curve. In such cases, a smooth curve should be drawn through the series of points to follow the general trend and thus represent an average. Suppose the plotted points show a straight-line trend such as shown in Figures C and D. To draw the straight line which best represents the relationship which produced the series of points, proceed as follows. First, cover the lower half of the points and draw a faint sharp cross in the centroid of the points in the top half of the series. Next cover the top half and mark a cross at the centroid of the lower group. Then draw a line straight through the two crosses and it will represent a true average.

are known. In general, the curve should fill most of the sheet if the data are known to three significant figures. If the data are known to only two significant figures, a large graph gives a false impression of the precision of the measurements.

Note the range of values of the independent variable (X quantity), and the number of spaces along the X axis. Choose a scale for the main divisions on the graph paper that are easily subdivided and such that the entire range of values may be included. Subdivisions such as 1, 2, 5, and 10 are best, but 4 is sometimes used; *never* use 3, 7, or 9, because these make it very difficult to read values from the

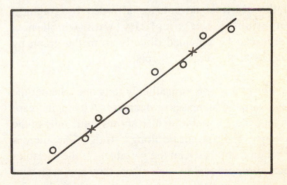

C

Figure C Incorrect Method of Fitting a Straight Line to a Series of Points.

D

Figure D Correct Method of Fitting a Straight Line to a Series of Points.

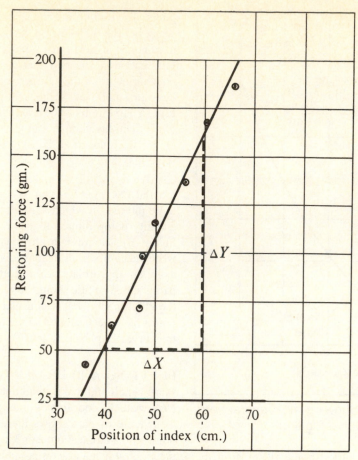

Figure E Analysis for Stiffness Constant of a Spring.

If the series of points appears to represent a function which is not a straight line, the points should lie on both sides of the curve along *all* parts of the curve.

When more than one curve is drawn on a graph, and it is desirable to distinguish the points associated with one curve from those associated with another, crosses (×), triangles (▽), squares (□), and circles (○) may be used.

ANALYSIS AND INTERPRETATION OF GRAPHS.
One of the principal advantages afforded by graphical representation is the simplicity with which new information can be obtained directly from the graph by observing its shape and intercepts.

The shape of a graph immediately tells one whether the dependent variable increases or decreases with an increase of the independent variable. It also shows something of the rate of change. If the points lie along a straight line, there is a linear relationship between the variables. If the variables are directly proportional to each other, they approach zero simultaneously, and the line passes through the origin. Curves which are straight lines and do not pass through the origin do not indicate direct proportion.

SLOPE.
In discussing the slope of a graph we must distinguish between physical slope and geometric slope. The geometric slope is usually the angular inclination of the line with respect to the X axis. In plotting physical data there may be an enormous difference in the size of the units on the two axes. The physical slope is found by dividing Δy by Δx (see Figure E), using for each the scales and units that have been chosen for those axes. The unit of the slope will be the ratio of the units on the respective axes. With the physical slope, one is not usually concerned at all with the angle the line makes with the X axis; the tangent of the angle has no meaning.

INTERCEPTS.
Significant information is often revealed by the intersections of the graph with the coordinate axes. This is true for other types of curves as well as for straight lines. The true interpretation of the intercept can be obtained only if the scales used begin at zero. In many cases there are no data available for drawing the curve to the axes. If the plotted points indicate the trend of the curve, one may be justified in *extrapolating* the curve to the intercept desired. *Extrapolation* is accomplished by extending the curve in the desired direction by a dotted line,

rather than a solid line, thus indicating that data are not available for this portion of the curve. Intercepts obtained by extrapolation may serve as clues to aid in the theoretical interpretation of the phenomena being observed.

LOGARITHMIC GRAPH PAPER.   In a few of the experiments in this manual, the variables to be plotted may be in the form of logarithms to the base 10. There are two methods for plotting such quantities: (1) *prepare a table of the logarithms of the variables and plot as log x and log y on regular linear cross section paper;* (2) *plot the variables themselves on regular logarithmic graph paper.* In the second method, if only one variable is logarithmic, then use semilogarithmic graph paper. If both variables are in logarithmic form, use log–log graph paper. The use of logarithmic graph paper is discussed in Experiment 3.

IV. Units

A great deal of time and frustration can be avoided if care is taken to always use correct units in laboratory calculations. These calculations will give meaningful results only when all physical constants and measured quantities are in a consistent set of units. If you are uncertain of the various units associated with the quantities in a given experiment, consult your textbook and find out what they are before beginning.

A problem that frequently arises in beginning laboratory investigations is the use of force units. It is often convenient to use gravitational force units in various experiments. This is usually done by suspending some *weights* which are marked in mass units such as grams or kilograms. For experiments in which acceleration is involved the forces should be recorded in dynes or newtons, which means that the mass values must be multiplied by the appropriate value of the acceleration of gravity, g, so that the units will be correct and the answers meaningful.

PART ONE
MECHANICS

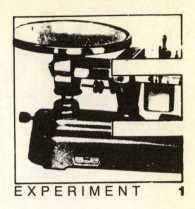

EXPERIMENT　　1

DETERMINATION OF LENGTH, MASS, AND DENSITY

NOTE TO INSTRUCTOR:

If it is desirable to speed up the pace of the early lab work, Experiments 1 and 2 may be combined into one by omitting Steps 1–4 in Experiment 1, and Step 3 plus the first graph in Step 9 of Experiment 2.

SPECIAL APPARATUS:*

Meter stick, vernier caliper, micrometer caliper, metal cylinder, copper wire (90–100 cm), set of masses, irregular solid, a 250-ml graduated cylinder, lightweight string, small sphere.

GENERAL APPARATUS:†

Laboratory balance.

THE PURPOSE OF THIS EXPERIMENT

is to get acquainted with laboratory measuring devices and their use in the determination of length, mass, and density.

INTRODUCTION

Before beginning this experiment the student should read Sections I and II of the Introduction dealing with laboratory procedures and the significance of errors and uncertainties in measurement.

VERNIER CALIPER.

When using a meter stick it is necessary to estimate the tenths of millimeters (the fractional parts of the smallest scale divisions). The vernier is a device which assists in the accurate reading of the fractional part of a scale division. The vernier caliper (Figure 1–1) consists of two scales: one is the fixed main scale of the instrument and the other,

called the vernier scale, is arranged to slide along the fixed scale.

In Figure 1–1, the principal divisions on the main scale represent centimeters, and are further divided into tenths of centimeters, or millimeters. The vernier, or movable scale, contains ten divisions, each of which is nine-tenths as long as the smallest main-scale division. Hence the ten divisions on the vernier scale have the same length as the nine divisions on the main scale (Figure 1–2). The scales in Figure 1–1 would appear as in Figure 1–2 if the jaws of the instrument were closed. It should be evident that the distance between the jaws of the instrument (Figure 1–1) is

---

*Items that will usually be checked out of the storeroom.
†Items of equipment that cannot be checked out of the storeroom, or items usually kept in the laboratory for general use.

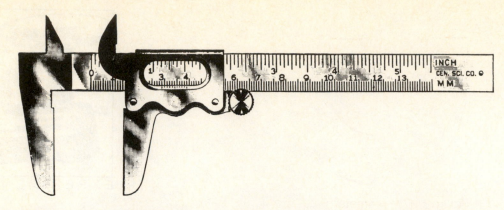

Figure 1–1 Vernier Caliper (Courtesy of Central Scientific Co.).

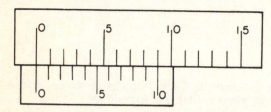

Figure 1–2 Vernier Scale Setting.

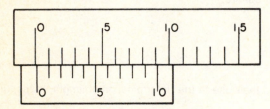

Figure 1–3 Vernier Scale Setting.

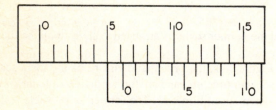

Figure 1–4 Vernier Scale Setting.

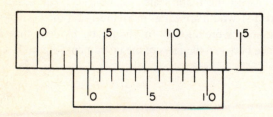

Figure 1–5 Vernier Scale Setting.

at all times the same as the distance between the zero marks on the two scales.

In Figure 1–2 it may be seen that both the 0 and the 10 mark on the vernier coincide with some mark on the main scale. The first mark beyond the zero on the vernier is one-tenth of a main-scale division short of coinciding with the first line beyond the zero on the main scale. This difference between the lengths of the smallest division on the two scales represents the least amount of movement that can be made and read accurately and is called the *least count* of the instrument. If the vernier scale is moved $1/10$ mm, the next mark beyond the zero coincides with a main-scale mark (Figure 1–3). If moved until mark 2 coincides with a main-scale division, the total distance moved is 0.2 mm. Thus it is seen that we can get an accurate setting for tenths of a millimeter and not have to estimate them as on a meter stick.

In Figure 1–4 we see that the zero on the vernier lies between the 6 and 7 mm marks, and since the second mark on the vernier scale makes coincidence, the fractional distance is 0.2 mm, and the complete reading is 6.2 mm, or 0.62 cm. By similar reasoning, the reading in Figure 1–5 may be seen to be 3.7 mm or 0.37 cm.

Nearly all vernier calipers have the length of n divisions on the vernier scale equal to the length of $n - 1$ divisions on the main scale, and the method of determining the reading is similar to that described above. The least count is always $1/n$ of the length of the smallest main-scale division. For example, in Figure 1–1 the least count is one-tenth of $1/10$ cm, or 0.01 cm.

MICROMETER CALIPER.

The micrometer is an instrument designed for the accurate measurement of small distances, such as the diameter of a wire or the thickness of a thin sheet. The jaw B (Figure 1–6) is the end of a screw passing through the cylindrical nut carrying the scale S. The object to be measured is placed between jaws A and B, and the head H which carries the screw is advanced toward the zero end of the scale until

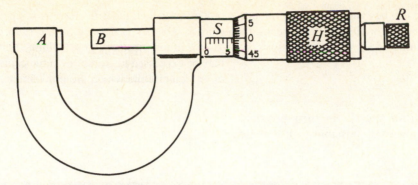

Figure 1–6 Micrometer Caliper.

contact is made with the object by both jaws. Correct pressure of the jaws is best determined by turning the ratchet R until it begins to slip. The distance the screw advances when turned through one revolution is called the *pitch of the screw*. The type most commonly used in the laboratory has a screw with a pitch of ½ mm. Hence if the divisions on the scale S are millimeter divisions, the head will make two revolutions while advancing between two marks, a distance of 1 mm.

From Figure 1–6, the position of the head on the scale indicates that the jaws have been opened by an amount greater than 6 mm. Since the scale on the revolving head indicates a reading of zero, we know that the head has been turned through one complete revolution since passing the 6 mm mark. Hence the reading is 6.50 mm, or 0.650 cm. Examination of the scale on the head shows that it contains 50 divisions, thus making the value of the smallest division on the head equal to one-fiftieth of ½ mm, or 0.01 mm. Careful examination of the reading and the position of the head H with respect to the space between two marks on the scale S will always make it possible to determine whether the head has made more than one revolution since passing the last mark visible on the scale S. If, when the jaws are closed against each other, the reading on the scale is not zero, a correction must be added to, or subtracted from, each scale reading. This is an example of a systematic error which will occur in each measurement unless a correction is made.

LABORATORY BALANCE.

The laboratory balance (Figure 1–7) carries this name because it is used to balance an unknown mass against certain standard masses. When balancing two masses in this way, we are comparing the gravitational pull on the two bodies, and we call this procedure *weighing*. The mass to be weighed is placed in the left pan, since it is seen that a movement of the rider R to the right is equivalent to the addition of weight to the right pan. Before you attempt to weigh anything, the rider should always be on the zero position and the pointer at the center of the scale, or the pointer should swing about equal distances on either side of the center of the scale. After the body has been placed on the left pan, a single mass near the mass of this body should be tried on the right pan. This saves the delay and annoyance caused by a large number of small masses and at last finding their sum to be too small. The final adjustment for balance is made by moving the rider R to the right. Its reading must be added to the masses in the right-hand pan.

Another commonly used balance is the triple beam form shown in Figure 1–8. In this balance the arms are unequal in length, but the triple beam with the movable masses is calibrated to indicate the correct weight on the pan. If the body to be weighed is too heavy for the indicated values on

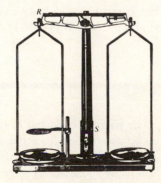

Figure 1–7 Laboratory Balance (Courtesy of Central Scientific Co.).

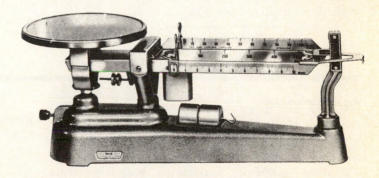

Figure 1–8 Triple Beam Balance (Courtesy of Sargent–Welch Scientific Co.).

the triple beam, the range can be extended by hanging one or more of the auxiliary masses on the end of the beam. Its indicated value must then be added to the reading on the beam scale.

DENSITY.

Different substances differ greatly in their physical properties, a fact that makes it possible to use these properties in their identification. Among these properties is density. *Density* is defined as the ratio of the mass of a sample of the substance to its volume; that is, $D = M/V$. It may be expressed in grams per cubic centimeter (gm/cm^3), or any other units which are convenient.

PROCEDURE

Steps 1–4 of the procedure are classified as preliminary measurements. They serve as an aid in understanding the principles of the vernier and micrometer calipers as measuring devices.

1. Using the metric scale of the vernier caliper, make three measurements of the diameter of a coin and record on the data form. Practice on other objects until you are sure you know how to read the instrument.

2. Now examine the micrometer caliper and watch how the scale reading changes as you carefully close the jaws. *Be sure you do not force the screw.* If your instrument should have a rachet R attached to the head H, it will slip when the jaws are touching, and thus protect the screw threads. Determine the zero correction and record it as $+$ or $-$, depending on whether it is to be added to, or subtracted from, actual instrument readings.

3. Measure the diameters of samples of human hair with the micrometer caliper. For variety, select samples from your own head and the heads of other class members. Record as male, female, brown, blond, and so on. In recording the values, to be sure of the precise reading, take care to apply the zero correction.

4. Measure the thickness of a sheet of paper in your book with the micrometer caliper. Make three trials by using different sheets. Record and determine the average. Now measure the total thickness of about all of the numbered sheets in your textbook with the vernier caliper (three trials); determine and record the average thickness of a single sheet in millimeters.

5. Now measure the diameter and length of the cylinder with the vernier caliper, about three trials for each; then estimate the uncertainty of measurement for each. Record the averages in the form $m \pm d$, and compute the percent uncertainty in the measurement of each dimension of the cylinder.

6. Measure the diameter of a copper wire with the micrometer caliper, and measure the length with a meter stick. Record both dimensions in centimeters. Why?

7. Place the small sphere between the end of the cylinder and the edge of your lab book, and measure its diameter in centimeters with a meter stick. Make three trials, using different reference points, and record on your data form in the column for length. Now make three trials of the diameter measurement using the micrometer caliper and record. Then record the average for each set with the uncertainty indicated.

8. Determine the masses of the cylinder, the wire, and the sphere by weighing on the laboratory balance. Compute the volume and density of each. The volume of a cylinder is given by $V = \pi r^2 h$, and that of a sphere by $V = \frac{1}{6}\pi d^3$.

9. Determine the mass of an irregular solid, weighing it on the laboratory balance. The volume can be determined by the water displacement method. Partially fill the graduated cylinder with water to some easily read level. By means of a lightweight string, lower the irregular solid into the water until completely submerged; then carefully read the new water level. Record the volume of the solid as that of the displaced water. Then compute the density and compare with the value listed in Appendix B (Table 1).

QUESTIONS

The questions below, and those following all experiments in this book, are designed to help the student analyze the results of the experiment and to draw some conclusions concerning the findings. In order that these analyses and calculations may be smoothly woven into your report, it is suggested that you not write out the questions, but that you answer them as separate paragraphs stated in such a way that the question can be ascertained from the answer. In general, these answers will have to be placed on one or more separate sheets of paper and attached to the Record of Data and Results.

1. Of the two methods used to determine the thickness of a sheet of paper, which do you think is the best? Explain.

2. How many significant figures did you have in the measurement of the total of the sheets? How many significant figures in the number of sheets used? If you had measured only one-tenth as many sheets, how would the number of significant figures associated with the number of sheets have been affected? Think carefully and then explain.

3. Would measuring the diameter of the cylinder with a micrometer caliper have changed the number of significant figures in the computed volume? Explain.

4. If your vernier caliper has more than one scale, examine the scale you have not used, and determine and record the following: (a) the size of the smallest main-scale division, (b) the number of divisions on the vernier and the number of main-scale divisions which they cover, and (c) the least count of the instrument for this scale. Measure and record the diameter of the coin, using this scale.

5. In the measurements on the cylinder, which do you consider was measured with the greatest precision, the length or the mass? Explain why.

6. In reality, does the laboratory balance give you the mass or the weight? Explain. If you did the experiment on the top of a mountain, how would your result be affected? Why?

7. In the measurements of the sphere, which instrument measured with the greatest precision. Explain how one chooses the most appropriate instrument for a particular measurement.

8. Would the zero correction of the micrometer be of greater significance in the measurement of the thickness of a sheet of paper or the diameter of the cylinder used in this experiment? Explain.

9. Estimate the uncertainties in the measurements of the length and diameter of the wire; compute the percent of uncertainty in each measurement. In which case do you have the most precise measurement? Explain.

10. Do the procedures in this experiment suggest any other method you could have used to obtain the volume of the wire? If so, discuss its accuracy.

11. Do you think the density determination of the irregular solid was as reliable, or accurate, as that of the cylinder? Explain.

12. Would the water displacement method of measuring the volume of an irregular solid be suitable for measuring the volume of the wire? Explain.

13. You made three trials in measuring the diameter of the wire at different places along its length. What assumption did you make in assuming that these three trials give an accurate value for the diameter?

WITHDRAWN

Learning Resources Center
Collin County Community College District
SPRING CREEK CAMPUS
Plano, Texas 75074

NAME *(Observer)* _____ Date _____

  (Partner) _____ Course _____

RECORD OF DATA AND RESULTS

Experiment 1 – Determination of Length, Mass, and Density

PRELIMINARY MEASUREMENTS

Vernier caliper		Micrometer caliper	Combination of both	
		Zero correction _____	Thickness of paper	
Coin den.	Diameter	Diameter — Hair Samples	One sheet	Bundle
1.	1.　　cm	1.　　mm	1.　　mm	No. sh.
2.	2.　　cm	2.　　mm	2.　　mm	Rdg.
3.	3.　　cm	3.　　mm	3.　　mm	1 sh.

LENGTH AND DIAMETER MEASUREMENTS

Object measured	Length		Diameter	
	Trials	Average	Trials	Average
Cylinder	1. 2. 3.		1. 2. 3.	
Wire	1. 2. 3.		1. 2. 3.	
Sphere	*Diameter with meter stick* 1. 2. 3.		1. 2. 3.	

DETERMINATION OF DENSITY

Object	Kind of material	Mass (gm)	Volume (cm³)	Density calc.	Density (Table)	Percent discrepancy
Cylinder						
Wire						
Sphere						
Irregular solid						

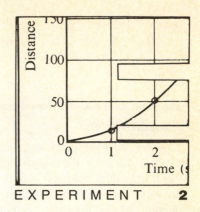

DETERMINATION OF π AND DENSITY BY MEASUREMENTS AND GRAPHICAL ANALYSIS

NOTE TO INSTRUCTOR:

See *Note to instructor* for Experiment 1.

SPECIAL APPARATUS:

Meter stick, vernier caliper, micrometer caliper, and about six circular disks having different diameters.*

GENERAL APPARATUS:

Laboratory balance.

THE PURPOSE OF THIS EXPERIMENT

is to determine π by two methods—(1) from geometrical measurements, and (2) by graphical representation of the data—and then to use graphical analysis to study relations of mass to radius of circular disks.

INTRODUCTION

Before beginning this experiment the student should read Section III of the General Laboratory Introduction dealing with graphical representation of experimental data, analysis and interpretation of graphs, and evaluation of the slope.

Since, by definition, π is the ratio of the circumference to the diameter of a circle, its value can be computed from measurements of these quantities. Handbooks often list the value of π to 10–15 significant figures that have been determined entirely by mathematical procedures. When π is determined from experimentally measured values of circumference and diameter, the resulting computed value can be only as precise as the measurements.

Experience shows that the accuracy of the computed value can be improved by making several trials of each measurement and using an average in the calculation. Furthermore, the mean (or average) deviation of the individual readings from the mean is a measure of the probable uncertainty of the experimental measurement. Large deviations between trials indicate an unreliable result and one should try to discover the cause. Small deviations tend to establish confidence in the reliability of the instrument used and the method of measurement.

In order to show the probable uncertainty in the measurement, the value should be recorded in the form:

$$m \pm d \qquad [1]$$

where m is the mean value of the trials made, and d is the mean deviation without regard to the algebraic sign of each d value. In order to show this uncertainty as a percent, the value may be recorded as

$$m \pm \frac{d}{m} \times 100\% \qquad [2]$$

*The disks should be of uniform thickness and of the same material, such as thin plywood or sheet aluminum.

If $m = 3.15$, and $d = 0.02$, the value would be written as $\pi = 3.15 \pm 0.02 = 3.15 \pm 0.6\%$.

In examining the series of circular disks, one might suspect that there is also a correlation between mass and radius over the series of disks; one of the objectives of this experiment is to determine if a relation does exist.

PROCEDURE

1. Study the equipment supplied and try to decide on a method of making the required measurements. If it seems that other equipment is needed, or would be better, perhaps the instructor can furnish the items you need. For the first method chosen, however, it is requested that all the circular disks be measured by the same method.

2. Measure the circumference and diameter of each disk, listing them in order of size, beginning with the smallest. On the data form supplied with this experiment, record these measurements along with the kind of material of each disk. In recording the data, follow the suggestions in Section II of the Introduction—dealing with significant figures. The unit to be recorded is given in the heading of each column on the data form and need not be repeated below.

3. Now select one of the medium-size disks, make at least six independent measurements of circumference and diameter, and record in the appropriate table. In order that the circumference measurements will be independent, start the measurement at a different place on the meter stick each time and let one partner read the position of the beginning point while the other reads the end point. The diameter measurements may be made independently by placing the measuring scale across the circle in a different direction each time. Be sure that in each measurement you record the reading as accurately as the measuring instrument will allow.

4. Compute the value of π, from the relation $\pi = C/D$, for each set of measurements made, and then record the average of the trials used.

5. Now find the difference between the mean (or average) and each individual value of π, and record the numerical values of these deviations from the mean without regard to algebraic sign. Then find the average of the deviations. This average deviation for each of the two methods is an indication of the uncertainties in the measured quantities used to compute the value of π. In order to show this probable uncertainty in your determination of π, record the value in the form indicated in Equation [1].

6. The standard value of π is usually written as 3.1416. In order to compare your value with this commonly used standard, it will be desirable to review Section II of the Introduction. When you have decided on the proper number of significant figures to use, find the percent discrepancy in each of your determinations of π.

7. Suppose we now check our computed value of π by graphical analysis. Following the directions in Section III of the Introduction, plot your diameters on the X axis and the circumference values on the Y axis. Carefully draw the straight line that best fits the trend of the series of plotted points. Now select two well-separated points *on the line* to provide values for Δy and Δx (or ΔC and ΔD). Then referring to Figure E in the Introduction, and the discussion there about slope, determine the slope of your graph and make it a part of your report. Also, interpret its meaning.

8. If one examines the shapes of the disks carefully, one may observe that they are cylinders, with radius r and very small height h. With this in mind we may express the mass of each disk in the form

$$\text{mass} = \text{volume} \times \text{density} \qquad \text{or} \qquad M = \pi r^2 h \rho \quad [3]$$

where ρ is the density of the material of the disk. Measure and record the height of each disk, and then determine the masses by weighing on the laboratory balance. Record the values of r and r^2 along with h and M in the table on the reverse side of the data form. The density ρ is an unknown quantity, but Equation [3] indicates that the mass is related to r, h, and ρ. Suppose we study this relation by graphical analysis.

9. Plot two graphs, (1) mass versus radius and (2) mass versus (radius)^2, in each case plotting mass on the Y axis as the dependent variable. Through each set of points, sketch a line that seems best to fit the trend of points in each of the two graphs. Do you get a straight line in both cases? Explain.

10. Examine the nature of both graphs and then find the slope of the one that results in a linear relation, following the hints given in Step 7 for computing the slope.

You have already determined π and have measured h, which should be about the same for all the disks. This leaves ρ as the only unknown in Equation [3]. From your slope and other known quantities, compute the density ρ. Rearrange Equation [3] so as to accommodate the substitution of your slope. Compare your value of ρ with values in Table 1 of Appendix B.

QUESTIONS

The questions below, and those following all experiments in this book, are designed to help the student analyze the results of the experiment and to draw some conclusions concerning the findings. In order that these analyses and conclusions may be smoothly woven into your report, it is suggested that you not write out the questions, but that you answer them as separate paragraphs stated in such a way that the question can be ascertained from the answer. In general, these answers will have to be placed on one or more separate sheets of paper and attached to the Record of Data and Results.

1. In the first case where you measured several different sizes of circular objects, did you notice any correlation between the accuracy of the computed values of π and the size of the circle measured? If so, give an explanation for the relationship. Since π is defined as the ratio of the circumference to the diameter of a circle, would you say that its value is independent of, or dependent on, the size of the circle one measured? Explain.

2. In which case did you get the best value of π, with the variety of sizes or by making several trials with one size? Why was this true?

3. Compare, by observation, the percent discrepancy of one of your average values of π and the percent uncertainty of the measurements which gave this average value. What is the significance of the percent uncertainty? Do you think your percent discrepancy was within the limits of expectations for the method you used? Explain why.

4. In the case where you made several trials on the same circular object, what explanation can you give to account for the fact that all of the circumference measurements were not the same? Also explain the variations in the diameter measurements.

5. Briefly describe the method you used in measuring the circumference and diameter. Suggest some changes in the method of determining the value of π that might result in less percent error.

6. In Step 5 of the procedure, you were told to disregard the algebraic sign in finding the average of the deviations. Why not use the algebraic signs to compute the average? Try it for the case of several trials with the same circle.

7. Suppose you are assigned the task of determining the value of π to five significant figures, a value near 3.1416, with meter sticks as the measuring device. What minimum size circular object would you request from the storeroom? Explain.

Questions 8–13 are related to interpretation of graphs.

8. What physical quantity does the slope of your first linear graph represent? How is this slope related to the definition of π? What unit does the slope have? Explain.

9. Does the graphical method of determining π have any advantages over the first method used? Explain.

10. In the graphs involving the masses of the disks, what name (type of curve) is associated with the one that is not a straight line? Is the mass related to the radius in this case? Explain how you know.

11. Could you have determined the density of any of your disks without using the graphical method? If so, explain how. What advantage does the graph method have?

12. What units does the slope of your graph of M versus r^2 have? Are they units of density? Explain.

13. A student measures the width of a room and records the result as 5.375 ± 0.1 meters (m). This means the uncertainty of measurement was ± 0.1 m. Discuss the recorded result and state what you think should have been recorded.

14. A builder has a 50-m steel tape on which the smallest divisions are centimeters. The tape is used to measure the width of a door and the length of a house. If the door is about 1 m wide, and the house is about 16 m long, what would be the uncertainty in each measurement, based on the estimation of fractions of a centimeter? What would be the percent uncertainty in each measurement?

RECORD OF DATA AND RESULTS

Experiment 2—Determination of π and Density by Measurements and Graphical Analysis

MEASUREMENT ON A VARIETY OF SIZES OF CIRCULAR OBJECTS

Description of object	Circumference (cm)	Diameter (cm)	Computed value of π	Deviation from mean
Average values				

Experimental value of π =

Percent discrepancy =

DIFFERENT TRIALS ON THE SAME OBJECT

Trial	Circumference measurements			Diameter (cm)	Computed π	Deviation from mean
	Initial rdg	Final rdg	Circum (cm)			
1						
2						
3						
4						
5						
6						
Average values						

Experimental value of π by measurement =	Percent discrepancy =
Slope of graph $= \dfrac{\Delta Y}{\Delta X} = \dfrac{C_2 - C_1}{D_2 - D_1} =$	Percent discrepancy =

DATA FOR RELATION OF MASS TO RADIUS

Kind of material Disk No.	Thickness of disk h (cm)	Mass of disk M (gm)	Diameter (From table 1) D (cm)	Radius of disk r (cm)	Square of radius r^2 (cm^2)
1					
2					
3					
4					
5					
6					
Slope of graph = =			ρ (calculated) =		
Handbook value of ρ =			Percent discrepancy =		

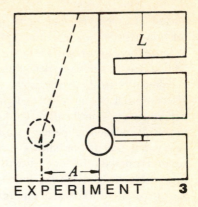

EXPERIMENT **3**

THE PERIOD OF A PENDULUM—AN APPLICATION OF THE EXPERIMENTAL METHOD

SPECIAL APPARATUS:

Two-meter stick, pendulum clamp, three pendulum bobs of different materials, protractor, two rulers, string or thread (about 2 m long), watch with sweep second hand, graph paper (several sheets).

GENERAL APPARATUS:

Table clamp, long rod, and pendulum clamp.

THE PURPOSE OF THIS EXPERIMENT

is to illustrate the experimental method of scientific investigation by finding how the period of a pendulum depends on various factors.

INTRODUCTION

A simple pendulum consists of a relatively small bob the mass of which is concentrated at the end of a string (or thread) suspended from a support so that it can oscillate back and forth. The time required for the bob to make one complete round trip of its motion is called the *period*, and the round trip is called a *vibration*.

As one examines the pendulum and its motion, one immediately wonders what factors determine the period—the time for one complete vibration. It may be that several factors are involved. As one holds the bob to one side and releases it, the fact that it moves to a lower position, indicates that gravitational attraction must be associated with the motion (Figure 3–1). Since such an attraction seems to be involved, the next question that comes to mind is, ''How will the period of the motion be affected by changing the material or size of the bob?'' What happens if its distance above the earth is changed?

It may be further noted that the pendulum bob can swing through various sizes of arcs, or amplitudes. Hence, one may ask, ''What effect does the size of the arc have on the period?'' Then again, if one changes the length of the supporting string, it will be noticed that the value of the

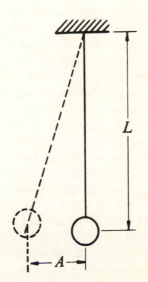

Figure 3–1 Simple Pendulum.

period changes. Hence, there must be some definite relation between the length of a pendulum and its period.

Since there are about four factors to which we must give attention in the study of a pendulum, we must perform what is called *a controlled experiment* in order to determine the effect of each. A controlled experiment is one in which the experimenter controls all of the factors which may affect the result and then deliberately varies one factor at a time. It is obvious that if one allows simultaneous variations in several relevant factors, it will be difficult, perhaps impossible, to sort out the individual contributions each variation made toward changing the result of the experiment.

Newton's law of gravitation tells us that, for two given masses, the gravitational force of attraction is inversely proportional to the square of the distance between their centers. Since the distance from the pendulum bob to the center of the earth is about 4000 miles, the small change in it that we could make by moving our pendulum up or down a few feet at a given place on the earth would not measurably affect the force of attraction. Hence, within the laboratory, we can assume that the factor associated with gravitational attraction remains constant, and we are left free to study the other three factors.

The variations related to the mass and size of the bob and the length of arc through which it swings will be mentioned in the procedure below. A very casual observation reveals that the period of the pendulum becomes shorter (or longer) as the length of the string is made shorter (or longer). This would indicate that the period T may be related to the length L by some relation such as

$$T = kL^n \qquad [1]$$

where n is an exponent and k is a proportionality constant. We can say that T varies directly as L^n and, if n is known, T plotted as a function of L^n would result in a straight line graph.

In most real situations, exponents for this type of relation are either simple fractions or simple whole numbers, such as one-half, three-fourths, 1, 2, 3, and so forth. The required value for n is the value that will give a straight line when T versus L^n is plotted on graph paper.

PROCEDURE

The factors which appear to be associated with, and may have some effect upon, the period of a simple pendulum are (1) the distance of the pendulum from the center of the earth, (2) the material and mass of the body, (3) the length of arc through which it swings, and (4) the length of the supporting string.

1. Since, for a given pendulum, we cannot within the confines of the laboratory essentially change our distance from the center of the earth, this factor can be assumed to remain constant in our study, and we can turn our attention to the other factors which we can control.

2. If the period of the pendulum is dependent on the mass of the bob, then one could determine this fact by finding the period for bobs of different masses or materials. In doing this, we must keep all other factors, such as length of arc and length of string, constant.

Measure the length from the lower part of the pendulum clamp to the *center* of the spherical bob, and use an arc of about 5 cm on either side of the center position. Select some particular bob and, using a length of 100 cm, time about 50 vibrations with your watch. Select some fixed point or line as a reference point from which to start the count, and let one student count while the other watches the time. Let the one who is to count begin by counting time, two, one, *go*, one, two, three, and so on. The countdown gives the timekeeper a warning about when to expect the ''go'' signal, so that the timing can be done to a split second. The counter should count to himself or herself until within three or four counts of the end of the counting period, and then count aloud again, indicating the last count by the word ''stop,'' or by a tap on the table.

Repeat with about two other bobs of different mass, being careful to keep the length, the arc, and the size of ball the same as before. Compute the period, the time for one complete vibration, and record along with other information about the pendulum.

3. In order to study the effect of the amplitude (arc) on the period, it is best to choose a heavy bob and use a length of about 150 cm, both of which must be kept constant while the arc is varied. Since it is desirable to use several sizes of arcs, it might be best to consider angular amplitudes of the bob, rather than linear amplitudes. Time 50 vibrations for angular amplitudes of about 5, 10, 15, and 25°, and record. Then time 20 vibrations with the much larger amplitudes of 60 and 80°, respectively. Compute and record the periods for all amplitudes used.

4. You already have the periods for lengths of 150 and 100 cm and you may record the results in the section for length effects. Use a small amplitude and time about 50 vibrations for a length of 120 cm, and then time 100 or more counts for each of the lengths of 80, 65, 50, 36, and 25 cm. In no case should the total counting time be less than 60 sec for any length, preferably about 100 sec.

5. The effect of changes in length can best be ascertained by plotting graphs of T versus L^n as discussed in the Introduction. However, before attempting to plot any graph read Section III on curve plotting in the General Laboratory Introduction, and then prepare your curves accordingly.

At this point, either of two methods may be used to determine the value of n. The method of plotting the graph is different. Ask your instructor which method you are to use.

Plotting of Graph—Method A: Steps 6 and 7

6. As a starting point, assume $n = 1$, and plot a graph with periods (T) as ordinates (Y axis) and L^1 as abscissas (X axis). Sketch a line (curve) through the points and examine it. If it is not a straight line, then the value of n is some value other than 1. If other assumptions for n seem necessary, then next try $n = 2$, and plot T versus L^2. This graph undoubtedly will have a different curvature than the previous one. Does this trial seem to be in the right direction to approach a straight line? If not, then make another try in the other direction with $n =$ one-half and plot T versus $L^{1/2}$, or T versus $\sqrt{L}$. Assign values to n and plot until a reasonably good straight line is obtained.

7. An examination of Equation [1] indicates that $k = T/L^n$, and since you have plotted T versus L^n, then the slope of your straight line graph will give $\Delta T/\Delta L^n$, *which is equal to k*. Determine the value of k from your slope and rewrite Equation [1] in your data record with the proper values of k and n supplied, so that T and L are the only variables. Show the steps of these computations below the data table.

Plotting of Graph—Method B: Steps 8, 9, and 10

8. By the use of logarithms we can arrive at the value of k and n in a manner more direct than Method A. If we take the logarithm of both sides of Equation (1), the result may be written

$$\log T = \log k + \log L^n \qquad [2]$$

because the log of a product equals the sum of the logarithms of the factors. But we can further simplify the above by writing it as

$$\log T = n \log L + \log k \qquad [3]$$

Now we have an equation with all terms of the first degree, and if $\log T$ versus $\log L$ is plotted as illustrated in Figure 3–2, a straight line should result.

If the coordinates $\log L_2$ and $\log T_2$, for point P_2, and $\log L_1$ and $\log T_1$, for point P_1, are sequentially substituted into Equation [3], we have

$$\log T_2 = n \log L_2 + \log k$$
$$\log T_1 = n \log L_1 + \log k$$

If these two equations are subtracted, the result becomes

$$\log T_2 - \log T_1 = n \log L_2 - n \log L_1$$
$$= n(\log L_2 - \log L_1)$$

and we have

$$n = \frac{\log T_2 - \log T_1}{\log L_2 - \log L_1} \qquad [4]$$

An examination of the illustrated graph in Figure 3–2 reveals that the right member of Equation [4] is the slope $\Delta y/\Delta x$ of the line. Hence, the slope of the graph provides the value of n needed to give a straight line from the original relation $T = kL^n$. If you have studied analytic geometry, you will recognize Equation [3] as having the straight line form $y = mx + b$.

9. Determine the values of $\log T$ and $\log L$ for plotting such a graph from your data. The coordinates of the intersection of the X and Y axes need not be (0,0), and you may be able to better utilize the graph paper by beginning on each axis with some number slightly below your smallest values, rather than (0,0).

If logarithmic graph paper is available you may plot this graph without determining all the logarithms. The logarithmic scale automatically "takes the logarithms" for you. Each cycle of logarithmic paper covers a range of values spanning exactly one factor of 10. The cycles on an axis occupy equal lengths on the paper and should be labeled as powers of 10 such as 1, 10, 100,

10. Select two points on your line (generally not any of the plotted points), well separated, and designate them as P_1, and P_2, respectively. Then substitute their values into Equation [4] and solve for n. Substitute this value of n into Equation [1] along with any corresponding pair of values for T and L and solve for k. Now rewrite Equation [1] in your data record with the proper values of k and n supplied so that T and L are the only variables. Show the computation as a part of your report.

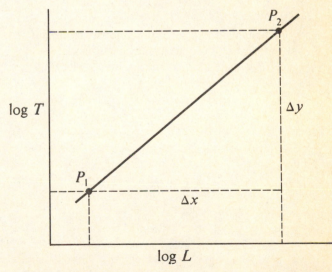

Figure 3–2 Illustration of log T Versus log L Graph

QUESTIONS

1. With all other factors remaining constant, what effect does changing the mass of the bob have on the period of a simple pendulum? Justify your answer from the results of your experimental investigation.

2. What other experimental arrangement could you have used to study the effect of different masses on the period?

3. Discuss what your findings show about the effect of the amplitude, or arc, on the period. Do you think your results are realistic, or accidental? Why? How do the values for different amplitudes compare with the percent uncertainty of your measurements?

4. When using the very large amplitude you were asked to time only 20 vibrations. Would the period as obtained for a large amplitude be more accurate if 50 or 100 vibrations were timed? Explain.

5. Why is it necessary to count more vibrations for a short pendulum than for a long one to obtain the same accuracy?

6. What effect would the expansion of the pendulum rod in a clock have on the rate of the clock?

7. From a study of your graphs, what relation do you find to exist between the period and the length of a pendulum?

8. Summarize the results of your investigations in a single sentence relative to the effects of the three factors which you were able to control during the course of the experiment.

9. Rewrite Equation [1] with your values substituted for k and n and test the validity of the equation with some random pair of values of T and L taken from your data. Do the results of your experimentation give you confidence, or doubt, concerning the assumption made in Equation [1]? Explain the reasons for your answer.

10. From the straight line graph determine the length of a pendulum having a period of 1.5 sec. Make the same determination using one of the nonlinear curves. Do you have confidence in the value of L thus obtained? What is the basis for your judgement?

11. The period of a simple pendulum is usually written in the form $T = 2\pi \sqrt{L/g}$, where g is the gravitational acceleration. From this equation you will note that g is a constant in a given locality, and thus becomes a part of the constant k in Equation [1]. Using your value of k, compute the value of g in the above equation. *Hint:* As a first step, rearrange the above equation so that all of the constant quantities are grouped together.

NAME *(Observer)* _____ Date _____

(Partner) _____ Course _____

RECORD OF DATA AND RESULTS

Experiment 3 — The Period of a Pendulum — An Application of the Experimental Method

NOTE: The last two columns of the data table may be adapted to Method A or Method B.

Observation	Material of bob	Angular amplitude (deg)	Elapsed time (sec)	Number of vibrations counted	Period T (sec)	Length L n = 1 (cm)	Method A n = ____ L^n	Method A n = ____ L^n
Effect of material of bob							Method B log T	Method B log L
Effect of amplitude or arc								
Effect of length								

COMPOSITION AND RESOLUTION OF FORCES—FORCE-TABLE METHOD

NOTE TO INSTRUCTOR:

A very convenient method of handling the force-system assignment is to issue to each student a card showing some definite arrangement of forces as to magnitude and direction. The cards are to be checked in again with the force-table equipment.

SPECIAL APPARATUS:

Set of slotted weights (weights ranging from 1 to 200 gm), four weight pans, ring with four strings attached, center pin for force table, two rulers, two protractors, assignment card.

GENERAL APPARATUS:

Force table with four pulleys attached (see Figure 4–4).

THE PURPOSE OF THIS EXPERIMENT

is (1) to determine graphically and analytically the resultant of several coplanar forces acting through a point, (2) to test the accuracy of the computation by setting up the forces on a force table, and (3) to represent graphically a system of forces in equilibrium.

INTRODUCTION

If a number of nonparallel forces are acting at the same point on a body, it can be shown that they may be replaced by a single force which will produce the same effect on the body. Such a force is called the *resultant* of the original forces. The process of finding this resultant is called the *composition of forces*. The single force which will hold a system of concurrent forces (forces acting through a common point) in equilibrium is called the *equilibrant* of the system. It is equal in magnitude to the resultant, but opposite in direction.

The part of the force effective in some particular direction is called a *component* of the force. The process of finding components of forces in specified directions is called the *resolution of forces*. The processes of composition and resolution may be performed by either of two methods, graphical or analytical. If f (Figure 4–1) is the

force being considered, the *analytical method* of finding the components consists of applying the proper trigonometric relations to the triangles formed by the force f and its components. As shown in Figure 4–1, the forces f_x and f_y are the horizontal and vertical components, respectively, of the force f.

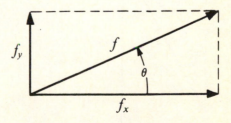

Figure 4–1 Force Components.

PROCEDURE

1. A portion of your apparatus consists of a small ring attached to four strings and held in place on the force table by the center post (see Figure 4–4). Set one pulley at the 0° position and suspend a 300-gm weight from the end of the string. *Note:* The weight of the hanger constitutes a portion of the suspended weight. In like manner, suspend a 400-gm weight at the 90° position. If this system is to be held in equilibrium, a third force of the proper amount must be applied at the required direction. Pull on one of the other strings until you determine the required direction, and then mount a third pulley at this position. Now suspend sufficient weight from this position to center the ring around the post. This force, expressed in grams, is the equilibrant of the other two forces. To check for equilibrium and to minimize the effects of friction in the pulleys, raise the ring a short distance above the table and release, noting the new position it takes. Record force and angle.

2. Since the two original forces are at right angles to each other, it is a simple matter to compute the resultant by the Pythagorean theorem. Set the computed value of the resultant on the force table at a position directly opposite to that of the equilibrant, remove the original two forces, and then check to see if the computed resultant balances the equilibrant. At this point review the definition of a resultant. Record force and angle and also $E - R$.

3. Obtain from the instructor an assignment of a concurrent-force arrangement of three forces. Choose some point on your report sheet as the origin and, after laying off the coordinate axes, draw to scale a vector

diagram of the force system, labeling the forces f_1, f_2, and f_3, and indicate their directions (see Figure 4–2).

4. Construct the horizontal and vertical components of each force. Measure their magnitudes and record on the diagram as shown in Figure 4–2. Also record on the data form.

5. Determine the vector sum of the horizontal components ΣF_x and the vector sum of the vertical components ΣF_y. Consider these sums as single forces; lay them off on a separate set of axes, as shown in Figure 4–3. Then construct the resultant R of these two force vectors, measure its magnitude and the angle θ that it makes with the positive direction of the X axis, and record.

6. Now place pulleys on the force table (Figure 4–4) at the positions of the three assigned forces and suspend the proper amount of weight for each. Then, while pulling on the fourth string in the direction required to produce equilibrium, set the pulley at the required position and add weights to the hanger until equilibrium is established. Remember to include the weights of the hangers in recording the forces. Test the system for equilibrium by the method suggested in Step 1. Also check the uncertainty in the magnitude of the experimental equilibrant by finding how much weight must be added to or removed from the weight hanger before a movement of the ring position is observed. Likewise, determine the uncertainty in the angular position by seeing how many degrees you can move the equilibrant pulley to each side before the ring moves off center. In the space provided for data, record the magnitude and direction of the experimental equilibrant, each accompanied by a ± value for the uncertainty. Now compute and record the percent uncertainty in the experimental value of E. Then compute and record the percent difference between the values of E determined experimentally and analytically.

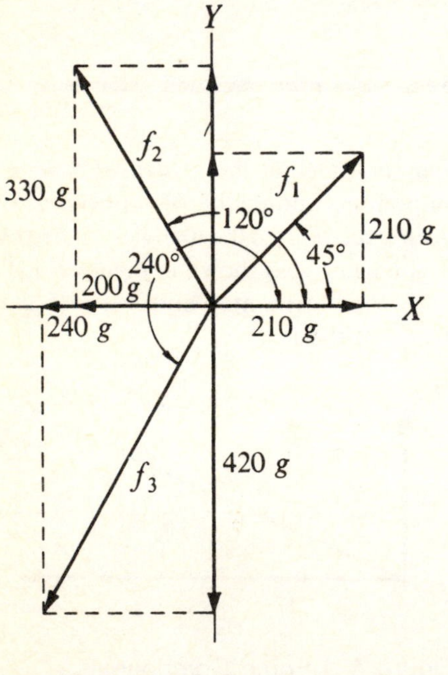

Figure 4–2 Method of Resolving Forces into Components.

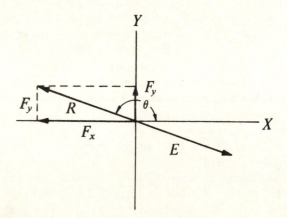

Figure 4–3 Relation of the Resultant and the Equilibrant.

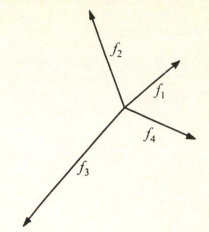

Figure 4–5 A System of Forces in Equilibrium.

Figure 4–4 Force Table (Courtesy of Central Scientific Co.).

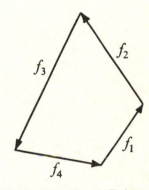

Figure 4–6 Force Polygon.

7. By applying the proper trigonometric relations to each of the forces in the system assigned to you, determine and record the horizontal and vertical components. Next find the sums of the horizontal and vertical components ΣF_x and ΣF_y, respectively, and record. Note that the resultant vector is the diagonal of the right triangle and determine its magnitude and direction analytically and record. How do these values compare with those found graphically? If materially different, check by balancing on the force table.
8. Figure 4–5 represents a system of four forces in equilibrium such as you had on the force table in step 6. By using the forces which you had balanced, with their

respective directions, sketch a diagram similar to Figure 4–5 for your system. Then construct a force polygon such as is represented by Figure 4–6, being very careful to determine the angle measurements correctly. If the vector polygon does not close, measure the amount of the discrepancy and record it on the figure.

QUESTIONS

1. State what part of this experiment best verified the definition of a resultant of a system of forces. Explain just how the definition was verified.
2. State clearly the relationship between the resultant and equilibrant of a system of forces.
3. Can the effectiveness of any single force be truly represented by two components of the force? Qualify your answer by the results of this experiment.
4. If the weights of the weight hangers are exactly the

same, could their weights have been neglected? Explain.
5. Was the ring in equilibrium when it was not centered around the post on the force table? Explain why it was necessary for it to be centered in this experiment.
6. If the ring had weighed considerably more, what would have been the resulting effect on the system?
7. By what amount, if any, did your polygon fail to close? Can this amount of discrepancy be justified by the uncertainty in the experimental value of the equilibrant? Explain.

8. To which did the equilibrium of the system seem to be more sensitive, small changes in force or in angle? From an examination of the setup, what explanation can you give for this?

9. Two forces of 200 and 300 gm, respectively, make an angle θ of 50° with each other. Find the resultant analytically by applying the cosine law, namely, $R^2 = A^2 + 2AB \cos \theta + B^2$, where A and B are the two forces and R is the resultant.

10. If the two forces in Question 9 make an angle of 90° with each other, find the resultant. To what form does the cosine law reduce in this case?

11. If the two forces in Question 9 make an angle of 0° with each other, find the resultant. Reduce the cosine law equation to the simplest form.

12. Does the percent difference between the experimental and analytical values of E seem to be justifiable for the experimental setup you used? Explain your reasoning.

13. In this experiment you used mass units on force quantities. In most cases this would be considered incorrect and would lead to incorrect results when used in computations. Why were you able to do this in this experiment? Show a sample calculation that indicates the use of correct units and demonstrate why this was not necessary for these computations.

RECORD OF DATA AND RESULTS

Experiment 4—Composition and Resolution of Forces—Force-Table Method

PRELIMINARY MEASUREMENTS — STEPS 1 AND 2

Force	f_1 (gm)	f_2 (gm)	Equili-brant, E	Resultant R	Difference E vs R
Magnitude	300	400			
Direction	0°	90°			

COMPOSITION AND RESOLUTION OF ASSIGNED FORCES

Assigned force problem			Force components, graphical method		Force components, analytical method	
Force	Magnitude	Direction	Horizontal	Vertical	Horizontal	Vertical
f_1						
f_2						
f_3						
ΣF_x; ΣF_y						
Resultant, R	Magnitude		— — — — — —		— — — — — —	
	Direction		— — — — — —		— — — — — —	
Equilibrant, E			Experimental		Analytical	
	Magnitude		— — — — — —		— — — — — —	
	Direction		— — — — — —		— — — — — —	
Percent uncertainty in magnitude of E (Exp) _____			Percent difference Experimental vs Anal. _____			

Show force diagram below for assigned problem.

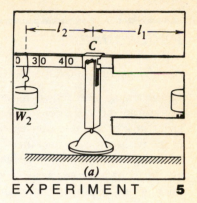

BALANCED TORQUES AND CENTER OF GRAVITY

SPECIAL APPARATUS:

Meter stick, balance support (see Figure 5–2), weight for loading the meter stick, unknown weight, set of weights, meter-stick clamp (with stirrup), spring balance.

GENERAL APPARATUS:

Laboratory balances, supply of thread or string.

THE PURPOSE OF THIS EXPERIMENT

is to study the conditions of equilibrium of a rigid bar and to locate the center of gravity of a nonuniform bar.

INTRODUCTION

If a body should be caused to move from its position of rest in either a straight-line direction (translation) or be set into rotation, its equilibrium condition would be disturbed. Then we may ask ourselves, "What conditions are necessary to maintain equilibrium?" *Translation* is produced by the action of an unbalanced force, while *rotation* is produced by the action of an unbalanced torque. Rotation involves an axis, and a *torque* may be defined as the turning effect about that axis due to some applied force. The perpendicular distance from the axis to the line of action of the force is called the *lever arm*. The torque is then defined as *torque = force × lever arm*.

The rotational effect of a body is often associated with the *center of gravity*, which is defined as the point at which a single upward force can balance the gravitational attraction on all parts of the body for any position of the body. The center of gravity would then be the point of action of the resultant of all of the gravitional forces. It may also be defined as the point about which the algebraic sum of all the gravitational torques is equal to zero for any orientation of the body.

Figure 5–1 shows an irregular-shaped bar balanced at the axis A with W_1 and W_2 suspended as shown. The weight of

the bar W_B is the gravitational pull which acts downward at the center of gravity C. The condition for rotational equilibrium states that the sum of the clockwise torques must equal the sum of the counterclockwise torques about any arbitrarily chosen axis. The condition for the above system is

$$W_1(l_1) = W_2(l_2) + W_B(l_3)$$

$$\text{(clockwise)} = \text{(counterclockwise)} \qquad [1]$$

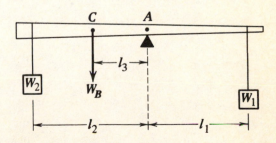

Figure 5–1 Balanced Torques.

PROCEDURE

A. A Study of Conditions Needed for Equilibrium

1. The arrangement shown in Figure 5–2a will be used to investigate the conditions of equilibrium. Balance the meter stick in the knife edge clamp and record the reading as the position of the fulcrum.

2. Select two different size weights (50 gm or more) and, by means of a thread, suspend one on each side of the fulcrum and adjust at convenient distances so that the system is balanced. Record the weights used and their positions at the time of balance.

3. What would you have to do to the setup to produce translation without rotation? Try two or three ways to disturb the rotational stability of the system and make a mental notation of them. Is there more than one arrangement of the two weights that will produce equilibrium?

4. Now suspend two weights at different positions on one side of the fulcrum and one on the other side, and record the weights and their positions when balanced. Do not place the weights too near the fulcrum. Does equilibrium depend on the number of forces pulling on either side of the axis?

5. Do the torque relationships needed in the above steps reveal a method for finding the value of an unknown force tending to produce rotation? Using the arrangement in Step 2 balance the meter stick with one of the forces being supplied by an unknown weight, and record all other amounts and positions of suspended weights.

6. From the torque relationships needed for equilibrium compute the value of the unknown weight and compare it with that obtained by weighing on the laboratory balance. The percent difference is a measure of the comparison of two such experimental values.

B. Weight of Bar and Center of Gravity

7. With only one weight, W_1 (100–200 gm), suspended near one end of the meter stick, slide the meter stick through the clamp at the fulcrum until a balance is obtained without a suspended weight on the other end (Figure 5–2b). What do you think is producing the torque to balance the torque due to weight W_1? Do you have an unknown weight in this case? Is there more than one force involved? If so, where is the position of the effective resultant responsible for the counterclockwise torque action? Review the definition of center of gravity and record all needed weights and their positions along with that of the new fulcrum (or axis).

8. By the application of torques about the new axis, compute the weight W of the meter stick and compare with the weight obtained by weighing on a laboratory balance. Show the method of computation on your report.

Now suppose we do additional study on the condition for translational equilibrium. With the knife edge clamp at the above axis position, fasten the stirrup on the knife edge part, and by means of a spring balance, determine the total upward force needed to balance the system used in Step 8. Compare this force with the total force pulling down. Examine the system to see if you have included all downward forces involved. If not, make the correction and

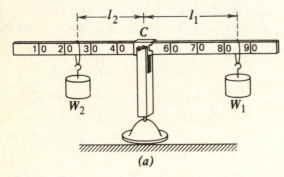

(a)

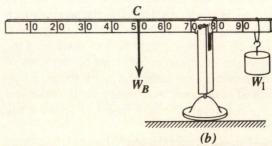

(b)

Figure 5–2 Balanced Meter Stick.

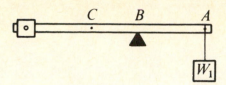

Figure 5–3 Loaded Meter Stick Balanced by a Single Weight.

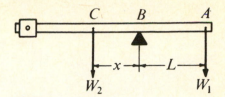

Figure 5–4 Forces and Lever Arms Illustrated for Loaded Meter Stick in Figure 5–3.

then compare. What is your conclusion about the condition necessary for translation equilibrium?

9. Now tie a mass, such as a strip of lead or a small cylinder, to the lower edge of the meter stick near one end. This arrangement is equivalent to an irregular bar with one end heavier than the other. Find the weight of the loaded bar on a laboratory balance, and keep the attached mass fixed in place for the remainder of the experiment.

10. The center of gravity of this loaded bar is now at some point C (Figure 5–3) other than the center of the meter stick. Replace the knife edge clamp, balance the loaded meter stick alone, and record the position of the fulcrum.

11. Suspend the weight W_1 from a point A near the free end of the bar. Adjust the position of the knife edge until a

balance is obtained. Record the position to the fulcrum and the weight W_1.

12. Repeat the above observation with the suspended weight W_1 at another position which is at least 5 cm from the first.

13. In Observations 11 and 12 the weight W_2 of the loaded meter stick may be considered as acting at some point C at a distance x from the fulcrum, its torque W_{2x} balancing the torque W_1L (Figure 5–4).

Compute and record the distance x for each of the positions of the weight W_1, and then determine the position of C as read from the zero end of the stick. Find the average of the two trials, and compare with the position found by balancing the loaded stick alone.

QUESTIONS

1. How much should the upward force against the knife edge be in any case where the meter stick is balanced in a horizontal position? What experimental evidence did you find from which to state the condition for translational equilibrium? Cite data to justify your statement.

2. What did you find to be the necessary condition to produce equilibrium with respect to rotation? Cite one result from your data record to justify your conclusion.

3. What determines the position of the center of gravity of a body? Does its position depend upon the position of the fulcrum or the amount and position of some balancing weight? Point out how the results of the experiment support your answer.

4. Make a sketch for the setup of Observation 4, showing the meter-stick readings and values of weights suspended. Assume the upward force to be unknown but acting at the fulcrum, and considering the zero end of the bar as the axis of torques, calculate the value of this upward force, and compare it with the sum of the downward forces. Consider the weight of the meter stick as one of the forces.

5. Repeat the above problem, using the 70-cm mark as the position of the axis of torques. What does this calculation reveal about the choice of axis?

6. In Procedure A what advantage, if any, was obtained by keeping the fulcrum at the center of gravity of the meter stick?

7. In the process of making calculations from the law of moments (or torques) what advantage results by choosing the axis of rotation at the position of one of the forces? Explain.

8. If, in any part of the procedure, the meter stick were balanced at an angle, rather than in a horizontal position, would the calculations be in error? Assume a 30° angle with the horizontal in Step 2 of Procedure A and make the calculations called for in the experiment. Explain the results.

9. When only the front wheels of an automobile are on a platform scale level with the ground, the scale balances at 2100 lb, and when only the rear wheels are on the scale it balances at 1800 lb. The wheelbase of the automobile is 130 in. Find the center of gravity of the automobile with respect to the front axle.

10. In this experiment you used mass units on force quantities. In most cases this would be considered incorrect and would lead to incorrect results when used in computations. Why were you able to do this in this experiment? Show a sample calculation that indicates the use of correct units and demonstrate why this was not necessary for these computations.

NAME *(Observer)* _____ Date _____

(Partner) _____ Course _____

RECORD OF DATA AND RESULTS
Experiment 5—Balanced Torques and Center of Gravity

A.   THE PRINCIPLE OF MOMENTS

Step	Position of fulcrum	Force (gm)	Position on bar	Lever arm	Clockwise torque	Counter-clockwise torque	Percent difference
(1) (2)							
(4)							
(5)		Unknown, *x*					
(6)	Unknown wt. calculated _____		Unknown wt. by weighing _____			Percent difference _____	

B.   WEIGHT OF BAR AND CENTER OF GRAVITY

Step 7	Step 8	Step 8	Step 8	Step 9
Wt. of W_1 _____	Wt. of meter stick, W, by balance _____	Wt. of clamp & stirrup _____	Total upward force, F _____	Wt. of loaded bar, W_2 _____

Step	Position of fulcrum	Position of W_1	Lever arm of W_1	Torque of W_1	Lever arm of W	W or x calculated	Position of center of gravity
(7)						W	
(10)							
(11)						x	
(12)						x	
(13)							

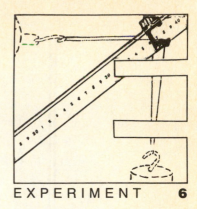

EQUILIBRIUM OF A CRANE

NOTE TO INSTRUCTOR:

Students will be able to complete either Procedure A or Procedure B but not both in a single laboratory period.

SPECIAL APPARATUS:

Meter stick, two rulers, two protractors, angle measuring caliper, set of weights, two spring balances (2-Kg range), 1-Kg load. (For Figure 6–6, substitute one 15-Kg-range spring balance and one 10-Kg load.)

GENERAL APPARATUS:

Crane setup with necessary rods and clamps (see Figures 6–3, 6–4, and 6–5).

THE PURPOSE OF THIS EXPERIMENT

is to study the conditions of equilibrium of a rigid body by comparing the computed values of the forces acting on a crane with those measured experimentally.

INTRODUCTION

One frequently sees a crane in operation around construction sites. This device is one example of the action of a system of coplanar forces. Other examples include a painter's ladder leaning against a wall, the A-frame supporting a child's swing, etc. All such force systems must satisfy two very basic laws if they are to support their loads and remain in static equilibrium. The problem involved in this experiment is also very pertinent to the work of structural engineers.

For any system of coplanar forces to be in equilibrium two conditions must be satisfied: (1) the algebraic sum of all force components in any one direction must equal zero, and (2) the algebraic sum of the torques exerted by these forces about an axis perpendicular to the plane of the forces must equal zero. In general, both conditions must be applied to solve a problem involving static equilibrium, and the equations used to express the equilibrium conditions are usually written as

$$\Sigma F_y = 0$$
$$\Sigma F_x = 0 \qquad \qquad [1]$$

and

$$\Sigma M_A = 0 \qquad \qquad [2]$$

If the forces are concurrent (acting through a common point), no torques are present, and only the first condition need be satisfied. Another way of stating this condition is, the vector sum of all of the forces must equal zero, or the polygon formed from the vectors must close. Procedure A is for this type configuration.

If the forces are not concurrent, both conditions must be satisfied. Using these two conditions, it is often possible to compute several unknown forces if a few of them are known. There is an infinite number of ways to write the second condition since it is valid for any axis perpendicular to the plane of the forces. However, if the position of the axis is judiciously chosen, the analytical solution can be simplified. This choice should be a point through which forces are acting since they will produce zero torque in this case and thus not appear in the equation. Procedure B is for this configuration.

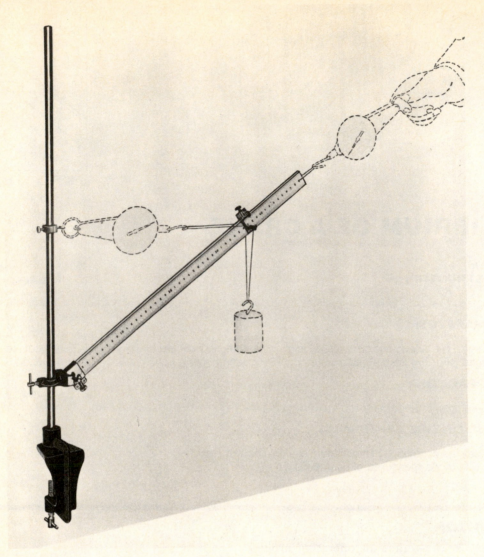

Figure 6–1 Crane Apparatus (Courtesy of Sargent–Welch Scientific Co.)

DESCRIPTION OF APPARATUS

Figures 6–1 and 6–2 show two cranes which are commercially available. Figures 6–3, 6–4, and 6–5 show several different setups for experimentally studying the crane. Figure 6–3 is used to look at the case of concurrent forces. The pivot blade at point A rests in a groove fastened to the frame. Point C is in equilibrium under the action of three forces: the vertical force due to W pulling downward, the tension T in the string pulling in the direction CB, and the outward thrust S of the boom, acting approximately parallel to the boom. The weight of the boom itself must be small in comparison to the other forces in order for the thrust to be approximately parallel to the boom, thereby allowing one to consider the forces as concurrent at point C.

Figure 6–4 is also used to study concurrent forces. For this setup the lower end of the boom is attached so that it can rotate about point A. The direction of the cord CB is varied by raising and lowering the pulley at B. The thrust is indicated by a calibrated spring built into the boom.

Figure 6–5 is the most general arrangement and is used to study coplanar nonconcurrent forces. The pivot blade at A rests in a groove attached to the frame. The horizontal and vertical components of this force are determined by the hanging weights S_y and S_x.

If an angle measuring caliper, similar to that shown in Figure 6–6, is available, hold it in contact with the members forming the angle, being careful not to disturb the

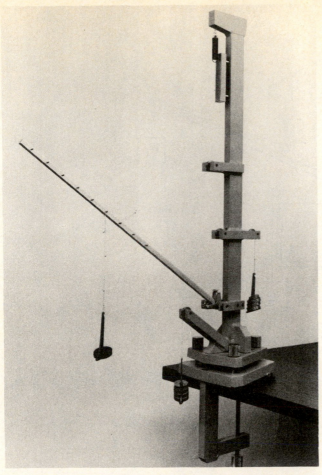

Figure 6–2 Crane Apparatus (Courtesy of Bernard O. Beck Co.).

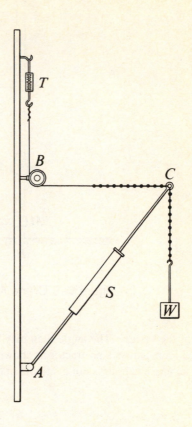

Figure 6–4 Simple Crane Compression Spring in Boom.

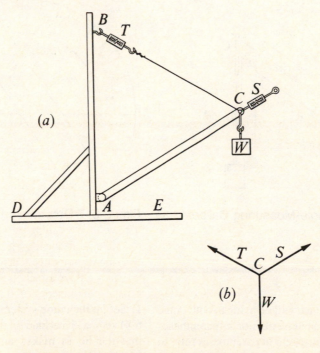

Figure 6–3 Simple Crane Rigid Boom.

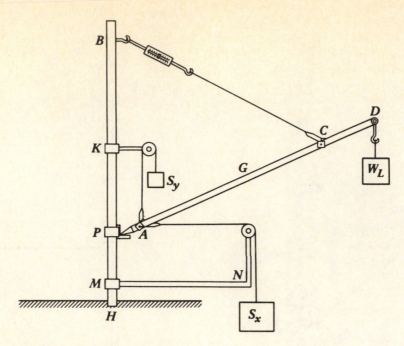

Figure 6–5 Crane Arrangement.

equilibrium of the system. The angle then may be read by placing the calipers across a protractor as shown in Figure 6–7. If such an angle caliper is not available, a fairly good estimate of the angles can be made by holding a 6-in. protractor in the proper position.

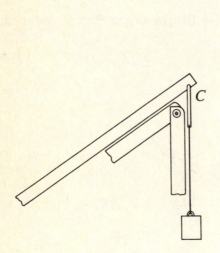

Figure 6–6 Setting Angle-Measuring Caliper.

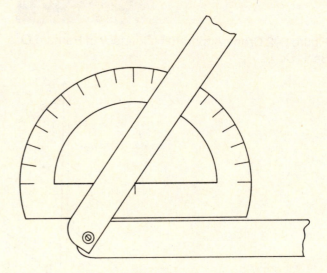

Figure 6–7 Determining Angle of Measuring Caliper.

PROCEDURE A

In the following steps, reference will be made to both crane types. Numbers, letters, or statements not in parentheses refer to Figure 6–3; those that appear in parentheses refer to Figure 6–4.

1. Set up the crane system as shown in Figure 6-3 (Figure 6-4) and, with no load on the boom, adjust the apparatus so that the boom makes an angle of 30 to 40° with the horizontal.

Read and record the no-load reading of the spring balance T. Still, with no load on the crane and with the spring balance S *parallel* to the boom, pull on the end of the boom until the pin A begins to move away from its support. (Pulling is not necessary for Figure 6-4 since the spring scale in the boom gives the force.) Read and record the no-load reading of spring balance S.

2. Now hang a 1-kg (10-kg) mass from the end of the boom and observe the approximate change in the angle at A. If it changes more than 5°, it will be desirable to adjust the length of the cord BC until the angles are within 5° of the no-load angles.

3. Measure carefully the distances AB, AC, and BC, and also the three interior angles of the triangle ABC. Use the angle measuring caliper if available. Record the values on the data form. Check the correctness of your angle measurements by finding their sum. These values which you have just observed will be called trial 1.

4. Record the value of the load W hanging on the end of the crane and the readings of the spring balances T and S while the crane is supporting the load. Pull on the end of, and parallel to, the boom (or read spring balance in boom) as directed in Step 1. Subtract the no-load readings from these readings and record the results as tension T and thrust S. In Figure 6-3 the pull S replaces the thrust and is therefore a measure of that force.

5. There might be situations where the load to be lifted by a crane is known but there are no devices in other members of the crane for measuring the tension in the cable and the thrust by the boom. Hence, suppose we consider T and S as unknown forces and compute their values from the observations which have been made. There are two methods by which this may be done: (1) a graphical solution based on the angles measured, and (2) an analytical method based on the dimensions of the components of the crane.

6. *Graphical solution:* Since point C is the point through which the forces act, it serves best as the starting point for drawing a vector diagram of forces. Represent C by some point on a large sheet of paper and, using some convenient scale, say 1 cm = 1 N, draw to scale a vector representing the force W from point C (Figure 6–8). By making use of the angles determined from your apparatus, construct vectors S and T from the lower and upper ends of W, respectively, and let them meet wherever they will at some point P. Now measure their lengths and, from your chosen scale, determine the magnitude of the forces and record as the calculated values. Record these values also on the diagram.

7. *Analytical solution:* Draw triangle ABC (Figure 6–9) to represent the directions of parts of the crane shown in Figure 6–3. These need not be drawn to scale. Write the measured lengths along these lines. Now draw from C the triangle CDE to represent the general shape of the force triangle. Triangles ABC and CDE are similar; hence, the values of T and S can be calculated by proportion using W as a known quantity. Make these calculations near the drawing for Figure 6–9 and record the results in the table.

8. In this case, the actual values of T and S are known from the observed readings of the spring balances used to measure them. Hence, if we consider these observed readings as the correct values of the forces, the validity of our two methods of computation can be checked by computing the percent discrepancy in each case.

9. If time permits, arrange the crane boom in a horizontal position and make the same measurements as in the above step. However, as an analytical method of solution for values of T and S, resolve the forces at point C into horizontal and vertical components and apply the first condition of equilibrium, namely, $\Sigma F_x = 0$ and $\Sigma F_y = 0$. Record these data as trial 2.

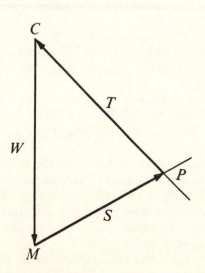

Figure 6–8 Force Triangle.

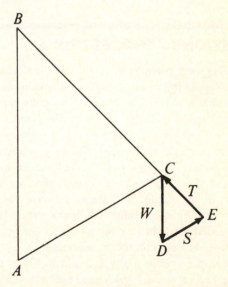

Figure 6–9 Force Triangle Proportional to Distance.

PROCEDURE B

1. Weigh the boom together with any wire hooks and string that may be attached to it. Also, locate the center of gravity by balancing the boom across a small rod and record its distance from the pivot point (or axis) at the lower end of the boom. Measure and record the distances from the pivot point to the other end of the boom D and to C where the tie rope is attached. Record the zero-load reading of the spring balance.

2. Assemble the boom as illustrated in Figure 6–5 or as indicated by your instructor. For the most general arrangement, the tie rope *should not be* attached either at the center of gravity or at the upper end where the load is suspended. The pulleys used for the forces at the lower end should be arranged so that the strings pull vertically and horizontally on the pivot point at A.

3. Suspend a load of about 10 N from the upper end of the boom and then suspend masses from the strings attached to A and passing over the pulleys until the end of the boom is just free from its support. The boom should now be suspended in space under the action of the forces at points A, C, D, and G (the center of gravity). Record all of the forces and then measure and record the inside (acute) angles at C and D. From these angles all other angles that may be needed can be computed.

4. Draw a free-body diagram of the boom as indicated in Figure 6–10, showing the directions of all of the forces acting on it. Indicate the lengths and angles which you have measured along with the value of the load suspended at D and the weight of the boom as a force at G. Label the vertical and horizontal forces at A as S_y and S_x, respectively, and the tension in the tie rope as T, but do not

indicate any values for them on the diagram. These are to be considered as unknown forces and will be computed by applying the conditions of equilibrium for a rigid body.

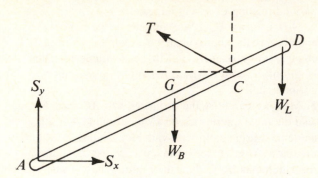

Figure 6–10 Free-Body Diagram of Forces on Crane Boom.

5. *Calculations from the conditions of equilibrium:* After making a judicious choice of axis from which to compute the lever arms, set up a torque equation in harmony with the *second condition* of equilibrium and compute the tension T in the tie rope. Then, by using this calculated value of T, set up equations in harmony with the *first condition* of equilibrium for vertical and horizontal directions and compute the components of the thrust at the pivot point A. Show these equations and their solutions in a neat form as part of your report.

6. Compare the above calculated values of T, S_y, and S_x with the experimentally measured forces by finding the percent difference.

QUESTIONS—PROCEDURE A

1. Sketch a force triangle on your report similar to that in Figure 6–8, label W as a known force (the value used in the experiment), and consider T and S as unknown forces. Label the angles with the values you measured and, by applying the sine law (see Appendix A), compute the forces T and S as a check on the other methods you have used.

2. How much uncertainty would you estimate to be in your readings of the spring balances? How much uncertainty, in degrees, in the angle measurements? Are the percent discrepancies in your computed forces greater than the percent uncertainties in the readings? If so, can you suggest other sources of error in the measurements?

3. By using the interior angles at C and M (Figure 6–8) and the computed values of T and S, find the sum of the vertical components of these two forces and compare with the value of W. These values are $T \cos \theta_1$ and $S \cos \theta_2$, where θ_1 and

θ_2 are the interior angles at C and M, respectively. Why should this sum be equal to W? *Hint:* Review the conditions for equilibrium.

4. By using the relations $T \sin \theta_1$ and $S \sin \theta_2$, find the horizontal components of T and S. What relation is indicated by their values? Why would you expect this relation?

5. If the angle between the boom and the horizontal is increased, how will the value of T be affected? If you are not sure, try it out.

6. What characteristic must a force polygon exhibit if the forces are in equilibrium? Do your data verify this condition of equilibrium? Explain.

7. Would you obtain the same reading of spring balance T if the balance were supported with the movable hook upward? Why?

8. What must be the position of the light boom (Figure 6-3) in order for the vertical component of T to balance exactly the load W. What balances the horizontal component of T in this case?

9. Suppose an actual industrial crane with the same angular relations as your experimental crane was being used to lift a 4000-lb automobile. Compute the forces T and S in the cable and the boom, respectively.

10. How does the weight of the boom in your equipment compare with the magnitude of the forces acting at the end of the boom? About what portion of this weight is pulling down at the end of the boom C. Do your data indicate that the subtraction of the no-load reading of the spring balances compensated for this force? How?

QUESTIONS—PROCEDURE B

1. Would you obtain the same reading of spring balance T if the balance were supported with the movable hook upward? Why?

2. In the torque method, some of the lever arms were zero in the equation used. What, in the apparatus arrangement makes these lever arms zero?

3. From the results, what conclusions can you draw concerning the validity of the two conditions for equilibrium of a rigid body, such as a crane boom, under the action of several coplanar forces? Refer to your data in justification of your answer.

4. Did the observed and calculated values of the thrust S as well as those of T check? If not, can you give a reason? (Reexamine the free-body diagram, Figure 6–10.)

5. Compute the actual magnitude and direction of the thrust S at the pivot pin by finding the resultant of the components found in Step 5 of the procedure.

6. One might find the weight of the boom by using either a laboratory balance or the spring balance. How will the accuracy of the comparison of the computed and observed values of the tension T be affected by the choice of weighing method used for the boom? Explain.

7. You probably found the stability of the setup to be quite sensitive to the proper alignment of the forces. If the forces are not coplanar, what effect would you expect this to have on the stability? Did any of your observations indicate such an effect? Explain.

8. Let us suppose that each of the computed forces differed from the corresponding observed forces by 0.5 N. Compute the percent difference in each case and compare with the percent difference obtained in Step 6 of the procedure. What percent uncertainty do you think might really exist in the observed values of T, S_x, and S_y? Comment on your reasoning.

9. If the forces S_x and S_y, as you used them, were removed from the boom at the point attached, would the system still remain in equilibrium? Explain the reason for your answer.

10. If the boom were raised by making the string shorter between point C and the spring balance at B, which forces acting on the system would change in value? If not all forces, why not?

NAME *(Observer)* _____ Date _____

(Partner) _____ Course _____

RECORD OF DATA AND RESULTS

Experiment 6—Equilibrium of a Crane—Procedure A

OBSERVED DATA

Reading of spring balances with load: T _____ S _____

No-load reading of spring balances: T_0 _____ S_0 _____

Trial	Length of BC	Length of AB	Length of AC	Angle A	Angle B	Angle C	Load W	Tension T	Thrust S
1									
2									

TABULATED RESULTS

Method	T, Observed	T, Calculated	Percent discrepancy	S, Observed	S, Calculated	Percent discrepancy
Graphical trial 1						
Analytical trial 1						
Graphical trial 2						
Analytical trial 2						

RECORD OF DATA AND RESULTS
Experiment 6—Equilibrium of a Crane—Procedure B
OBSERVED DATA

Reading of spring balance with load: T_1 _____

Zero reading of spring balance: T_1 _____

Length measurements from A	Position of center of gravity: AG	
	Effective length of boom: AD	
	Distance to tie-rope: AC	
Angle measurements	Interior angle at C	
	Interior angle at D	
Force values observed from crane setup	Weight of boom	
	Load suspended at D	
	Corrected tension T	
	Vertical thrust: S_y	
	Horizontal thrust: S_x	
Force values calculated	Comparison: Calc. *vs* Obs.	Percent difference
	Tension $T =$	
	Thrust $S_y =$	
	Thrust $S_x =$	

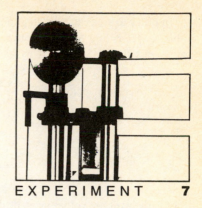

UNIFORMLY ACCELERATED MOTION

NOTE TO INSTRUCTOR:

Select either Procedure A or Procedure B depending on equipment available.

SPECIAL APPARATUS:

Graph paper, ruler, protractor.

GENERAL APPARATUS:

See Description of Apparatus.

THE PURPOSE OF THIS EXPERIMENT

is to study the nature of accelerated motion by observing the position versus time path of a uniformly accelerated object.

INTRODUCTION

When the velocity of an object is changing it is being accelerated. The average acceleration is defined as

$$\text{average acceleration} = \frac{\text{change in velocity}}{\text{time for change}}$$

The standard symbol for change is a $\triangle$ which means the previous definition can be written as

$$\bar{\mathbf{a}} = \frac{\triangle \mathbf{v}}{\triangle t} = \frac{\mathbf{v}_f - \mathbf{v}_i}{t_f - t_i} \qquad [1]$$

where the bar over the **a** indicates average value and the subscripts i and f indicate initial and final, respectively. Recall that displacement, velocity, and acceleration are vector quantities. In this discussion it is assumed that the object is constrained to move along a line. While vector notation is not used in the equations it should be remembered that even for this case these three quantities are vectors and appropriate signs must be associated with each one. If the velocity is not changing the average acceleration is zero. To determine the average acceleration over a time interval it is necessary to measure the velocity at two points and the time for the object to travel between the two points. The average acceleration over a time interval approaches the instantaneous acceleration at the center of the time interval as the time interval becomes very small.

If the acceleration is constant, the average and instantaneous accelerations are constant and equal over any time interval. This type of motion can be studied in depth using fairly simple techniques and analysis. Figure 7–1 is a graph of acceleration versus time for this case.

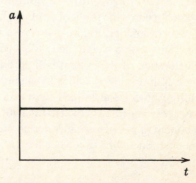

Figure 7–1 Constant Acceleration.

In order to study accelerated motion it is necessary to be able to determine the velocity as a function of time. The average velocity is defined by

$$\text{average velocity} = \frac{\text{change in displacement}}{\text{time for change}}$$

or

$$\bar{v} = \frac{\triangle s}{\triangle t} = \frac{s_f - s_i}{t_f - t_i} \qquad [2]$$

Thus, the average velocity can be determined by measuring a displacement interval and the corresponding time required to travel that interval. The average velocity over a given time interval approaches the instantaneous velocity at the center of the time interval as the time interval becomes very small.

If the acceleration is constant the change in velocity for a given time interval is always the same. This means the velocity is changing uniformly (at a constant rate) and a graph of instantaneous velocity as a function of time would look like Figure 7–2. The slope of this line is simply the acceleration, as can be seen from Equation [1]. Also note that the distance traveled in a given time interval is always given by

$$\triangle s = \bar{v} \triangle t \qquad [3]$$

For this case the average velocity over the interval shown in Figure 7–2 is given by

$$\bar{v} = \frac{v_f + v_i}{2} \qquad [4]$$

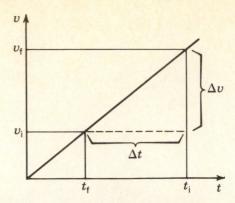

Figure 7–2 Constant Acceleration.

Thus, the distance traveled for the interval t_i to t_f is the area under the curve of Figure 7–2 provided the origin is at (0,0). This can be shown to be true for any velocity versus time curve.

Newton's second law implies that if the acceleration of a given object is constant, the force on that object is also constant. In general, forces that exist naturally in nature are not constant. In the laboratory, gravity is an exception to this. It is important to note that the earth's gravitational force on a given object is not constant but dependent on the distance the object is from the center of the earth. Since the laboratory is approximately 6400 km from the center of the earth, the gravitational force on an object varies a negligible amount from the top to the bottom of the laboratory. Thus we expect to be able to treat gravitation as a constant force in the laboratory. The magnitude of the acceleration of gravity, g, varies a small amount as you move over the surface of the earth but can be considered constant in a local region.

DESCRIPTION OF APPARATUS—PROCEDURE A

There are a number of different experimental setups that can be used to obtain the data needed to study accelerated motion. Figures 7–3 through 7–5 show three different types of laboratory setups which can be used for this type of experiment. Figure 7–3 shows a common free-fall apparatus which records the flight of an object falling freely under the action of gravity. Figure 7–4 is a variation of the free-fall apparatus with a weight and pulley added. In this case the acceleration of the system will be different from that of a free-falling object. Other available apparatuses have the falling object constrained by tracks which then allows the system to be tilted between 0 and 90°. In this case only a component of gravity actually accelerates the object along its unconstrained direction when the track is not in its vertical position.

Figure 7–5 shows a linear air track. On this system a glider floats on a cushion of air allowing it to move essentially friction free. Two techniques are commonly used to study accelerated motion on air tracks. The pulley arrangement shown in Figure 7–5 is one way. A second method is to tilt the track so that the glider moves down a friction-free inclined plane. Note that an air table with the puck constrained to move along a line could also be used. All of these setups have data collection systems which record the positions of the objects as functions of time. The most common system uses a spark timer. Spark timers are high-voltage spark devices which burn pinholes through paper at small regular time intervals. Figure 7–6 shows an example of what a spark timer record of an accelerated object would look like.

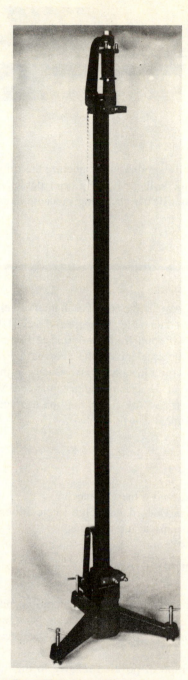

Figure 7–3 Free-Fall Apparatus (Courtesy of Central Scientific Co.).

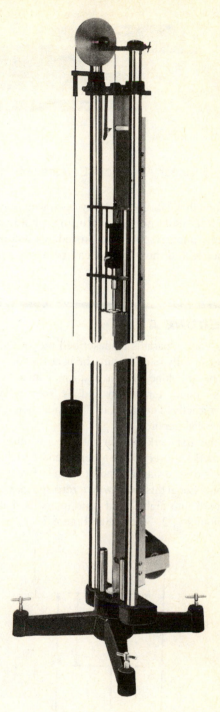

Figure 7–4 Acceleration Apparatus with Atwood Attachment (Courtesy of Sargent–Welch Scientific Co.).

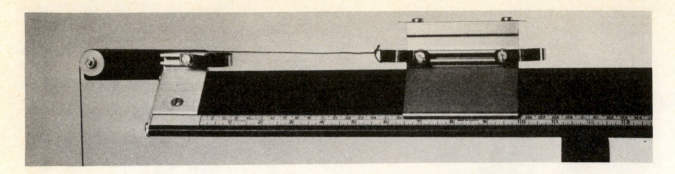

Figure 7–5 Linear Air Track (Courtesy of Pasco Scientific Co.)

There are other data collection systems which are used to study accelerated motion including photogates and stroboscopic photographs. The procedures will assume a spark timer record; however, it can be easily adapted to other records by appropriately labeling the timing points. Your instructor will give you a brief description of the apparatus you will use including precautions and special techniques.

PROCEDURE A

1. Set up your apparatus as described by your instructor. Inspect it carefully making sure the necessary leveling and calibrations are done. Try a dry run without your data collection system turned on and be sure everything is working correctly. Turn on your data collection system and make your data record.

2. Label the first mark on your data record with an ×. The first few marks are very close together and difficult to work with so it is best to start at a point three to five marks after the first one. Label this mark with a zero and then divide the entire record into eight equal time intervals. Label these interval points 1, 2, 3, etc. See Figure 7–6.

3. Record the time at the end of each interval relative to the mark 0 in your data table. Measure and record the distance of each interval point from mark 0. See Figure 7–6.

4. For constant acceleration the average velocity over a time interval is the instantaneous velocity at the center of the time interval. Determine the instantaneous velocities at times t_1, t_2, t_3, ..., by computing the average velocities about these times. Thus

$$\bar{\mathbf{v}}_1 = \frac{\mathbf{s}_2 - \mathbf{s}_0}{t_2 - t_0}, \quad \bar{\mathbf{v}}_2 = \frac{\mathbf{s}_3 - \mathbf{s}_1}{t_3 - t_1}, \quad \ldots$$

Compute and record these values.

5. Calculate and record the averge accelerations at times t_2, t_3, ..., by means of the relations

$$\bar{\mathbf{a}}_2 = \frac{\mathbf{v}_3 - \mathbf{v}_1}{t_3 - t_1}, \quad \bar{\mathbf{a}}_3 = \frac{\mathbf{v}_4 - \mathbf{v}_2}{t_4 - t_2}, \quad \ldots$$

6. Find the mean value of the acceleration and record the deviation of each value from the mean. Now find the mean deviation and record your value of the acceleration as $\mathbf{a} \pm \triangle\mathbf{a}$.

7. Make a graph of position relative to the mark 0 point versus the time relative to that point. Fit a smooth curve to the plotted points.

8. Make a graph of velocity versus time with velocity and time zero at the orgin. Do these points indicate a straight line? Draw the best line or curve fit to the plotted points. Determine the slope of the line and compare to your value of **a**. Compute the percent difference between these numbers. Which of the two values for **a** would you consider the better number?

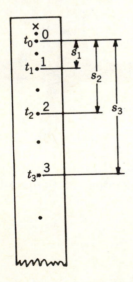

Figure 7–6 Spark Timer Data Record.

9. If you were measuring the acceleration of a free-falling object compare your results to the accepted value of g by computing the percent discrepancy. If you were not measuring a free-falling object, try to compute the theoretical acceleration of your system and compute the percent discrepancy in your experimental results.

10. The first plotted point in step 8 is at some time other than the beginning of the record. At the initial position of the record the velocity is zero. If we extrapolate the line to zero velocity it should intersect the time axis at the time corresponding to the beginning of the motion. Record this value. Look at your record and estimate the time from start to the mark 0 point. Compute the percent difference. Which of the two numbers is closer to the actual time? Why?

QUESTIONS—PROCEDURE A

1. Was there any variation in the separate values of the calculated acceleration? If so how do you account for them?

2. Suppose some mishap in obtaining the data caused one of the calculated values of the acceleration to be only one-half of the general trend of values. If all the values are used in obtaining the average, how much weight is given to this bad value? Would the graph method correct this situation? How?

3. Determine the velocity of the object at the mark 0 point. Do this from your velocity versus time graph and also analytically. Compare your results.

4. Is there any experimental evidence that indicates friction may have affected your results? Explain.

5. How did you obtain the accepted value of g for your location? Average your value for g with that of some other students. Is this value for g better than the accepted value you used? Explain.

6. Did your data collection system impair your results in any way? Explain.

7. In obtaining the slope of a curve, what are the advantages of taking the points far apart? If the difference in the abscissas of the points were increased by a factor of 5, what effect would this have on the accuracy of the slope determination?

8. What does the shape of the curve in Step 7 imply? Explain.

DESCRIPTION OF APPARATUS—PROCEDURE B

When electronic timing equipment is not available, it is possible to produce a position versus time record for a uniformly accelerated object by projecting a rolling ball across a tilted table. The Packard form of this apparatus is shown in Figure 7–7. This type of apparatus can easily be made simply by using a piece of smooth hard material such as glass or slate and arranging for it to tilt between 10 and 20°. A small incline with a guide is also required to project the ball in the horizontal direction and some type of level measuring device is helpful to insure that the plane is level in the horizontal direction.

Since the ball rolls without slipping and the rolling friction will be very small, we may assume that the horizontal component of velocity will be constant as the

Figure 7–7 Packard's Form of Inclined Plane (Courtesy of Sargent–Welch Scientific Co.).

ball marks the trajectory shown in Figure 7–7. Using elementary rigid body mechanics and the equations of projectile motion, one can show that the time required for the ball to roll down an incline plane of length *l* making an angle of θ with the horizontal is given by

$$t = \sqrt{2.8l/g \sin \theta} \qquad [5]$$

where *g* is the acceleration of gravity in the same system of units used for *l*. The ball will experience acceleration in the direction of the slope so the trajectory traced by the ball is a space plot of *y* versus *x* with both *y* and *x* being related in time. This trajectory traced out on graph paper provides a means for observing and analyzing the relation of the space coordinates to time.

PROCEDURE B

1. Place on the inclined plane a piece of coordinate paper with one set of coordinates exactly horizontal, and the left edge a little under the guard *G* (Figure 7–7). Start the ball rolling along the guide with a moderate speed so that when the ball rolls across the paper it will leave the paper near the lower right-hand corner. After an appropriate speed has been determined with a few trials, place a sheet of carbon paper over the coordinate paper and allow the ball to trace its own trajectory. If your trace is not a good one, make other trials and erase the undesirable traces. Before leaving your setup, measure the angle it makes with the horizontal. Then remove your coordinate paper to an ordinary laboratory table and make the following studies of the tracing. However, before beginning the next steps, read Section III of the Introduction on graphical representation of data and use it as a guide.

2. If you will now examine the trace of the ball's path, you will note that as the ball moved across successive squares in the *X* direction, the displacements down the plane successively increased. Since the ball travels equal distances in the *X* direction in equal time intervals, the *X* direction can serve as a convenient time axis. Pick two points on the trajectory, one near the top and one near the bottom. Carefully determine the *Y* axis distance between the two points which will be *l* in Equation [5]. Using the previously measured value of θ, determine the time for the ball to roll between the two points. Knowing the number of lines crossed, you can now determine the time required for the ball to roll horizontally across one large square of the graph paper and thus establish a time scale along the *X* direction. The time interval thus defined will be considerably less than a second and could be called a "minisecond."

3. The ball has plotted its own path, but in order to study the space–time relationships, we need to establish a set of coordinate axes. Since the ball begins its motion near the upper left corner of the graph paper, it appears convenient to use this corner as the origin. Since the purpose of projecting the ball horizontally was to establish a time unit, the displacement down the slope can be related to time if we plot time units along the *X* axis and displacement in centimeters in a downward direction on the *Y* axis, considering all numbers as positive. Label the axes and set

up your scale on each. The process of laying out the scale on the time axis is simplified if you use integral multiples of the calculated time interval for each major division of the graph paper. If we represent this time interval by t_m, then the units along the *X* axis become t_m, $2t_m$, $3t_m$, and so on. As an alternative, you can use arbitrary units for the time interval by calling each major division a minisecond. Then, the time scale on your graph is represented by simple integers 1, 2, 3, ... If the trace is not very clear, a dot placed on the curve at each integral value of *X* will be helpful in reading the values, *which should be estimated to a fraction of the smallest division on the scale*.

4. Note that the *Y* coordinate of the displacement of each point from the preceding point is the distance down the plane traveled in one time interval. This displacement divided by the time interval is the average velocity (displacement per unit time) in the *Y* direction in that time interval. Since these velocities are the average velocities, they represent the velocities at the midpoint of the time interval, rather than at the end. This is indicated on the data form where the values are to be recorded.

5. If we now find the differences in the successive average velocities, the values will represent the change in velocity in one time unit. Hence, these velocity differences divided by the time interval are the average accelerations (change in velocity per unit time) in the *Y* direction down the plane. Determine and record these values and examine the column of values obtained. What relationship do you observe? How do you account for it? Compute and record the average value for the acceleration. Also record the deviation of each value from the mean and the mean deviation. Record your acceleration as $\mathbf{a} \pm \triangle\mathbf{a}$.

6. Since acceleration is defined as the change in velocity per unit time, a graph of velocity versus time should reveal information about the acceleration. Choose the lower left-hand corner of a sheet of coordinate paper as the origin of this graph, plotting time on the *X* axis and average velocity on the *Y* axis. Spread the velocity scale out so that it will utilize two-thirds to three-fourths of the sheet of paper, and *remember that the velocity values are to be plotted at the midpoints of the time intervals*.

7. For a constant slope of the tilted plane, one would expect the acceleration to be constant, and that a plot of velocity versus time would produce a straight line, whose slope, $\triangle v/\triangle t$, is the average acceleration. Does your series of points indicate this? If so, sketch a straight line through them and determine the slope by following the instructions on slopes of graphs given in Section III of the Introduction. Compare this value with the average value computed in Step 5 above. Compute the percent difference.

8. It is quite likely that the first plotted point is at some time after the ball left the end of the guard G (Figure 7–7). Will this graph reveal to us the time the ball began its descent in the Y direction? This would be the time when the Y component of the velocity equals zero, or the intercept on the X axis. If we extrapolate our straight line until it intersects the X axis (see note on intercepts in Section III of the Introduction), we will have the time when $v_y = 0$, or the time when the ball first began to fall. Suppose we let T_g be the time the ball left the guard G. Examine the beginning portion of the trace of the ball and compare with the time found from the graph. Does your graphical analysis support actual observations and assumptions made from theory relative to the ball's trajectory?

QUESTIONS—PROCEDURE B

1. In this experiment we assumed that the vertical component of the velocity would be proportional to the time of fall. Do your experimental data support this assumption (examine the graph)? Explain.

2. How do you account for the variation in the separate values of the calculated acceleration?

3. Suppose some mishap in obtaining the data caused one of the calculated values of the acceleration to be only one-half of the general trend of values. If all the values are used in obtaining the average, how much weight (importance) is given to this bad value? Would the graph method correct this situation? How?

4. How does the time at which the ball began to fall compare with the time for zero velocity found from the extrapolated curve? If they are not equal, give a probable reason.

5. If the ball had been projected across a quite large inclined plane, do you think the horizontal component of the velocity could have been considered constant? Explain why.

6. Examine the general shape of your projectile trace. To what family of geometrical curves do you think it probably belongs?

7. In obtaining the slope of a curve, just what are the advantages in taking the points far apart? If the difference in the abscissas of the points was increased by a factor of 5, what effect would this have on the accuracy of the slope determination?

8. How might your results have been affected if the plane and guard for the ball had not been level in the X direction? Explain.

9. In the beginning, we assumed that the horizontal component of the velocity of the ball would remain constant. Do your data and results support this assumption? Cite experimental evidence to support your answer.

NAME (Observer) _____ Date _____

(Partner) _____ Course _____

RECORD OF DATA AND RESULTS

Experiment 7—Uniformly Accelerated Motion—Procedure A

$\triangle t$ of timing device _____ sec

Interval mark	Total time(sec)	Total distance(cm)	Velocity (cm/sec)	Acceleration (cm/sec^2)	Deviation from mean
0	$t_0 = 0$	$s_0 = 0$			
1	$t_1 =$	$s_1 =$	$v_1 =$		
2	$t_2 =$	$s_2 =$	$v_2 =$	$a_2 =$	
3	$t_3 =$	$s_3 =$	$v_3 =$	$a_3 =$	
4	$t_4 =$	$s_4 =$	$v_4 =$	$a_4 =$	
5	$t_5 =$	$s_5 =$	$v_5 =$	$a_5 =$	
6	$t_6 =$	$s_6 =$	$v_6 =$	$a_6 =$	
7	$t_7 =$	$s_7 =$	$v_7 =$		
8	$t_8 =$	$s_8 =$			

Mean value of **a** _____ Mean deviation _____ **a** $\pm$ $\triangle$**a** _____

Step 8

Slope _____

Percent difference _____

Step 9

Computed value of **a** if applicable _____

Percent discrepancy in **g** or **a** _____

Step 10

Time of mark $\times$ from mark 0:   From graph _____

From observation _____

Percent difference _____

NAME *(Observer)* _____ Date _____

(Partner) _____ Course _____

RECORD OF DATA AND RESULTS

Experiment 7—Uniformly Accelerated Motion—Procedure B

Angle of incline (θ) _____ **Length of incline (l)** _____

Time	Distance on Y-axis	Velocity $\dfrac{\Delta y}{\Delta t}$	Acceleration $\dfrac{\Delta v}{\Delta t}$	Deviation from mean Δa
			Average $\dfrac{\Delta v}{\Delta t}$ =	$a \pm \Delta a$ =

INFORMATION FROM GRAPH — STEPS 7 AND 8

Slope = $\dfrac{\Delta v}{\Delta t}$ =	*Avg.* $\dfrac{\Delta v}{\Delta t}$ (Step 5) =	Percent difference =
T_G from intercept =	T_G from trace =	Percent difference =

IMPULSE AND MOMENTUM

NOTE TO THE INSTRUCTOR:

It is considered desirable for the student to do both Procedures A and B if appropriate equipment is available. However, this may not be possible in a single laboratory period. Thus either two periods should be allowed or some discretion used to omit some parts of the procedures.

SPECIAL APPARATUS:

Set of weights, string, rubber band or compressible spring, pulleys and clamps, spark paper or photographic film, graph paper.

GENERAL APPARATUS:

Air track or table, with timing system (spark timer, photogates, or strobe photography), laboratory balance.

THE PURPOSE OF THIS EXPERIMENT

is to study impulse and momentum by observing the effects of different impulses.

INTRODUCTION

The most general form of Newton's second law is expressed as "Force is equal to the time rate of change of momentum." Momentum is a very useful physical construct defined to be the product of the mass of an object and its velocity at any time. Thus the momentum of an object can be expressed as

$$\mathbf{p} = m\mathbf{v} \qquad [1]$$

where the boldface type is used to denote vector quantities. If the net force on the object is constant, Newton's second law can be written as

$$\mathbf{F} = \frac{m\mathbf{v} - m\mathbf{v}_0}{t - t_0} = \frac{\mathbf{p} - \mathbf{p}_0}{t - t_0} \qquad [2]$$

where $\mathbf{v}_0$ is the velocity at the initial time t_0 and $\mathbf{v}$ is the velocity at some later time t. The majority of forces in nature are not constant so for these forces this expression is only approximately true over time intervals for which $\mathbf{F}$ undergoes very small changes.

When two bodies interact they exert forces on each other. If the force between the bodies were constant for a given period of time rising instantaneously from zero and dropping instantaneously to zero the graph of force versus time would be like that shown in Figure 8–1. A more typical force versus time plot which would be similar to the effect of a bat striking a ball is shown in Figure 8–2. The area under either of these curves is defined to be the

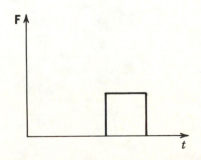

Figure 8–1 Constant Force Impulse.

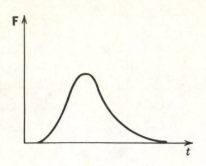

Figure 8–2 Collision Impulse.

impulse, $\mathbf{I}$. From Equation [2] and Figure 8–1 we can see that for a constant force

$$\mathbf{I} = \mathbf{F}(t - t_0) = \mathbf{p} - \mathbf{p}_0 \qquad [3]$$

Therefore, the impulse is equal to the change in momentum produced by the constant force. This can be shown to be true in general so that the impulse for a single force can always be expressed as

$$\mathbf{I} = \mathbf{p} - \mathbf{p}_0 = m\mathbf{v} - m\mathbf{v}_0, \qquad [4]$$

where $\mathbf{p}_0$ is the momentum before the force started acting and $\mathbf{p}$ is the momentum after the force becomes zero.

While impulse is not often investigated experimentally, the results of an impulse are frequently used. In this experiment the impulse of some special situations will be investigated to help make the concept of impulse more meaningful.

Once the interaction force drops to zero the momentum and thus the velocity remain constant until some other force acts on the object. Constant velocities are easily determined by measuring the time required to travel a known displacement. Then

$$\text{velocity} = \frac{\text{displacement}}{\text{time}} \qquad [5]$$

DESCRIPTION OF APPARATUS

In order to investigate the effect of a single force it is necessary to use equipment which shows very small friction effects. The linear air track or air table makes this situation possible. These devices allow gliders to travel on a cushion of air which eliminates surface contact and thus makes friction negligible.

The general experimental setup includes a track or table, air blower, gliders, miscellaneous components (pulleys, string, brackets, etc.), and a data collection system. The most common data collection system is a spark timer which is a high-voltage spark system designed to burn a pinhole in paper at fixed time intervals. This allows the experimenter

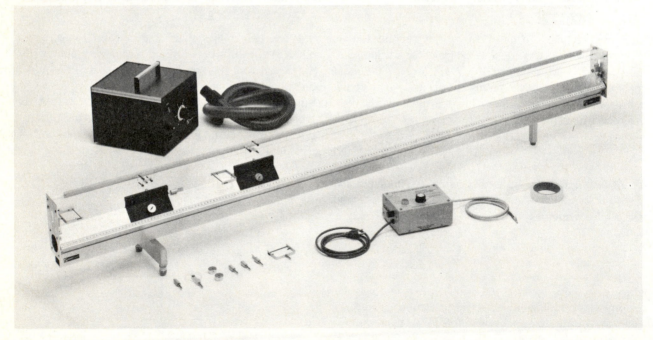

Figure 8–3 Air Track System (Courtesy of Pasco Scientific Co.).

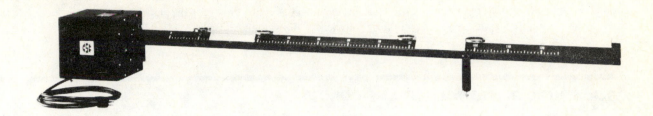

Figure 8–4 Air Track System (Courtesy of Central Scientific Co.)

to obtain a record of the glider's motion as a function of time. Other types of data collection systems exist including strobe photos and photocell gates. The strobe photos are made in a darkened room with an open camera lens and a strobe light illuminating the glider at fixed time intervals. The photocell gates consist of lamps shining in photocells which can be used to turn timers on or off when the beam is interrupted or made complete. Several air systems are shown in Figures 8–3 and 8–4. If an air table is used the puck will need to be constrained to linear motion.

Like any other experimental apparatus these systems require good techniques. Such things as leveling your system and proper alignment of pulleys or other components is important to obtaining good data. Your instructor will brief you on your specific setup before you begin.

PROCEDURE A—CONSTANT FORCE

Gravity will be used as the constant force for this procedure. The general experimental arrangement is shown in Figure 8–5.

1. Set up your air system as briefed by your instructor. Attach a string to the glider and align a pulley so that the string will pull horizontally on the glider (Figure 8–5). Arrange the string so that the weight will come to rest on the table or floor after having traveled a distance h of approximately 30 cm. Determine the weight necessary to move the distance h in approximately 5 sec. Record this force as **w** in newtons. Use a laboratory balance to determine the mass of your glider–weight system, including the string.

2. Determine a method of measuring the time, t, required for the weight to move through the height h. Use whatever is available (stopwatch, photogate, spark timer, etc.). You will also need to measure the time T for the glider to coast a distance d after the force drops to zero (hanging weight resting on table or floor). This will be used to determine the final velocity of the glider. The height h and the distance d should be carefully measured and recorded.

3. The technique used to release the glider–weight system is very important. The most common techniques use a light thread attached to the glider. The release may be accomplished by burning, cutting, or simply letting go of the thread. The timing needs to start at the instant of release. Practice a few releases and make sure everything is in order.

4. If electronic timing is used, only one or two runs are necessary. If a stopwatch is used, it is advisable to make three to five runs and use the average values. Record m, **w**, h, d, t, and T in your data table.

5. Now increase the mass of the glider system to m_2, which should be approximately $2m$, by adding weights to the glider. Then repeat Step 4 using the same force **w**, height h, and distance d. Be sure all data are recorded.

6. Make graphs of force versus time for Steps 4 and 5. Determine the area under each curve which gives the impulse, **I**, for each case. Are these impulses the same? Why?

7. Compute and record the coasting velocity of the glider after the accelerating force dropped to zero for both cases. Determine the momentum change produced by each

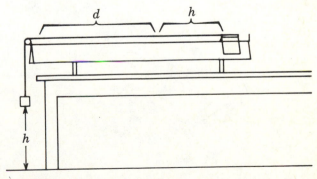

Figure 8–5 Air Track: Constant Force Setup.

impulse. Compare the momentum changes with the corresponding impulses. According to Equation [1] they should be equal. Compute the percent difference for each case. Is this within the expected accuracy of the equipment

and technique used? Explain.

8. Compute the kinetic energy of the glider during the coasting periods for the light and heavy gliders. Are they the same? Can you explain? See Question 6.

PROCEDURE B—NONCONSTANT FORCE

In this procedure the impulse due to a force which varies as a function of deformation is investigated. Electronic timing is necessary to obtain reasonable results. A rubber band, compressible spring, or metal band is required to supply the variable force. The setup should be similar to that shown for a rubber band in Figure 8–6. Note that the larger the distance over which the force acts the better the results. Thus, the rubber band or spring should not be too stiff and must allow three (preferably more) timed points while the force is acting.

1. Set up your system as briefed by your instructor. Use a laboratory balance to determine the mass of your glider. It is desired to measure the applied force as a function of time. This requires the position of the glider as a function of time while the force is acting. Also, the coasting velocity after the force drops to zero is needed. Thus, the time T required for the glider to travel a known distance d after the force is zero must be measured.

2. Refer to Step 3 of Procedure A for a description of release techniques. When all is ready release the glider but do not disturb the timing record at this point because some additional measurements are required.

3. Assemble a string, weight, and pulley system as shown in Figure 8–7 so that the deformation force can be measured as a function of time. Place weights on the string which cause the glider to rest at each time point. Record each weight (w_1, w_2, etc., in newtons) as a function of the corresponding time (t_1, t_2, etc.). Also record the distance of each point from the zero force position (x_1, x_2, etc.).

4. Measure the coasting distance d and the time T it took to travel this distance. Using these values compute the coasting velocity and record.

5. Plot a graph of force versus time. Note that you started with the force at its maximum value at $t = 0$. This means that the force versus time curve is a vertical line at $t = 0$. Draw a smooth curve through the points. By counting graph squares under the curve determine as accurately as possible the impulse **I** and record. Using the coasting velocity, compute the total momentum change the glider experienced and compare it to the impulse. Record the percent difference between these two numbers and discuss any factors which might have contributed to this difference.

6. Repeat Steps 2 through 5 using a heavier glider. Be sure the starting position is the same as before. Is the impulse the same? Why?

7. Make a graph of the force versus the displacement from the zero-force position. Plot both sets of points obtained from the two runs with different masses on the same graph. Forces which behave in this way obey Hooke's law which is given by $\mathbf{F} = -k\mathbf{x}$, where k is called the force constant. Determine k by measuring the slope of this line.

8. Compute the kinetic energy of the glider while coasting for each mass and compare. Try to explain your results. See Question 5.

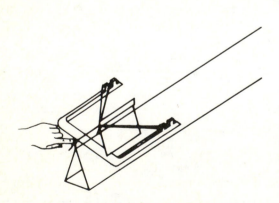

Figure 8–6 Air Track: Nonconstant Force Setup.

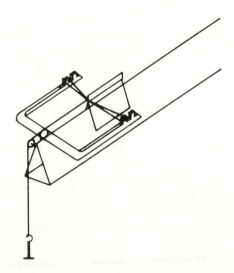

Figure 8–7 Force Measurement Setup.

QUESTIONS—PROCEDURE A

1. What factors determine the acceleration of the glider?
2. Was the acceleration the same for the light and heavy gliders? Why?
3. Was the accelerating force the same for the light and heavy gliders? Explain.
4. Was the acceleration of the glider uniform? If a graph of velocity versus time during the acceleration period were made what would you expect the plot to look like?
5. Compute the potential energy, relative to its final rest position, of the hanging weight before release. What is the potential energy change of the hanging weight during its fall?
6. What is the kinetic energy of the hanging weight just before it hits its rest position? Subtract this kinetic energy from the initial potential energy computed in Question 5 and compare this difference to the kinetic energy of the glider when coasting. Are they the same? Explain?
7. How would your results have been affected if your string had not been horizontal? Explain.

QUESTIONS—PROCEDURE B

1. What factors determine the acceleration of the glider?
2. Was the acceleration the same for the light and heavy gliders? Why?
3. Was the accelerating force the same for the light and heavy gliders? Explain.
4. Was the acceleration of the glider uniform? If a graph of velocity versus time during the acceleration period were made what would you expect the plot to look like?
5. It can be shown that the total energy, E, stored in an elastic system obeying Hooke's law is given by

$$E = 1/2kx^2$$

where x is the maximum displacement of the system from its zero-force position. Compute this total energy value and compare to the kinetic energy of the coasting glider.
6. If friction were not negligible how would it affect your results?

RECORD OF DATA AND RESULTS

Experiment 8—Impulse and Momentum—Procedure A

OBSERVED DATA

h = _____ d = _____ w = _____

Mass	Time	Trial 1	Trial 2	Trial 3	Trial 4	Trial 5	Mean
m_1	t						
	T						
m_2	t						
	T						

CALCULATED RESULTS

I_1 (area under curve) = _____ I_2 (area under curve) = _____

v_1 (coasting) = _____ v_2 (coasting) = _____

P_1 (mv_1) = _____ P_2 (m_2v_2) = _____

Percent diff. (I_1 vs P_1) = _____ Percent diff. (I_2 vs P_2) = _____

KE_1 = $1/2\ m_1v_1^2$ = _____ KE_2 = $1/2\ m_2v_2^2$ = _____

NAME (Observer) _____ Date _____

(Partner) _____ Course _____

RECORD OF DATA AND RESULTS
Experiment 8—Impulse and Momentum—Procedure B

OBSERVED DATA

Light glider			Heavy glider		
$m_1 = $ _____ $d = $ _____ $T = $ _____			$m_2 = $ _____ $d = $ _____ $T = $ _____		
t	F	x	t	F	x
0			0		

CALCULATED RESULTS

I_1 (area under curve) = _____ I_2 (area under curve) = _____

v_1 (coasting) = _____ v_2 (coasting) = _____

P_1 (mv_1) = _____ P_2 (mv_2) = _____

Percent diff. (I_1 vs P_1) = _____   Percent diff. (I_2 vs P_2) = _____

$k = $ _____

$KE_1 = 1/2\ mv_1^2 = $ _____ $KE_2 = 1/2\ m_2v_2^2 = $ _____

$E_1 = 1/2\ kx^2 = $ _____ $E_2 = 1/2\ kx^2 = $ _____

Percent diff. (E_1 vs KE_1) = _____   Percent diff. (E_2 vs KE_2) = _____

EXPERIMENT 9

WORK, ENERGY, AND FRICTION

SPECIAL APPARATUS:

Set of slotted weights, weight hanger, car, meter stick, protractor.

GENERAL APPARATUS:

Inclined plane with attached pulley, support rod, table clamp, small rod, right angle clamp, laboratory balance, supply of twine.

THE PURPOSE OF THIS EXPERIMENT

is to use the inclined plane in making a study of the relationship of work and energy and to determine the coefficient of friction.

INTRODUCTION

Energy as applied to a mechanical device may be defined as the capacity the device has for doing work. Various types of machines have been used to convert energy from one form into another, but both experimental and theoretical considerations indicate that the total energy associated with the operating system is conserved.

A very simple type of machine which may be used to study the principle of work and the conservation of energy is the inclined plane, such as that illustrated in Figure 9–1. As a body is pulled up the incline, the amount of work done is equal to the product of the pulling force and the distance the body is moved. Since friction is involved, some of this expended energy is used to do work against the frictional forces. Another portion of energy will be used to increase the potential energy of the body being lifted up the inclined plane.

If there were no friction between the plane and the body and it were being moved at uniform speed, the same amount of force would be required to move it up the plane, down the plane, or hold it still. Suppose we call this force F. Now let the opposing force of friction be designated f, and the force to pull the body up the plane F_u. Then for uniform motion up the plane we have both the component

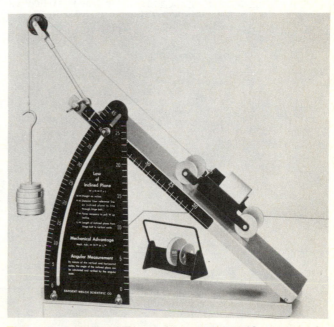

Figure 9–1 Inclined Plane Arrangement (Courtesy of Sargent–Welch Scientific Co.).

of the weight F parallel to the plane and the force of friction f to oppose the force F_u. Hence, we have

$$F_u = F + f \qquad [1]$$

and to allow the body to move down the plane with uniform motion will require a force F_d holding back equal to

$$F_d = F - f. \qquad [2]$$

From these two equations we get

$$F_u - F_d = 2f \qquad [3]$$

from which the force of friction can be computed.

If the body is lifted a height h as it moves up the plane a distance L, then we have information from which to find the following:

The work done to pull the body up $= w = F_u L$ [4a]
The work done against the forces of friction $= w_f = fL$ **[4b]**
The increase in potential energy $= PE = mgh$ [4c]

The *force of friction* is a force between the surfaces, acting parallel to the plane of the sliding surface and opposing any tendency for the surfaces to slide relative to each other. The ratio of the force of friction to the force perpendicular to the surfaces, pressing the bodies together, is called the *coefficient of friction*. When a body rolls by means of wheels instead of sliding, we are then concerned with rolling friction between the wheels and the plane and with sliding friction on the axle of the wheel. Hence, any

measurement of the coefficient of friction will involve one or both of these effects.

A very convenient method for determining the coefficient of sliding (or rolling) friction is to tilt an inclined plane (Figure 9–2) until the body on the plane will just slide (or roll) down at uniform speed. In this case, the component F of the weight of the body parallel to the incline is just balanced by the opposing force of friction f. If F_n is the component of the weight normal to the plane, the coefficient of kinetic (or sliding) friction $\mu = f/F_n$. But since the force triangle is similar to the triangle ABC, we may write

$$\mu_k = \frac{f}{F_n} = \frac{F}{F_n} = \frac{h}{b} = \tan \theta_k. \qquad [5]$$

This angle θ_k is the angle for which uniform motion occurs, and the relations in Eq. [5] are not true for any other angle.

When an attempt is made to push a body which is at rest on a flat surface, friction acts as an opposing force. This frictional force attempts to prevent the body from moving and, hence, is called *static friction*. If f_s is the force parallel to the surface needed to start a body which exerts a force of F_n normal to the surface, the coefficient of static friction, μ_s, is given by the relation

$$\mu_s = \frac{f_s}{F_n}. \qquad [6]$$

If the body is resting on an incline as in Figure 9–2, the limiting angle of static friction is the angle of incline, θ_s, required for the body to be on the *verge* of slipping. In this case, $\mu_s = \tan \theta_s$. In general, $\theta_s > \theta_k$ and thus $\mu_s > \mu_k$.

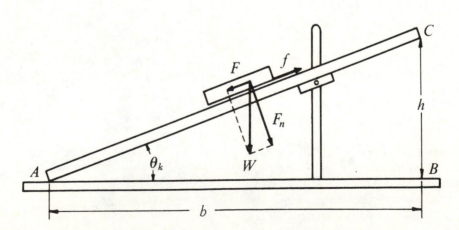

Figure 9–2 Illustration of Limiting Angle for Sliding Friction.

PROCEDURE

1. Weigh the car and arrange the inclined plane as shown in Figure 9–1 elevated at an angle of about 15° with the horizontal. If the car accelerates up the plane by the force of the weight hanger alone suspended over the pulley, place a weight of 200 gm or more in the car and then consider the total weight as the weight of the car. Also make sure that the string pulling the car is parallel to the plane.

2. Determine the force F_u required to pull the car up the incline at a slow uniform speed after the car is given a small push to get it started. Try moving it without the small push and see what happens. Then remove sufficient weights from the weight hanger to allow the car to roll down the plane at a slow uniform speed after being started. This gives you force F_d.

3. Now measure the length L of the inclined plane and the height h between the two ends. Although the arrangement of apparatus prevents you from moving the car the entire length of the plane, the work computed with reference to the entire length and height will be proportional to the work computed for any smaller lengths that might have been used.

4. Repeat the above steps for two other angles of the inclined plane, one about 30–40°, and the other about 60–70°.

5. Compute, and show as a part of your report, the work that would be required to pull the car up the entire length of the plane, the work done against friction, and the increase in the potential energy of the car. Since none of the work done on the car was used to increase its velocity, this work must have been transformed into other forms of energy. Examine your results and determine if energy was conserved in this operation. Examine the various parts of the entire setup to determine if any form of energy has been overlooked, or neglected. If so, what is its nature, and could it possibly account for any discrepancy in your conservation of energy check.

6. Detach the string from the car; with the same weight (if any) in the car, set the plane at a much smaller angle and adjust the angle until the car rolls down the plane at uniform speed after being given a small push. Do not allow the car to bump into any hard object at the bottom of the plane or to roll off the table. Measure the height h and the base b as indicated in Figure 9–2, and record. You now have information from which you can compute the coefficient of rolling friction. If you prefer, you may measure θ and use $\tan \theta$ to evaluate μ_k.

7. Remove the excess weight (if any) from the car and set it on the plane in an upside down position so that its sliding motion can be studied. Repeat Step 6 for the car sliding down the incline. This information now allows you to compute the coefficient of sliding friction between the surfaces of iron and wood.

8. Suppose we use a setup which is more directly associated with the definition of the coefficient than the inclined plane. Place the plane board flat on the table with the pulley end extending over the edge so that masses can be suspended from it. With the car on its back find the force needed to slide it at a slow uniform speed after being started. How much is the force of friction in this case? Use the definition of the coefficient of friction and compute its value from the observations you have just made for metal sliding on wood. Compare with the value found in Step 7.

9. With the board flat on the table, determine the force needed to just start the upside down car from a rest position. Record this force as the force of static friction μ_s. Now try to push the car with your hand and note the comparison of the force to start the motion and that required to maintain motion after the car starts.

10. Again with the car in the upside down position and the setup of Figure 9–2, very carefully adjust the angle so that the car is on the *verge* of slipping. Measure and record the angle θ_s and use $\mu_s = \tan \theta_s$ to compute the coefficient of static friction on the incline. Record both μ_s and θ_s.

QUESTIONS

1. What was the purpose of the small push on the car to get it in motion? Is this situation related to starting friction? Explain.

2. If the car had been allowed to accelerate up the plane, how would its motion at the top be related to its motion at the bottom? Make use of the acceleration and the time in your answer.

3. If the car had been allowed to accelerate up the plane, would any additional factors have to be included in considering the conservation of energy? If so, what and why?

4. Did the angle of incline seem to have any effect on the force of friction involved in pulling the car? If so, what? How can you account for it?

5. By making use of the definition of, and the value you obtained for, the coefficient of friction, compute the force of friction for the (loaded) car and compare with the value obtained from experiment and the application of Equation

[3]. Compare by finding the percent difference.

6. What do the results of this experiment indicate about the relative advantage of rolling, over sliding, a barrel up an incline?

7. Add Equations [1] and [2] and obtain a value for F, the force required to hold the car on the incline if there were no friction. Compute F from your values of F_u and F_d and compare with the component of the weight of the car parallel to the plane given by $W \sin \theta$.

8. At what angle would you set the inclined plane in order for the frictional force to be a minimum? At what angle for a maximum? Assume you are sliding the upside down car.

9. When one pulls a body along a level surface at uniform speed, what are the opposing forces? If the plane is elevated at an angle θ, what are the opposing forces?

10. If a body is accelerated up an inclined plane, do any opposing forces other than those mentioned in Question 9 have to be considered? If so, what are they, and what determines their magnitude?

11. The mechanism of friction is not well understood, and reliable laws are very hard to formulate. Do any of the results of your experiment indicate that this difficulty may exist? If so, what evidence indicates such?

12. What difference did you find between the forces of static friction and sliding friction? Is the effect of pushing by hand verified by experimentally measured values?

13. In driving an automobile, one is usually advised to avoid locking the wheels when a sudden stop is needed, especially if the road is slick. Explain the relationship of this advice to kinetic (sliding) and static friction.

RECORD OF DATA AND RESULTS
Experiment 9—Work, Energy, and Friction

OBSERVED DATA — WORK AND ENERGY

Weight of car and contents = _____ gm.

Angle of incline	F_u, gm	F_d, gm	L, cm	h, cm	f, gm

CALCULATED RESULTS — WORK AND ENERGY

Angle of incline	Work done to pull car up, F_uL	Work done against forces of friction, fL	Increase in potential energy, Wh	$fL + Wh$	Percent difference columns 2 and 5

THE COEFFICIENT OF FRICTION

Weight of car gm	Apparatus arrangement	Height, h cm	Base, b cm	Coefficient of friction
	Rolling on incline			μ_k =
	Sliding on incline			θ_k = μ_k =
	Sliding on flat surface	Force to pull car _____	Force of friction _____	μ_k =
	Static friction			μ_s =
	Verge of sliding on incline	θ_s =	tan θ_s =	μ_s =

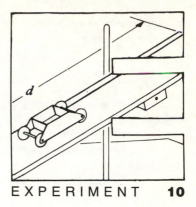

MECHANICAL ADVANTAGE AND EFFICIENCY OF SIMPLE MACHINES

SPECIAL APPARATUS:

Two pulley blocks, car, pulley cord, meter stick, set of weights, weight hanger, protractor, knife edge support, spring balance, wheel and axle.

GENERAL APPARATUS:

Table clamp, rod, small rod and clamp, inclined plane, laboratory balance, supply of twine.

THE PURPOSE OF THIS EXPERIMENT

is to learn how to measure the mechanical advantage and the efficiency of machines, and to study the conditions which determine their value.

INTRODUCTION

In many activities in offices, shops, and factories, we see machines of various kinds in operation. As we observe their operation, it becomes obvious that these machines are devices used to aid man in performing some particular task. In many cases, a machine is used to overcome (or exert) a large resisting force at one point by the application of a smaller force at another point. In other cases, a machine may be used to produce a greater velocity of some moving part than would be possible without the machine. Since machines move objects through some distance by the application of an applied force, they may also be defined as devices which aid man in doing work.

Since machines are often used to exert a large force by the application of a smaller force, we speak of the machine as having a force advantage or a mechanical advantage. The magnitude of the *actual mechanical advantage* is the ratio of the resisting force to the applied force. For a given resisting force the amount of the applied force will always depend on the kind of machine and the amount of friction present. Since friction is always present in the mechanism of machines, the work supplied to the machine is more than that done by the output part of the machine. Hence, we may define the efficiency of a machine as the ratio of the work

output to the work supplied (work input) to the machine. The work in each case is the product of the force and the corresponding distance through which it acts.

Let us assume the machine lifts a weight W through a height h by the application of a force F acting through a distance d. If there were no friction present, the work output, Wh, would equal the work input, Fd. Hence, we can write $Wh = Fd$, or

$$\frac{W}{F} = \frac{d}{h} \qquad [1]$$

Assuming no friction, the ratio W/F is the mechanical advantage under ideal conditions, and is called the *ideal mechanical advantage (IMA)*. Friction prevents us from measuring the ideal ratio W/F, but since the distance ratio is independent of friction, the ideal mechanical advantage may be obtained from the distance ratio d/h.

Most modern-day machines are quite complex in structure, but on careful analysis, it may be seen that most machines consist of combinations of two simple types of machines, the lever and the inclined plane. These two types will be studied in this experiment.

PROCEDURE

A. The Lever Type

1. You have no doubt used a pry pole to lift some heavy object too large to lift directly by hand. Set up a meter stick on a knife edge support such as shown in Figure 10–1. The forces may be applied by suspending masses from the meter stick with a string loop.

2. Suspend about 1 kg of mass as the load to be lifted and, by trying several positions for the axis (fulcrum), get a feel for the magnitudes of the required forces F by pushing with the hand for each case. What seems to determine the actual mechanical advantage for the various positions of the fulcrum? Select some one arrangement and record lever arms and forces. Then compute and record the mechanical advantage.

3. Now interchange the positions of the fulcrum and the load and compare the value of F with that in Step 2 for the same positions of forces otherwise. Is there a difference? Make a comparison by computing the actual mechanical advantage in the two cases. Use a suspended mass for F, or a spring balance if the force must act upward. Record forces and lever arms as in Step 1.

4. With the fulcrum near the left end of the bar, interchange positions of F and W from Step 3 and note the effect. What advantage do you observe for this arrangement of a lever? Measure lever arms and compute the mechanical advantage. Can you think of any common tool which utilizes this arrangement? Results are to be indicated in answers to questions.

B. The Inclined Plane

Various types of inclined planes are in use, but one of the simplest forms consists of a smooth board (Figure 10–2) supported by a small horizontal rod passing through (or under) a block attached to the underside of the board.

Suppose we let F_u be the force needed to pull the car up the incline at uniform speed, the motion being attained by adding weights to a string passing over the pulley. Then let F_d be the force (or weight) on the end of the string which will allow the car to descend uniformly. Let F represent the applied force in the absence of friction, which would be the same whether the load were ascending uniformly, descending uniformly, or standing still. If f represents the frictional forces, which always act opposite to the direction of motion, then when the car is ascending we have

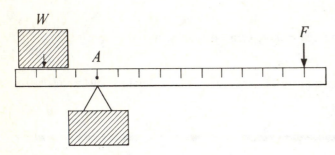

Figure 10–1 A Lever Arrangement.

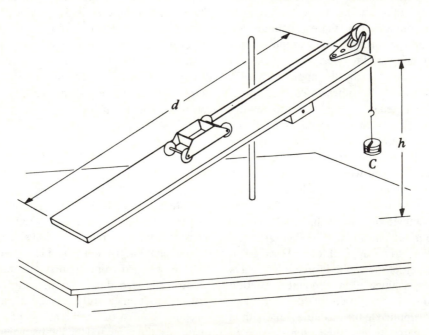

Figure 10–2 Inclined Plane as a Machine.

$$F_u = F + f$$

and when descending we have

$$F_d + f = F \quad \text{or} \quad F_d = F - f$$

If we add these two equations we get

$$F_u + F_d = 2F + f - f$$
$$\text{or}$$
$$F = (F_u + F_d)/2$$

Hence, the value of the applied force in the absence of friction is the average of F_u and F_d.

5. Adjust the inclined plane at an angle of about 30° with the horizontal and record the approximate angle. Weigh the car and attach a weight hanger to it by means of a string about the length of the inclined plane. If the weight hanger alone pulls the car up the incline, place a weight of 200 gm or more in the car and then consider this added weight as a part of the car. Be sure to include the weight of the hanger itself as a part of the pulling force.

6. By adding the proper amount of weight to the weight hanger, determine the force needed to pull the car up the incline at uniform speed and the force to let it roll down at uniform speed. Record these forces as F_u and F_d along with the load W being pulled.

7. The car probably does not move the entire length of the plane, but the magnitude of the forces F_u and F_d is independent of the distance the car moves. Suppose the car had moved the entire length of the plane, the distance d (Figure 10–2) through which the car rolled would be equal to the distance through which the applied force F_u acted, and the height h would be the distance through which the load W was lifted. Measure carefully with a meter stick the height h and the length d, and record.

8. Compute and record each of the following: the force F (as if there were no friction), the actual mechanical advantage W/F_u, the distance ratio d/h, the ideal mechanical advantage W/F, and the efficiency.

9. If time permits, repeat the above observations with the incline at a different angle. Make the angle at least 15° more than the first angle.

C. The Pulley

10. After weighing the lower pulley blocks, set up a pulley arrangement similar to that shown in Figure 10–3. (Blocks of three pulleys each may be substituted.) Examine the system to see if it has any resemblance to a lever. If so, what is the lever arm of the pulley, and how much is the mechanical advantage of one individual pulley wheel?

11. Suspend a 1-kg load from the lower pulley block and determine the amount of force F_u at the free end of the cord required to lift the load at a slow uniform speed. You may need to give it a start with your hand. Why? Note that the total load W is the sum of the lower pulley block and the suspended weight. Select some distance through which to lift the load and measure the corresponding distance required for the movement of the applied force F_u. Record these distances as h and d, respectively. Also record the total number of pulleys in both blocks as N_1.

12. From your measured values of forces and distances, compute the actual mechanical advantage, the ideal mechanical advantage, and the efficiency of the pulley system.

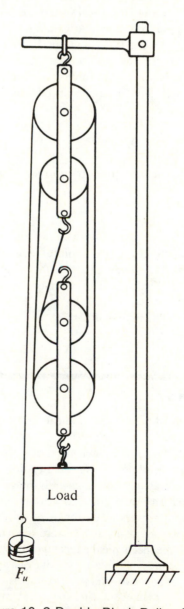

Figure 10–3 Double-Block Pulley System.

13. Now assume there is no friction and designate the force at the free end of the string as F. How much is the force on the other side of the upper pulley? Then how much force is pulling up on the left side of the lower pulley? Continue this process until you have determined the amount of force in each strand of cord actually lifting up on the lower block. What is the total of the forces needed to support the lower block? What is the ratio of the total to the applied force at the free end of the cord? Does this ratio compare with any of the quantities computed in Step 12?

14. If time permits, Steps 10–12 may be repeated with a different arrangement of pulleys, with the number of pulleys designated as N_2.

D. The Wheel and Axle

Various kinds of machines with belt and pulley arrangements, as well as gear wheel systems, employ some form of a wheel and axle device to gain a force or speed advantage. The system illustrated in Figure 10–4 is mounted on an axis at A and consists of a wheel of radius R and an axle of radius r. If the force F moves through a distance d as the load W is lifted a distance h, then, for ideal conditions, we have

$$Wh = Fd$$

However, under actual conditions friction is present and the work relation becomes

$$Wh = F_u d - fd$$

If the system makes the complete revolution, the distance through which the force acts is the circumference of the wheel, or the axle. Hence, we may write

$$W(2\pi r) = F(2\pi R),$$

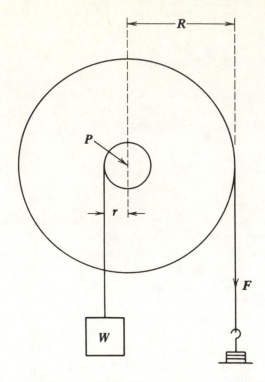

Figure 10–4 Wheel and Axle.

$$IMA = W/F = 2\pi R/2\pi r = R/r$$

However, the actual mechanical advantage is still W/F_u.

15. Apply the procedure used in the pulley system to determine the forces and distances to compute the ideal mechanical advantage and the actual mechanical advantage. Also compute and record the efficiency.

QUESTIONS

1. In what way did the lever setup indicate that the machine had a mechanical advantage? What factors seemed to influence the amount of the mechanical advantage?

2. In the interchange of positions made in going from Step 2 to Step 3, how was the mechanical advantage affected? Explain why. Make a drawing of each setup and name some everyday simple machines as examples of each.

3. When the positions of F and W were interchanged in Step 4, how was the mechanical advantage affected? Draw a diagram of the setup and explain the value of a machine of this type. Give one or two everyday examples.

4. State how the distance ratio, the ideal mechanical advantage, the actual mechanical advantage, and the efficiency were affected by a change in the angle of the inclined plane. Explain?

5. State how the distance ratio, the ideal mechanical advantage, the actual mechanical advantage, and the efficiency were affected by a change in the number of supporting cords in the pulley. Explain.

6. The percent difference between the actual mechanical advantage and the distance ratio is greatest in which machine? Give a reason for this.

7. Prove that the efficiency of a machine may be expressed as the ratio of actual mechanical advantage to distance ratio. Check the validity of this ratio by using one set of your data.

8. What are the limits (maximum and minimum values) of the distance ratio of an inclined plane? What arrangement would you use for each value?

9. Draw a diagram for a pulley arrangement whose ideal mechanical advantage is 5.

10. If the efficiency of the pulley of Question 9 is the same as for the pulley you used in this experiment, compute the applied force required to lift the same load.

11. At what angle would you set an inclined plane in order for the ideal mechanical advantage to be 3? At what angle for an IMA = 1.15? Show your work.

12. If two more pulleys were added to each of the pulley blocks, how would the distance ratio be affected? How do you think the efficiency would be affected? Why?

13. Can a pulley be classified as a form of lever? If so, explain why.

14. Describe a method for obtaining the IMA of a pulley system without actually measuring any distances or forces.

15. The principle of the wheel and axle is most closely related to which of the other devices used in this experiment? Explain the similarity.

16. If you desired a speed advantage from the wheel and axle, what change, from that shown in Figure 10–4, would you use to accomplish this advantage?

NAME *(Observer)* _____ Date _____

(Partner) _____ Course _____

RECORD OF DATA AND RESULTS
Experiment 10—Mechanical Advantage and Efficiency of Simple Machines

DATA AND RESULTS FOR THE LEVER

Procedure	Position of fulcrum	Position of load W	Position of force F	Lever arm of W	Lever arm of F	Mechanical advantage
2						
3						
4						

OBSERVED DATA FOR MACHINES

Machine	Arrangement	Load W	F_u	F_ϕ	F	d	h
Inclined plane	Angle = Angle =						
Pulley	$N_1 =$ $N_2 =$						
Wheel and axle							

TABULATED RESULTS FOR MACHINES

Machine	Arrangement	Distance ratio d/h	IMA W/F	AMA W/F_u	Efficiency $Wh/F_u d$
Inclined plane	Angle = Angle =				
Pulley	$N_1 =$ $N_2 =$				
Wheel and axle					

Show calculations for efficiency.

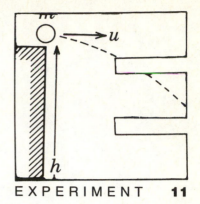

INELASTIC IMPACT AND THE VELOCITY OF A PROJECTILE

SPECIAL APPARATUS:

Two meter sticks, plumb line, set of masses, graph paper and protractor (for Procedure B).

GENERAL APPARATUS:

Ballistic pendulum, including auxiliary items, and laboratory balance.

THE PURPOSE OF THIS EXPERIMENT

is to study the momentum and energy relations involved in an inelastic impact and to investigate some properties of projectile motion.

INTRODUCTION

When two bodies collide, they exert equal and opposite forces on each other, and hence experience equal and opposite changes in momentum if no externally applied forces intervene. If such be the case, the total momentum is said to be conserved. If the bodies adhere after colliding, the collision (or impact) is said to be inelastic, and energy is dissipated in the inelastic deformation. In this type of impact, momentum is conserved but kinetic energy is not conserved.

The algebraic equations for the momentum and energy relations are derived as follows: If a mass m, moving with a velocity u, collides inelastically with a mass M, which was initially at rest, and the two bodies move off together with a common velocity v, (Figure 11-1) then in view of the conservation of momentum, we have

$$mu = (m + M)v. \qquad [1]$$

Since some of the energy is dissipated in the impact, only a fraction F of the kinetic energy of the impinging body is transferred to the system after impact. Hence, we have

$$F(1/2)mu^2 = (1/2)(m + M)v^2. \qquad [2]$$

If Equation [1] is solved for u and the resulting expression is substituted in Equation (2), it is easily shown that

$$F = \frac{m}{(m + M)}. \qquad [3]$$

This equation indicates that the fraction of the kinetic energy conserved in the collision may be predicted from only the masses of the bodies involved.

Equation [1] suggests a straightforward technique for measuring the velocity, u, of a projectile fired from a gun. The only requirement is to determine the velocity, v, after collision. If the mass M is suspended as a free-swinging

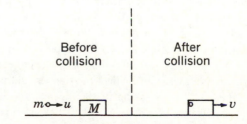

Figure 11–1 Inelastic Collision.

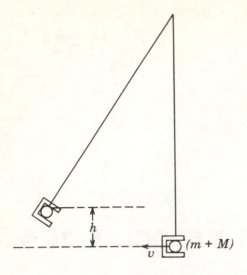

Figure 11–2 Diagram of Impact Pendulum.

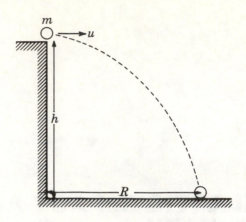

Figure 11–3 Trajectory of Horizontally Projected Object.

pendulum (Figure 11–2), the conservation of energy provides a simple way to find v. In this arrangement the kinetic energy of the combined objects immediately after impact is converted entirely to potential energy at the top of the pendulum swing. Conservation of energy then gives

$$1/2(m + M)v^2 = (m + M)gh. \quad [4]$$

Solving for v we have

$$v = \sqrt{2gh}. \quad [5]$$

Therefore, if we measure h, we can compute v and then by using Equation [1] we can find u.

Another method of measuring u if it is not too large, is to let the object of mass m be projected horizontally off of a table as shown in Figure 11–3. During free flight the object will be acted on only by the vertical force of gravity (neglecting air resistance). The horizontal velocity u will

remain constant during this flight so that the range R of the projectile is given by

$$R = ut. \quad [6]$$

The projectile will be accelerated by gravity in the vertical direction and since the initial vertical velocity is zero we have

$$h = 1/2gt^2. \quad [7]$$

Eliminating t from Equations [6] and [7] and solving for u give

$$u = R\sqrt{\frac{g}{2h}}. \quad [8]$$

Thus, by measuring R and h the value of the projected velocity can be computed.

DESCRIPTION OF APPARATUS

The apparatus to be used (Figure 11–4 or 11–5) is a combination of a ballistic pendulum and a spring gun. The pendulum consists of a large bob hollowed out to receive the projectile and is suspended by a light rod. The pendulum may be removed from its supporting yoke by unscrewing the shoulder screw at the yoke. The projectile is a brass ball which, when propelled into the pendulum bob, is caught and held by a spring in such a position that its center of gravity lies on the axis of the suspension rod. A

pointer on the side of the bob indicates the height of the center of gravity of the pendulum–ball system. When the projectile is shot into the pendulum, the latter swings in an arc and is caught at its highest point by a pawl which engages a tooth on the curved rack. The height of rise is determined by measuring the heights of the center *of gravity above the table before and after the impact. This gives a direct measurement of h.*

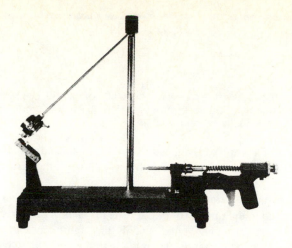

Figure 11–4 The Blackwood Ballistic Pendulum (Courtesy of Central Scientific Co.).

Figure 11–5 The Beck Ball Pendulum (Courtesy of Bernard O. Beck Co.).

PROCEDURE A—THE VELOCITY OF A PROJECTILE

1. Weigh and record the masses of both the ball and the ballistic pendulum.

2. When the pendulum has been replaced and aligned with the gun, place the ball on the firing rod and cock the gun by pulling back on the ball until the trigger engages and holds the rod against the spring. For the Beck gun (Figure 11–5), it will be necessary to hold the ball release knob down while pushing the ball ram back to the desired notch. There are three positions from which the ball can be fired. Fire the gun and record the number of the notch on the curved scale reached by the pawl. To remove the ball from the pendulum, push it out with the finger while holding up the spring catch.

3. Repeat the above procedure for four or five more trials, recording the number of the proper notch on the rack each time.

4. Compute the average of the numbers of the notches which you have recorded. Place the pendulum pawl on the notch number nearest to your computed average and, while it is in this position, measure and record the height of the center of gravity from the table. Also measure and record the corresponding height when the pendulum hangs in the prefiring vertical position.

5. By using Equation [5] and your measured heights, compute the velocity of the system immediately after collision. Then use Eq. [1] and compute the velocity of the projectile before collision.

6. The following procedure will provide another value of the projectile velocity. Set the pendulum upon the notched rack out of line of the firing direction. Fire the ball out across the floor and make note of the approximate spot where it hits the floor. Now place on this spot two sheets of paper, with a carbon between; fasten the sheets together so they will not be easily moved. The impact of the ball will leave a carbon mark on the bottom sheet. Make five trials by firing the projectile horizontally and measure the range for each shot.

7. With the ball placed on the firing rod, measure the vertical height from the bottom of the ball to the floor. Compute and record the initial velocity u of the projectile by using Equation [8].

8. Compute the percent difference in the values of the initial velocity of the projectile determined by the two methods.

9. Note that Equation [2] may be solved for F, thus giving the ratio of the two kinetic energies. Compute the kinetic energies after and before impact, and find the fraction F by computing the ratio of the two energies. Compare this value of F with the theoretical value predicted by Equation (3). (Find the percent difference between the two values.)

PROCEDURE B—RANGE VERSUS ANGLE FOR A PROJECTILE

NOTE TO INSTRUCTOR:

Due to the nature of this experiment and the confusion produced by several sets of equipment operating simultaneously, it is suggested that it be performed as a demonstration (or group) experiment. All of the students can participate in making measurements and then work individually on all remaining parts.

1. Clamp the apparatus to a support in such a way that it can be fired at various angles above the horizontal. The angles may be measured with a protractor.

2. Fire the projectile at angles of 10, 20, 30, 40, 50, 60, 70, and 80° (three trials for each) and measure the range for each angle. The range to be measured is the horizontal distance between the points of projection and impact. Due to the difference in elevation of these two points, the horizontal distance between them is not the true range of a projectile, but it will serve our purpose for this investigation.

3. Plot a graph of range values on the Y axis and angles on the X axis and draw a smooth curve through the series of points.

4. **Interpretations of the Graph.** (a) Determine the angle for maximum range. (b) Extrapolate the ends of the curve to angles of 0 and 90° and discuss the significance of the intercepts. Do you note any relation of the range at 0° to the range determined in Procedure A? Explain.

5. From the range equation $R = v_0^2 (\sin 2\theta)/g$, determine the angle for maximum range. If your experimental value differs from this, explain some possible reasons.

QUESTIONS

1. By using the suggestions given in the Introduction to this experiment, derive Equation [3].

2. A bullet of 40-gm mass is fired horizontally into a 10-kg block which is suspended as a ballistic pendulum. The impact causes the center of mass of the block to rise 10 cm. What was the velocity of the bullet at the instance of impact?

3. Did you find that the momentum was conserved in the impact? Justify your answer with data obtained in the experiment.

4. Does the force of gravity affect the horizontal component of the projectile's velocity? Why?

5. If the projectile should be fired with a higher initial (or muzzle) velocity, would the range be greater? Would the time of fall be different? Explain.

6. If the spring gun were carried to a place where $g = 900$ cm/sec^2, and fired with the same initial velocity, would the range be different? Would the time of fall be different? Explain.

7. In an actual situation, such as a military artillery gun, where both the time and range are of considerable size, what factors might alter the theoretical values of the time and range? Using the same origin for both, sketch the trajectories for the theoretical and actual flights to illustrate the comparison.

8. If you were interested in dropping an explosive-type projectile over a hill at the enemy, but not interested in the maximum range, what change in gun orientation would you make? What change would result in the shape of the trajectory?

NAME *(Observer)* _____   Date _____

        (Partner) _____   Course _____

RECORD OF DATA AND RESULTS

Experiment 11—Inelastic Impact and the Velocity of a Projectile

PROCEDURE A—THE VELOCITY OF A PROJECTILE

Trial	Notch number of pawl pos.	Range of projectile
1		
2		
3		
4		
5		
6		
Avg.		

Mass of ball _____

Mass of receiver _____

Height of center of gravity

    (a) after impact _____

    (b) before impact _____

Height of rise of C.G. _____

Horizontal projectile data

    Vertical fall distance _____

    Average range _____

        Calculated velocities of the projectile and system

            Velocity v of system after impact _____

            Velocity u of projectile before impact _____

            Velocity u of projectile from range data _____

            Percent difference between the two values of u _____

        Results of energy computations

            Kinetic energy of the system after impact _____

            Kinetic energy of the projectile before impact _____

            Fraction F of energy lost from KE ratio _____

            Fraction F of energy lost from mass ratio _____

            Percent difference between the two values of F _____

PROCEDURE B—RANGE VERSUS ANGLE FOR A PROJECTILE

Angle of gun	Average range of projectile		
		Angle for maximum range (graph)	_____
		Intercept at 0° angle	_____
10°			
20°		Intercept at 90° angle	_____
30°			
40°		Angle for max range (step 5)	_____
50°		Angle for maximum height	_____
60°		Vertical component of velocity at highest point of trajectory	_____
70°		Point on trajectory when above component is greatest	_____
80°			

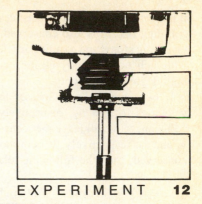

ELASTIC COLLISION—MOMENTUM AND ENERGY RELATIONS IN TWO DIMENSIONS

NOTE TO THE INSTRUCTOR:

Select either Procedure A or Procedure B depending on equipment available.

SPECIAL APPARATUS:

Procedure A—Two steel balls and one glass ball approximately 1.3 cm in diameter, large sheet of paper, two sheets of carbon paper, ruler, protractor, meter stick. Procedure B—Spark paper or photographic film, ruler, weights.

GENERAL APPARATUS:

Procedure A—Collision apparatus, plumb line, table clamp. Procedure B—Air table, pucks, spark system or strobe photo system.

THE PURPOSE OF THIS EXPERIMENT

is to study the momentum and energy relations involved in a two-dimensional elastic collision.

INTRODUCTION

When two bodies collide, they exert equal and opposite forces on each other and, hence, experience equal and opposite changes in momentum, if no externally applied forces intervene. If the only forces involved are those due to the interactions resulting from the collision, the total momentum of the colliding bodies is said to be conserved.

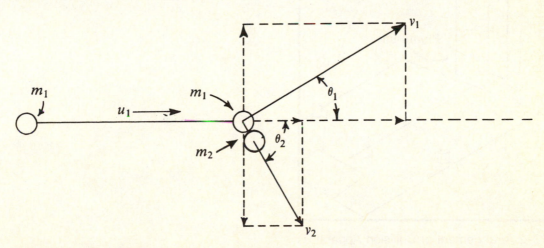

Figure 12–1 The Elastic Collision of Two Balls.

If the collision is perfectly elastic, the kinetic energy of the colliding bodies is also conserved. Many of the collisions involving atoms and atomic particles are of the elastic type. Although not perfectly elastic, steel and glass possess a high degree of elasticity.

If two steel balls of masses m_1 and m_2, traveling at velocities u_1 and u_2, respectively, collide in a head-on collision, they will leave the scene of action with velocities v_1 and v_2, respectively. If both momentum and energy are conserved, then the following relations must be true.

$$m_1u_1 + m_2u_2 = m_1v_1 + m_2v_2$$
(conservation of momentum) [1]

$$1/2m_1u_1^2 + 1/2m_2u_2^2 = 1/2m_1v_1^2 + 1/2m_2v_2^2$$
(conservation of energy) [2]

However, many collisions are not head-on. Consider the case illustrated in Figure 12–1 where m_1 strikes m_2 at a glancing angle, and after the collision the two balls leave at angles of θ_1 and θ_2, respectively, with the original direction of m_1. We now must give consideration to the conservation of momentum in two dimensions: along the line of the incident ball, and in the direction at right angles to this line. This is best done by considering the components of the momenta along these two directions. Now if we consider m_2 at rest before the collision, $u_2 = 0$, and the law of conservation of momentum then requires that

$$m_1u_1 = m_1v_1 \cos \theta_1 + m_2v_2 \cos \theta_2 \quad \text{(x direction)} \quad [3]$$
$$0 = m_1v_1 \sin \theta_1 + m_2v_2 \sin \theta_2 \quad \text{(y direction)} \quad [4]$$

In Equation [4], the numerical values of $\sin \theta_1$ and $\sin \theta_2$ have opposite algebraic signs.

Since energy is a scalar, and not a vector, Equation (2) will apply to either type of collision, in considering the conservation of energy.

DESCRIPTION OF APPARATUS—PROCEDURE A

The apparatus shown in Figure 12–2 consists of a sloping track, shaped at the lower end so as to project a ball horizontally as it leaves the track. If a second ball is mounted just beyond the end of the track of the proper height, this target ball will also be projected horizontally after collision with the incident ball. The point on the floor directly below the target ball is located by a plumb line suspended from the target ball support. Final positions on the floor of the projected balls are indicated by carbon paper placed in the area of impact. The alignment of the balls is made by an adjustable support which holds the target ball in place.

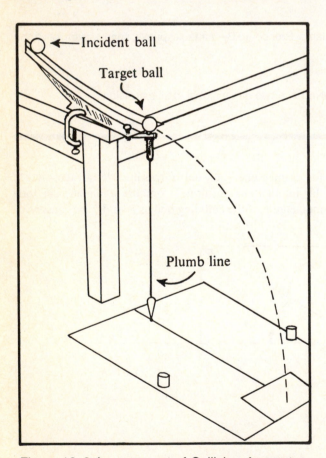

Figure 12–2 Arrangement of Collision Apparatus.

PROCEDURE A

1. Allow one of the steel balls to roll down the incline and observe the manner in which it leaves the track. Adjust the position of the target support so that the *center* of the target ball, when resting on the support, will be at a distance equal to the radius of the ball beyond the end of the track and directly in the path of the projected incident ball. The height of the support should be such that the incident ball will just clear it on leaving the track. Now with a large sheet of paper securely held to the floor, suspend the plumb line from the target support so the bob almost touches the paper, and mark the spot as point Z. This marks the position of the incident ball for all future collisions.

2. Select some position near the top of the incline from which to release the incident steel ball. As it is projected horizontally and then falls to the floor, note the approximate position of the impact. Make two or three trials to see if the ball is going to hit at nearly the same place each time. Then place a carbon paper face down over this area and cover it with a sheet of notebook paper to prevent tearing the carbon paper at impact. By releasing the steel ball at exactly the same position each time, near the top of the track, make four or five trials for the position of impact on the floor. Mark these points 1, 2, 3, and so on, draw a circle around the group, determine an average position, and label as position A for future identification. The length of the line ZA is proportional to the horizontal velocity of the incident ball at the point of collision.

3. Now let us arrange the apparatus for a head-on collision with a target ball of the same size and mass on the adjustable support. For this arrangement, it will be best to adjust the support so the target ball will be a distance of three radii from the end of the track. (Why is this arrangement desirable?) Work with the adjustment until the incident and target balls are aligned for a true head-on collision and are also at the same height at the instant of impact. Now mark the new position of the plumb bob as point X, and make four or five trials of the head-on collision, locating the final position of the projected ball with the carbon paper as before. Circle this group of points and label the average position as B. Note again that X is the starting point for the target ball and the line XB is proportional to the velocity of the target after collision. What happens to the incident ball in these cases?

4. Since the times of fall were the same in all cases and the masses of the colliding balls were equal, the horizontal distances traveled after collision are proportional to the velocities at the instant after collision. Hence, the length of a vector from Z to A is proportional to the velocity u_1 of the incident ball, and the length of the vector X to B is proportional to the velocity v_2 of the target ball after collision. Measure these vectors and record on the large sheet of paper. Also record on the data form.

As an *optional exercise* at this point, and to supply information for some of the questions at the end of experiment, replace the target steel ball with a glass ball of the same size and examine the magnitude and directions of the two balls as a result of a collision in two dimensions between a steel ball and a glass ball. What changes in the situation do you notice? Now use the steel ball as the target ball and observe what happens. (See Question 8.)

In order to determine whether momentum was conserved in this experiment it will be necessary to compare the total momenta before collision with the total momenta after collision. Since Equation [1] is designed for this purpose, the comparison can be made by substituting experimental values in this equation. Since $m_1 = m_2$, the masses will divide out, and the vectors from your data sheet can be substituted for the velocities. Show this computation in your report.

Also check the energy relationship by substituting experimental values in Equation [2].

5. At this point let us arrange the position of the target ball so as to approximate the type of collision illustrated in Figure 12–1. This is accomplished by moving the target slightly to one side, being careful to allow enough distance from the track so the incident ball will be over point Z at the instant of collision. Mark this new position of the plumb bob as point Y. Make a few preliminary trials to determine the approximate position of impact on the floor for the two balls. It would be best if the angles were between 25 and 65° with the incident direction. When all adjustments are made,

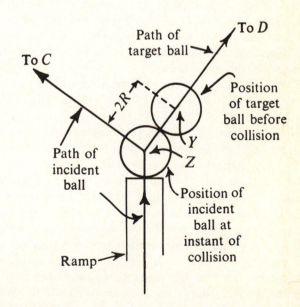

Figure 12–3 Relative Position of Two Balls at Moment of Impact.

make four or five trials of the collision and label the impact positions of the incident and target balls as C and D, respectively. From the splatter of impact points estimate the percent uncertainty in the magnitude and direction of these vectors.

6. Draw a vector from point Y to D, the final position of the target ball. Now note from Figure 12–3 that the incident ball was projected from point Z instead of Y. Hence, the vector which represents the velocity of m_1, after collision is drawn from point Z to point C. Use the idea illustrated in the diagram to locate the point Z from the position of Y which is known. Note that the distance ZY does *not* constitute a part of either vector.

7. In order to have the direction of the incident ball represented on your data sheet draw a dotted line from point Z across the data sheet parallel to and equal in length to, the vector ZA, and label this vector as ZA. Now measure and record the angles θ_1 and θ_2 which the vectors ZC and YD, respectively, make with the incident velocity vector ZA.

Consider the incident direction as the $+ X$ direction and distances and angles to the left of incident direction as $+ Y$ direction. At this point perhaps your instructor will want a smaller diagram of the data sheet, not necessarily to scale, sketched on a regular sheet of paper for your report on this experiment. From the splatter of impact points estimate the percent uncertainty in the magnitude and direction of these vectors.

8. With the magnitudes and directions of the velocity vectors known for the two-dimensional collision, it is now possible to check the momentum relations with the aid of Equations [3] and [4], and the energy relations by means of Equation [2]. Show these calculations as part of your report, and compute the percent difference between the momentum (and energy) before collision and the momentum (and energy) after collision.

9. If experimental measurements were made with balls of unequal masses, make the necessary calculations for checking the conservation of momentum and energy.

QUESTIONS—PROCEDURE A

1. What two initial experimental conditions were employed to reduce Equations [1] and [2] to the simplest possible form before actually substituting values from your data sheet?

2. What additional information would be needed to obtain the actual velocities of the projected balls rather than using vectors which are only proportional to the velocities?

3. What changes in the experimental setup would have to be made so that vectors which you measure will be numerically equal to the projected velocities of the balls?

4. In the case of the *head-on* collision, how did you arrive at a value for the vector v_1 which represents the velocity of the incident ball after collision? What actual observation suggested this value for v_1?

5. Why is it very important that the heights of the incident and target balls be the same before collision? Give two reasons.

6. For the head-on collision you were advised to set the target ball a distance of three radii beyond the end of the track. Explain why this was desirable.

7. In the case of the two-dimensional collision, the target ball did not always hit the floor at exactly the same spot. If the incident ball were released from the same height each time, how do you account for this scatter of impact points?

8. In connection with the collision between a steel ball and a glass ball of the same size, answer the following questions. (a) Which ball should be used as the target? Give an explanation based on your observations. (b) What differences did you observe in the motion of each ball after collision as compared to those for balls of equal masses?

9. Based on the observations which you have made on the effect of a collision between balls of unequal mass, make a

sketch similar to that of Figure 12–1 for each of the following cases: (a) a proton collides with a stationary nucleus (not head-on); (b) a carbon nucleus collides with a stationary proton (not head-on).

10. When using balls of different masses, explain how you would proceed in making computations to determine if momentum is conserved.

11. In checking the conservation of momentum and the conservation of energy, which seemed to be most nearly conserved in your experiment? Use your data and results as evidence. Was this to be expected? Explain.

12. In the case of the two-dimensional collision between balls of equal masses, divide out the masses in Equation [2] and examine the equation that results. Does it have any similarities to the Pythagorean theorem? If so, what should be the value of the angle between v_1 and v_2? Measure this angle on your data sheet and compare. How is this comparison related to the conservation of energy in the collision?

13. How did the Y components of the momentum for your two-dimensional collision compare? What is the sum of these two components? Explain how the value of this sum supports the law of conservation of momentum. Base your answer on experimental observations.

14. In the two-dimensional collision, compare the total kinetic energy before collision with that after collision by finding the percent difference. Do you think this percent difference is within the percent uncertainty of v^2 for any one of the velocity vectors? (*Hint:* What was the percent uncertainty in determining v from the splatter of impact points?) If so, what is your evidence? If not, what reason can you give for the discrepancy?

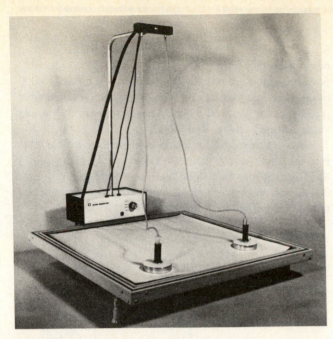

Figure 12–4 Air-Table System with Camera (Courtesy of Sargent–Welch Scientific Co.).

Figure 12–5 Precision Air Table (Courtesy of Daedalon Corp.).

DESCRIPTION OF APPARATUS—PROCEDURE B

Air tables shown in Figures 12–4 and 12–5 provide a good way to study momentum and energy relations in two dimensions. The pucks move on a practically friction free cushion of air and thus forces in the plane of motion are essentially zero during coasting motion. The two most common types of data collection systems for air tables are spark timers and strobe photos. The procedure will accommodate either type of data record.

PROCEDURE B

1. Set up your system as described by your instructor. Make sure the system is level. Practice a few collisions and learn to gauge your release technique of the incident puck for speed and direction. The speed should be such that you get a good set of data points to determine the velocities of the two pucks before and after the collisions.

2. Using identical pucks obtain a record of a collision as close to head-on as possible. Repeat for a collision which changes the direction of the incident puck between 25 and 65°.

3. Repeat Step 2 with the mass of the incident puck approximately twice that of the target puck. Then repeat Step 2 with the mass of the target puck approximately twice that of the incident puck. You should now have six records. Record the masses of both pucks for each case.

4. Before and after the collisions the velocities of the pucks are constant. These velocities can be determined by measuring the straight-line distance s the puck travels in a known time t. The velocities are then given by

$$v = \frac{s}{t}. \qquad [5]$$

Measure a free coasting distance for both pucks before and after collision and record these values along with the times it took the puck to travel between the points. Compute and record the velocities for all cases.

5. On your data records, define the X directions as the directions of the velocities of the incident pucks. Measure and record the scattering angles θ_1 and θ_2 defined in Figure 12–1 for each case. Note that θ_2 should be recorded as a negative number.

6. Using your data, compute and record both sides of Equations [3] and [4]. Is momentum conserved? Compute and record the percent differences. Did some cases appear to work better than others? Can you explain why?

7. Using Equation [2] determine the kinetic energy of the pucks before and after collision for all cases. Compute the percent differences. Were these collisions elastic? Explain.

QUESTIONS—PROCEDURE B

1. What differences did you note between the three head-on collision cases? Can you explain these observations based on conservation of momentum and energy?

2. Do your experimental results show that within experimental error the momentum was conserved in all six collisions? Cite evidence.

3. Did some cases show better momentum conservation than others? Explain why.

4. Do your experimental results show that within experimental error the energy was conserved in all six collisions? Cite evidence.

5. Did some cases show better energy conservation than others? Explain why.

6. Suppose it were possible to make the target mass 2000 times more massive than the incident mass. Describe what type of scatter you would expect to see for various limiting values of the incident direction.

7. Answer Question 6 for the case where the incident mass is 2000 times heavier than the target mass.

8. Why is it important that your table is level?

9. Did you observe any experimental evidence of forces on the pucks during coasting periods? Cite evidence.

NAME *(Observer)* _____ Date _____

(Partner) _____ Course _____

RECORD OF DATA AND RESULTS

Experiment 12—Elastic Collision—Procedure A

OBSERVED DATA

The Vector measured	Head-on collision		Two-dimensional collision	
	Magnitude	Direction	Magnitude	Direction
Vector representing u_1				
Vector representing u_2				
Vector representing v_1				
Percent uncertainty v_1				
Vector representing v_2				
Percent uncertainty v_2				

CALCULATED VALUES OF MOMENTUM AND ENERGY RELATIONS

	Percent difference
1. Head-on collision—momentum & energy—Step 4	
(a) Total momentum *before* collision—equation [1] (b) Total momentum *after* collision—equation [1]	
(c) Total energy *before* collision—equation [2] (d) Total energy *after* collision—equation [2]	
2. Two-dimensional collision values—Step 8	
(a) Total momentum *before* collision—equation [4] (b) Total momentum *after* collision—equation [4]	
(c) Total energy *before* collision—equation [2] (d) Total energy *after* collision—equation [2]	

NAME (Observer) _____ Date _____

(Partner) _____ Course _____

RECORD OF DATA AND RESULTS

Experiment 12—Elastic Collision—Procedure B

OBSERVED DATA

Case	Puck	Head-on collision					
		Mass	s	t	v	θ_1	θ_2
1	Incident						
	Target						
2	Incident						
	Target						
3	Incident						
	Target						
Case	Puck	Two-dimensional collision					
		Mass	s	t	v	θ_1	θ_2
4	Incident						
	Target						
5	Incident						
	Target						
6	Incident						
	Target						

COMPUTED DATA

Case	Momentum equation [3]			Momentum equation [4]		
	Left side	Right side	Percent diff.	Left side	Right side	Percent diff.
1						
2						
3						
4						
5						
6						

Case	Kinetic energy before collision	Kinetic energy after collision	Percent diff.
1			
2			
3			
4			
5			
6			

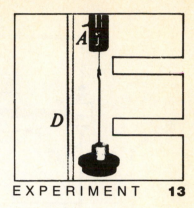

CENTRIPETAL FORCE

SPECIAL APPARATUS:

Set of weights, two springs (different force constants), meter stick, a bob (mass) supplied with hooks and a pointer, vernier caliper for Figure 13–2.

GENERAL APPARATUS:

Centripetal force apparatus (see Figure 13–1 or 13–2), two table clamps, ringstand for Procedure B (Figure 13–3). (Spring and bob listed above may be attached to the apparatus.)

THE PURPOSE OF THIS EXPERIMENT

is to determine the centripetal force required to keep a mass moving in a horizontal circle with uniform speed, and to study the effect of some of the factors upon which the force depends.

INTRODUCTION

Newton's first law of motion states that a body, once in motion, tends to continue in motion in a straight line unless acted on by some external force. If an external force is applied, Newton's second law states that the change in motion is proportional to the force and in the direction of the force. If a body is made to move in a circular path at uniform speed, it can be shown that the external force acting on the body is directed toward the center of the circle. Such an external force, which causes a moving body to travel in a circular path, is called *centripetal force*.

The magnitude of the centripetal force on a body moving at uniform speed v in a circle of radius r is given by

$$F = \frac{mv^2}{r}. \qquad [1]$$

If the moving body makes one revolution in T sec or f revolutions per second, the velocity, v, is given by

$$v = \frac{2\pi r}{T} = 2\pi rf. \qquad [2]$$

Equation [1] can then be written as

$$F = 4\pi^2 f^2 mr. \qquad [3]$$

The units of F in each of these equations will depend on the units assigned to the other quantities in the equation.

An examination of the above equation indicates that the centripetal force F is directly proportional to the square of the frequency, the mass m, and the radius r of the circle. In order to study the effect of each of these quantities on the force, only one at a time should be varied. However, the nature of the equipment available does not permit easy control of some of the factors. Hence, some of these effects may have to be measured in terms of more than one variable.

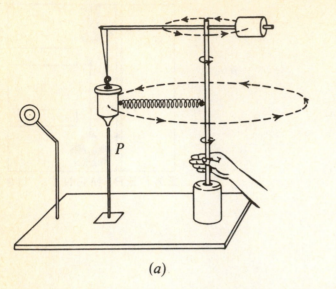

(a)

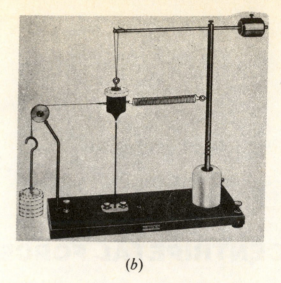

(b)

Figure 13–1 Hand Operated Centripetal Force Apparatus (Courtesy of Sargent–Welch Scientific Co.).

DESCRIPTION OF APPARATUS

Two forms of apparatus are commercially available for this experiment, the hand operated type (Figure 13–1) and the motor driven type (Figure 13–2). In either type of centripetal force equipment, the general procedure is the same: the determination of the angular rate of rotation required to hold a given mass in a circular path of predetermined radius for a given centripetal force, provided by the tension in a stretched spring.

The Hand Operated Centripetal Force Apparatus. In Figure 13–1*a*, a bob of mass *m* is suspended by a double string from the cross pin at the end of a horizontal arm which is mounted at the top of a vertical rotating shaft. The supporting arm is counterbalanced to provide smoother operation, but the position of the counterbalance is not

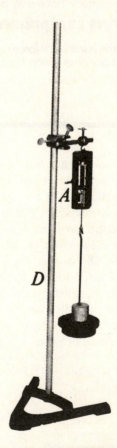

Figure 13–2 Motor Driven Centripetal Force Apparatus (Courtesy of Central Scientific Co.).

Figure 13–3 Measurement of Force Constant of Spring (Courtesy of Central Scientific Co.).

critical. An adjustable vertical pointer mounted on the base serves as a radius indicator. The mass of the bob may be varied by attaching slotted masses to it by means of a screw at the top. A pulley mounted near one end of the base is used to make a direct measurement of the force exerted by the spring when coupled to the bob (Figure 13–1b).

The Motor Driven Centripetal Force Apparatus. In the apparatus shown in Figure 13–2, a vertical shaft is rotated at variable speeds by the proper positioning of a friction ring against a motor driven disk. The mechanism at the top of this vertical shaft provides the means for measuring the centripetal force. As the speed of rotation increases, the mass *m* slides along its guides until the stretched spring exerts enough centripetal force to hold the mass in a circular path. The desired tension for the spring can be obtained by the proper position of the adjustable screw *K*.

When the rotating mass approaches the end of the guide, it touches a lever which causes a pointer arm to rise. When the pointer is opposite its index, the mass is at the desired measurable radius. The arrangement shown in Figure 13–3 provides a method for directly measuring the force exerted by the spring for the radius used.

PROCEDURE A (see Figure 13–1)

1. If the position of the pointer **P** is adjustable, adjust it for the smallest possible radius and then measure and record the radius of the circle of rotation, i.e., the distance from the tip of the pointer to the center of the vertical rotating shaft. Remove the bob and determine its mass on a laboratory balance. With the bob hanging freely from the string attached to the cross-arm, adjust the position of the cross-arm until the bob is suspended directly over the pointer, the spring being unattached.

2. By rolling the knurled section of the rotor between thumb and finger, practice rotating the bob until you are able to maintain uniform motion with the bob passing over the pointer each time. In this rotation process, it is obvious that the rotating mass and the radius of the circle are being kept constant. In order to determine the frequency of the rotation, it will be necessary to count from 50 to 100 revolutions, depending on the speed of the rotor. In any case, the counting interval should not be less than 100 sec.

In preparation for timing the revolutions, it is desirable to use the countdown technique. When you have the speed adjusted, begin counting backward by saying five, four, three, two, one, go, one, two, etc. This countdown gives warning to the person with the watch so that he can spot the exact position of the second hand when he hears the word *go*. Most of the counting can be done silently, but when you get within five revolutions of the predetermined number to be counted, warn the timekeeper by counting aloud, six, seven, eight, nine, *Stop*. By this procedure the timekeeper can estimate fractions of seconds on the watch dial.

Repeat for three additional trials, with the partners changing duties. Record all of the trials and use the average values in making computations. You now have one set of values for *f*, *m*, and *r*, from which *F* can be computed by the application of Equation [3].

3. In the course of the rotation a stretched spring was exerting a force on the bob to hold it in the circle. This force can be measured directly by measuring the force required to pull the bob out over the pointer while attached to the spring. This is accomplished by attaching a string to the outside of the bob, passing it over the pulley, and suspending the required mass from the other end (see Figure 13–1b). Determine the force exerted by this suspended mass and compare with the value computed from Equation [3] by finding the percent difference. Be sure you use the same units for both.

4. Now suppose we determine the effect of a change in mass of the bob. This is accomplished by unscrewing the nut on top of the bob and inserting a slotted mass (100 gm or more) under it. Place the open end of the slot outward and tighten the screw against the added mass. Time 50 or more revolutions and determine the average frequency from two or more trials. Which of the other quantities in Equation [3] do you think has been caused to change as a result of a change in the mass of the bob? Determine the force by calculation and by direct measurement and record.

5. Remove the added mass in order to return the bob to its original mass but move the pointer farther out to provide for a larger radius. Time 50 to 100 revolutions for two or more trials and compute the force from Equation [3], this time with a change in radius but with the original mass. Measure the force directly and compare with the computed value. What quantities changed as a result of a change in radius?

6. Replace the spring which you have been using with another of a different stiffness. Now with the same mass and radius as in Step 2, time 50 to 100 revolutions and find the average frequency from two or more trials. Compute the centripetal force and compare with the measured value.

PROCEDURE B (see Figure 13–2)

It is desirable that both partners work actively as a team on this experiment and acquaint themselves with the operation of all parts of the apparatus before attempting to record any data. Note the following particularly: how the speed of rotation is varied, how the revolution counter operates, how the tension in the spring is adjusted, and how the pointer works.

Caution:

It is dangerous to use excessive speeds, so *do not* go beyond the needed values. You may find it impossible to control the speed precisely enough to keep the pointer exactly opposite the index. If such is the case, you will need to watch the pointer continuously and keep the speed adjusted to reduce the oscillation of the pointer to a minimum.

1. Adjust the threaded collar on the spring so that the tension is near the minimum (zero on the scale), record the initial reading of the revolution counter, and decide on some time interval to use, in the range 100–120 sec.
2. Read the description of the countdown technique in Step 2 of Procedure A of this experiment, but do the countdown on the watch instead of the revolutions. As the timekeeper counts down from 55 sec to **Go** (60 sec), the partner will engage the revolution counter as he hears the word **Go**. Disengage the counter on the word **Stop** after a 5-sec count

warning from the timekeeper. Read the revolution counter and record the number of revolutions and the time.
3. Repeat the above procedure for about three more trials, with the partners changing duties. Record all of the trials and use the average values in making computations. When finished with the rotator, retract the driving disk to prevent permanent distortion in the shape of the rubber disk. If desired, other trials may be made with different tensions in the spring.
4. Remove the rotor assembly and suspend it on a support in the manner shown in Figure 13–3. Attach a weight hanger to the eye on the mass *m* and add masses until the pointer position is opposite its index. Compute the weight of the suspended masses and record as the force by direct measurement. While the masses are suspended from the rotor, measure with a vernier caliper the distance of the center of mass *m* from the center of rotation, and record as the radius of the circle.
5. Use Equation [3] and average values of the recorded data to compute the centripetal force and record as the calculated force. Compute the percent difference in the values of the centripetal force determined by the two methods.
6. Other trials, with some variation of conditions, may be made by adjusting the spring tension to other values, say 10 and 20, on the scale. Record these spring settings in the data column designated Spring No.

QUESTIONS

1. In the equipment you used, on what body does the centripetal force act? What is the direction of this force?
2. If the centripetal force on a moving body is zero, what is the nature of the motion of the body? Explain.
3. With a given radius, what effect does Equation [1] indicate that an increased velocity would have on the centripetal force required to maintain uniform motion in the same circle? How does your experiment bear this out? Explain.
4. Assume you have made one trial with some particular arrangement. If you then increase the frequency of the rotation, what changes take place in other quantities? Try it.
5. From Newton's second law $F = ma$, what quantity in Equation [1] represents the acceleration? What is the direction of this acceleration? Use the statement of Newton's second law, given in the Introduction, and the actual observations made in this experiment to explain what evidence you have from which to determine the direction of the acceleration.
6. Select one set of your data and compute the acceleration of the bob.

7. When the mass of the bob was increased with the radius being kept constant, how were the frequency and force affected? Explain the reasons for the resulting effect on each of these quantities.
8. When the radius was increased with the mass of the bob remaining the same, what quantities were found to change? What was the nature of the changes? Could you have predicted these changes? Explain why.
9. By assuming the velocity in one set of your observations to be doubled, but the radius to remain the same, compute the centripetal force and compare it with the original value. How do you account for this amount of difference?
10. Does the centripetal force acting on the rotating mass do any work? Explain.
11. In Procedure B, the spring was in a vertical position for the direct measurement and in the horizontal position while in rotation. Does this introduce an error? Explain. How would you propose to avoid this situation?
12. With the arrangement mentioned in Question 11, the weight of the spring acting on itself was neglected in the measurement of the gravitational force. Is a correction needed? Explain. (Examine the spring as you pull on it with your hand.)

RECORD OF DATA AND RESULTS

Experiment 13—Centripetal Force

Spring no. & other initial conditions			Rotation information				Computed quantities		
Spring no.	Mass of bob	Radius of circle	Trial	Number of rev. counted	Total time, sec	Number of rev. per sec	Average f	Computed F	Measured F
							Percent diff.		
							Percent diff.		
							Percent diff.		
							Percent diff.		

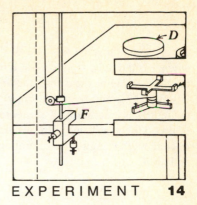

MOMENT OF INERTIA

SPECIAL APPARATUS:

Two rod pulleys, two right-angled clamps, 2-meter stick, set of slotted weights, weight hanger, vernier caliper, stop watch (or student's watch with sweep second hand).

GENERAL APPARATUS:

Rotating support with ring and disk (or cylinders), table clamp, support rod, supply of string, heavy duty laboratory balance (5-kg capacity), spirit level.

THE PURPOSE OF THIS EXPERIMENT

is to determine experimentally the moment of inertia of a body about an axis, and to compare this with the theoretical value computed from the mass and dimensions of the body. Energy relations for rotating bodies will also be studied.

INTRODUCTION

If a body conforms to some simple geometrical configuration, its moment of inertia about an axis can usually be computed from its mass and dimensions. If, however, the body is of irregular shape, the moment of inertia I must be determined experimentally.

If a torque τ acts on a body to give it an angular acceleration α, by Newton's second law for rotational motion τ and α are related by

$$\tau = I\alpha \qquad [1]$$

where I is the moment of inertia about the axis of rotation. If torque and resulting acceleration can be measured, the moment of inertia can be computed.

The apparatus in Figure 14–1 provides for measuring torque and angular acceleration. A cord wound on the drum of a metal cross passes to the lower pulley, then over the upper pulley. The driving force of mass m attached just below the upper pulley provides a torque to rotate the metal cross. The cross is mounted on a vertical axis and the cord pulls toward the lower pulley.

As mass m descends, the net force accelerating the mass is $mg - F$, where F is the tension in the cord. The equation of motion of the mass m is $mg - F = ma$, or

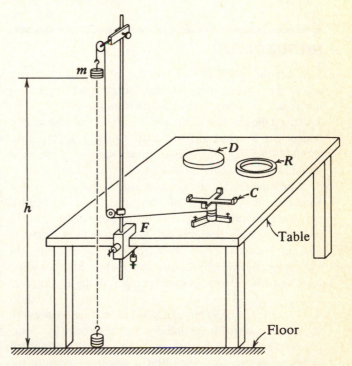

Figure 14–1 Moment of Inertia Apparatus (Courtesy of Central Scientific Co.).

$$F = mg - ma. \qquad [2]$$

The torque acting on the axle of the rotating cross is Fr, where r is the radius of the axle. If the linear acceleration of the mass m can be measured, the force F can be found from Equation [2], and then the torque can be computed. The tangential acceleration of the cord on the surface of the drum (axle) is the same as the acceleration of mass m, and is related to the angular acceleration α of the axle by the equation

$$a = r\alpha. \qquad [3]$$

Hence, when a is determined, both I and α are easily calculated, and I is thus determined experimentally.

Another method of measuring I employs the law of conservation of energy. Consider the elevated mass m and the metal cross (Figure 14–1) as a system through which energy can be transferred without loss from one part to another. As the mass m accelerates downward a distance h, the potential energy which it loses during the descent is transformed into kinetic energy associated with the moving mass m and the rotational speed gained by the metal cross. The law of conservation of energy then requires that

$$mgh = 1/2mv^2 + 1/2I\omega^2 \qquad [4]$$

where ω is the angular velocity acquired by the cross. The moment of inertia I can be computed from Equation [4].

Under the action of a constant driving force $mg - F$, the mass m will descend the distance h in a time t with constant acceleration. Hence, we have the relations

$$h = 1/2at^2 \qquad [5]$$

and

$$h = \bar{v}t = \frac{v_0 + v}{2} t \qquad [6]$$

from which both a and the final velocity v can be determined if the time of fall t is measured. When v is known, then ω can be determined from the relation

$$\omega = v/r. \qquad [7]$$

A review of the foregoing discussion will reveal that a measurement of the time t for a known mass m to descend through the height h, together with the value of the axle radius r, will furnish all of the needed information for an experimental determination of the moment of inertia of the rotating cross.

If any other body, such as the ring or disk shown in Figure 14–1, is then set on the metal cross, the same quantities can be measured for a rotating system whose moment of inertia is I'. The moment of inertia I_1 of the added body is then given by

$$I_1 = I' - I. \qquad [8]$$

PROCEDURE

1. Level the metal cross by means of a spirit level and adjust the lower pulley so that the cord will pull in a horizontal direction as it unwinds from the axle.

2. Adjust the position of the upper pulley so that the weight hanger can descend a distance of at least 1 m. The cord should be long enough so that about a meter of it can wind around the axle of the rotator.

3. When the cord is all in place with about a meter of it wound around the axle of the metal cross, suspend a sufficient amount of mass from the free end of the string to produce slow uniform motion. The weight of this mass is just the amount to counterbalance the frictional forces. In later steps, this weight remains as a part of the descending load, but it is *not* to be counted as a part of the accelerating force.

4. Now add about 50 gm of additional mass to serve as the mass m and measure the time for it to descend to the floor. Measure and record the distance to be used. (If a stopwatch is not available, you may find it easier to measure the distance traveled in some exact number of seconds as indicated on the sweep second hand of a wrist watch.)

Make at least three trials of these measurements of time and corresponding distances of fall. Record the trials and the average value of the time of fall.

5. Place one of the regularly shaped objects on the metal cross so that its center of mass is on the axis of rotation. Add a new set of small masses to determine the forces of friction for this arrangement. Then add sufficient mass m (perhaps 150 to 200 gm) to produce a sizable acceleration, and determine the time and corresponding height as before.

6. Repeat the measurements of Step 5, placing other regularly shaped objects on the rotating cross as directed by the instructor. Any object will work if symmetrical with the axis of rotation.

7. Measure and record the diameter of the axle around which the string was wound, and also the diameters of all other objects placed on the cross. If a ring was used measure both inside and outside diameters. If any smaller bodies were used at some distance away from the axis of rotation, measure the distance from the axis to the center of the bodies.

8. After you have determined the torque and angular

acceleration, use Equation [1] to compute the moment of inertia of the cross alone and also for each system used. Show computations as a part of your report. Then for at least one case, determine the linear and angular velocities of the cross, and use Equation [4] to obtain the moment of inertia from energy relations. Also show these computations on the report.

9. By the application of Equation [8], determine the moment of inertia for each of the regularly shaped bodies placed on the cross. These are experimentally determined values. Compute the moment of inertia of each of the objects used with the metal cross from the appropriate theoretically derived formula and compare with the experimentally measured value. The formulas for a solid disk and ring are $1/2MR^2$ and $1/2M(R_1^2 + R_2^2)$, respectively. For bodies displaced from the axis of rotation, the parallel axis theorem, $I = I_c + Md^2$, must be used, where I_c is the moment of inertia of the body about its center of mass and d is the distance from its center of mass to the actual axis of rotation.

QUESTIONS

1. Why is the mass determined in Step 3 not included as a part of the mass m used in the equation for the moment of inertia?

2. What other frictional forces are present besides the friction in the bearing of the metal cross?

3. Which will influence the value obtained for the moment of inertia most: an error of 5 cm in the measurement of h, or an error of 1 sec in the time measurement? (Justify your answer from the data obtained in Step 4 of the procedure.)

4. What is the origin of the additional friction when the ring (or disk) is placed on the cross?

5. In Step 4 of the procedure you attempted to counterbalance the frictional forces by adding a small mass to the end of the string. Did you observe any increase in velocity as this small mass reached the lower part of its path? If so, explain why.

6. Are the moments of inertia of the ring and disk directly proportional to their respective masses? Explain why.

7. If the cord, wrapped around the axle, were wound with more than one layer, would any measured quantities be affected? If so, which one and why?

8. In making your theoretical calculation, do you think a correction should be made for the hole in the center of the disk? Why?

9. By employing Equations [2], [3], and [5], solve Equation [1] for the moment of inertia, and thus derive an equation in terms of the quantities you actually measured.

10. By employing Equations [6] and [7], solve Equation [4] for the moment of inertia I, and thus derive an equation in terms of the quantities you actually measured.

Date _____
Course _____

RECORD OF DATA AND RESULTS
Experiment 14—Moment of Inertia

OBSERVED DATA

System used	Diameter	Mass of rotating body	Friction counter-balance	Mass on string	Average time	Height of fall
Drum and cross		✕			1. 2. 3. Avg. _____	
					1. 2. 3. Avg. _____	
					1. 2. 3. Avg. _____	

SUMMARY OF COMPUTED RESULTS

System used	Radius of rotating system r	Accel. of descending mass a	Angular accel. of system α	Torque producing rotation τ	Moment of inertia (Exper.) I_1	Moment of inertia (Calc.) I_1	Percent diff.
Metal Cross						✕	✕

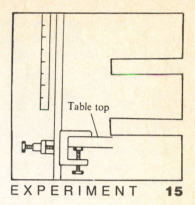

Table top

ELASTICITY AND VIBRATORY MOTION

SPECIAL APPARATUS:

Set of slotted masses with hanger (or set of hooked masses), two rulers, one 2-meter stick, steel spiral spring, brass spiral spring, unknown weight, four or five rubber bands (four lengths of size 32R rubber bands looped together in chainlike fashion work quite well; other sizes may also be used), graph paper, timer.

GENERAL APPARATUS:

Support rod, table clamp, right angle clamp, small rod, laboratory balance.

THE PURPOSE OF THIS EXPERIMENT

is to study the elastic properties of different materials and to investigate the vibratory motion of a spiral spring.

INTRODUCTION

Elasticity is that property of a body which determines the extent to which the body returns to its original size and shape after being distorted by some force. If the recovery of size and shape is complete, the body is said to be perfectly elastic. Casual observation indicates that bodies, such as a spiral spring or rubber band, stretch and become longer when a force is applied. If the distortion is not too great, these bodies appear to return to their original lengths when the distorting force is removed.

It is quite obvious that the elongation y of a stretched spring increases as the force increases. Since $y = 0$ when the force $F = 0$, one might assume that the relation of force to elongation could be expressed by an equation of the form

$$F = -ky. \qquad [1]$$

The minus sign is included because the direction of the force is opposite the direction of the displacement. If F versus y is plotted on graph paper, the result should be a straight line whose slope $= k$. The constant k depends on the shape and elastic properties of the spring and is called the force constant of the spring. Equation [1] is one form of a mathematical statement of Hooke's law, the quantitative condition for a body to be elastic.

Another common characteristic of elastic materials is that they set up vibratory motion when deformed and released. Vibratory motion is characterized by an object periodically repeating a to-and-fro path. The time required for one complete vibration is called the *period* of the motion. The number of vibrations per unit time is the *frequency* and the maximum displacement from the equilibrium position is called the *amplitude*.

A very common and important type of vibratory motion is called simple harmonic motion. It is produced by any elastic substance obeying Hooke's law. A spring vibrating with a suspended mass is an excellent example of this type of motion. The period of vibration of a mass m, suspended from a spring having a force constant k_1 is given by

$$T = 2\pi\sqrt{m/k}. \qquad [2]$$

When suspended alone, the spring stretches to some extent due to the weight of the lower portion pulling on the coils above. In reality, any part of the spring is equivalent to a mass suspended from those parts above, tending to stretch, not the entire spring, but those portions above the section in question. This fraction, called the effective mass m_1, must be considered as a portion of the total mass involved in maintaining the motion of the spring. It can be shown to be one-third of the mass of the entire spring.

If we now represent the total mass by $m + m_1$ Equation [2] takes the form

$$T = 2\pi\sqrt{(m + m_1)/k}. \qquad [3]$$

All of these quantities are easily measured except m_1. A rearrangement of Equation [3] may give a clue to a method for measuring m_1. If both sides of the equation are squared and the terms rearranged it can be written in the form

$$T^2 = \frac{4\pi^2}{k}(m) + \frac{4\pi^2}{k}(m_1). \qquad [4]$$

The terms T^2 and m are the only variables in the above equation. A close examination will reveal that the expression has taken on the slope-intercept form of a linear equation. Hence, a plot of T^2 versus m will give a straight-line graph with slope $S = 4\pi^2/k$ and intercept $I = (4\pi^2/k)(m_1)$. Since both S and I can be read from the graph, m_1 can be computed from the relation $I = S(m_1)$.

PROCEDURE

A. The Elastic Properties of Materials

1. Support the brass spring as shown in Figure 15–1 and determine how much mass suspended from the end of the spring will stretch it to three or four times its original length. This is about the maximum load that should be put on it. Select some maximum load that can be divided easily into six to eight equal units (such as 0.1-kg units) with sizes that will fit the set of masses available. The sizes of the units will depend on the stiffness of the spring.

2. Select some place near the bottom part of the system as a reference point from which to make readings. It is usually desirable to suspend the first mass unit and make the initial (or zero) readings with this mass on. This will be called the no-load reading. Parallax in making readings may be reduced by sighting along the bottom of this first mass for all of your readings. Note and record the initial reading on the scale while the first mass unit is suspended. Add one unit of mass at a time until all of the six or eight units selected have been used, taking a reading after each one is added. The suspended load should be recorded in newtons, which requires one to multiply the mass units by the acceleration of gravity. Record all these readings in a column opposite the weights used.

3. While the maximum load is still suspended, displace it a small amount and record the reading after it is released and comes to rest. Now remove the masses one unit at a time, and record the readings opposite the corresponding ones taken while adding the masses.

4. Compute the average scale reading from the two columns of scale readings and from these determine the corresponding elongation for each load used.

5. Plot a graph with loads in newtons as ordinates and elongations in meters as abscissas. The slope of this graph is the constant k in Equation [1] expressed in newtons per meter. Determine this slope and record.

6. Repeat Steps 1–5 for the steel spring, and then repeat for a chain of rubber bands. Loop about four rubber bands together to form a chain of four links and use them as you did the springs. Do you observe any shift in the scale reading while a particular mass is suspended? Observe one of the readings for a period of 4 or 5 min. Plot graphs for these two sets of data to the same scale and on the same axis on which you have already plotted data for the brass spring.

Figure 15–1 Arrangement for Studying the Elasticity and Vibration of a Spring.

B. Vibratory Motion of a Spiral Spring

NOTE TO INSTRUCTOR:

The latter part of the procedure is designed to provide the choice of method that best fits the background of the student. After completing Step 11, you may choose to do either Step 12 or 13, or to omit them and do Steps 14 and 15. The methods are different but the end result is essentially the same.

Suppose we begin our investigation of vibratory motion by observing the motion of a vibrating spring such as shown in Figure 15–1.

7. Weigh the spring and record as the mass, m_s. Measure the length of the unloaded spring as it hangs vertically and compare it with the length measured by laying it horizontally on a meter stick. Record your observations. Is there a difference in length? Why? Do you think the weight of the spring itself could be associated with this change in length when suspended vertically? If so, would you say the whole weight of the spring is involved? What pulls on the unloaded spring?

8. Now hang a 0.2-kg mass on the spring and set it into vibration in a vertical (up and down) direction. Observe all phases of the motion and write down your description of the various aspects. Consider such factors as direction, acceleration, velocity, displacement, amplitude, etc. Questions 9, 10, and 11 may serve as a guide in formulating your description of observations made. Record a brief description of each observation on the data form.

9. With a 0.2-kg mass on the spring, and using the *countdown* procedure described in Experiment 3, time about 40 vibrations and compute the period both for a steel spring and for a brass spring. Any difference in the periods is due in part to the difference in force constants resulting from differences in material. What else do you see that might have a bearing on the comparison?

10. With the 0.2-kg mass suspended from the brass spring, check the effect of the amplitude on the period by timing about 50 vibrations, first with a small amplitude (say, 2 cm) and then with a larger amplitude (say, 10 cm). Record your data and compute and record the period for each case. Does amplitude affect the period?

11. Now replace the 0.2-kg mass with a 0.1-kg mass and by simple observation determine if a change in mass affects the period. If so, time 50 or more vibrations and compute the period. Repeat for masses of 0.3, 0.4, and 0.5-kg (unless other values are specified by your instructor). Record the above value for 0.2-kg with this group. Does mass affect the period?

12. Equation [2] shows the general relation between the period of a vibrating spring and the mass involved in maintaining motion. When m is expressed in kilograms, the force constant k must be in newtons per meter. Check the validity of this equation by substituting one of the masses (say, 0.2-kg) used in Step 11 above for m and the force constant k from the graph. Compare this computed value of T with that measured by a watch. Do they check within the limits of your experimental uncertainty? If not, do you have an explanation?

13. Now review the results of Step 7 of the procedure and decide if you have maybe overlooked anything in computing T from Equation [2]. Use one of your observed values of T in Equation [2], compute the value of m needed to yield this value of T, and designate this computed mass as m_2. The differences of these two masses, $m_2 - m$, represents the effective mass of the spring m_1. If m_s is the actual mass of the spring determined in Step 7 by weighing, compute and record the ratio m_1/m_s which is the fraction of the total mass represented by the effective mass. Record in the last three blocks of the final data form.

●●●●●●●●●●●●●●●●●●●●●●●●●●●●●●●●●●

14. You have measured all of the quantities for computing the effective mass of the spring m_1 described in the Introduction and expressed in Equation [4]. Compute and record the values of T^2 from the observed values of T. Then, from the different masses used in Step 11 plot a graph of T^2 versus m, with T^2 expressed in seconds squared on the Y axis and m in kilograms on the X axis.

15. Determine the slope of your line and express it in units taken from the graph. Using the slope $S = 4\pi^2/k$, check the units on each side to see that they are equivalent. Note that the intercept I on the Y axis is the point where $m = 0$ and $T^2 = I$. Hence, at this intercept point, Equation [4] reduces to $I = 0 + S \cdot m_1$, and we have $m_1 = I/S$ as the value of the effective mass of the spring. Compute the ratio m_1/m_s, where m_s is the mass of the entire spring determined in Step 7 of the Procedure. Record the results of Steps 14 and 15 in the last section of the data form.

●●●●●●●●●●●●●●●●●●●●●●●●●●●●●●●●●●

QUESTIONS

1. What relationship between force and elongation for steel and brass is indicated by your graphs? What is the evidence?

2. Which has the higher force constant, steel or brass? To what factors do you attribute this difference? Do you have enough evidence to determine which of the two materials is the more elastic. If so, what is it?

3. Generally, did you find the graph for rubber to be any

different from that for steel and/or brass? If so, describe the difference.

4. Do you think the same equation, with a different value of k, can be used to represent the force versus elongation relationship for steel as for brass? Explain the reason for your answer.

5. Can the same form of equation be used for the force–elongation relationship for rubber as for brass? Explain. If the equation should be modified, what type of modification would you suggest?

6. Which do most of the general populace consider to be the more elastic, rubber or brass? In view of the definition of elasticity, do the results of this experiment support the above opinion? Use experimental evidence to justify your answer.

7. Would you say the distortion in the stretched spring is a streching or a bending process? Examine the spring and explain.

8. Do you think the weight of the spring plays any part in the stretching when hanging vertically? If so, is it due to the entire weight of the spring? What evidence do you have for your answers?

9. As a suspended mass on the spring vibrates:

a. At what point in the path did the velocity appear to be the greatest?

b. As the mass descended below the midpoint, was it accelerated (or decelerated)? If it was accelerated, what was the direction of the acceleration?

10. a. At what point in the path do you think the spring exerted the greatest force on the suspended mass?

b. At what point do you think the acceleration was greatest?

11. As you further observed the vibratory motion:

a. What did you notice about the relative magnitudes of the amplitudes of the successive vibrations?

b. What did your timing of the period for vibrations indicate? In other words, how did the damping of the motion affect the period?

12. In comparing the periods of springs of brass and of steel, which had the greatest frequency? In which case did the suspended mass have the greatest acceleration? On what evidence do you base your answer?

13. What effect did you find different amplitudes to have on the period? Justify your answer from your data.

14. What general effect did increasing the mass have on the period? Do you think the period and suspended mass are directly proportional? Check your answer by comparing the ratio of two of the periods with the ratio of the corresponding masses. Include one-third of the mass of the spring in your comparison. Make another check by comparing the ratio of the *squares* of two of the periods with the ratio of the corresponding masses. Now make your statement.

15. Review the paragraph on extrapolation of curves in the section on graphs in the Introduction to this book; then extrapolate your load versus elongation curve until it intersects the Y axis. If it does not go through the origin, what is the significance of the Y intercept?

16. In the plot of Equation [3], extrapolate the line to the X axis. This intercept is the point on your graph where $T^2 = 0$. Set $T^2 = 0$ in equation [3] and solve for m. Now read the value of the X intercept and compare with the value of m_1 determined in Step 15. Explain the comparison.

17. Based on your observations would you say that a stretched spring possesses potential energy? Explain. In the vibrating spring did you observe any type of energy transformation? Explain. At what point in a single vibration was the potential energy the greatest? Where was the kinetic energy the greatest?

RECORD OF DATA AND RESULTS
Experiment 15—Elasticity and Vibratory Motion

A. ELONGATION OF MATERIALS AS A FUNCTION OF THE APPLIED FORCE

Kind of material being stretched	Suspended load (N)	Scale reading		Average scale reading (m)	Elongation for each total load (m)
		Mass being added (m)	Mass being removed (m)		
Brass					
Steel					
Rubber					

SLOPES OF GRAPHS AND CONSTANTS OF SPRINGS

Kind of material	Slope of graph, S	Force constant, k

B. VIBRATORY MOTION OF A SPIRAL SPRING

MEASUREMENT OF THE LENGTH OF THE SPRING

Position of spring for measurement	Scale Readings		Length of spring	Change in length	Mass of spring m_s
	One end	Other end			
Hanging vertically					
Lying horizontally					

DESCRIPTION OF THE NATURE OF VIBRATORY MOTION

Direction	
Velocity	
Acceleration	
Amplitude	
Displacement	
Force	
Damping	

PERIOD OF A VIBRATING SPRING

Effect under consideration	Conditions imposed on spring	Suspended mass (kg)	Number of vibrations counted	Elapsed time sec	Observed period T (sec)	T^2 (sec)^2 for steps 14 & 15
Material of spring	Steel spring					
	Brass spring					
Amplitude	2 cm					
	10 cm					
Amount of vibrating mass	Load suspended from spring (step 6)					

DATA RELATED TO EFFECTIVE MASS OF SPRING

Slope of T^2 vs m graph	Intercept I (sec)^2	Effective mass, m_1	Mass of spring m_s	Mass ratio m_1/m_s

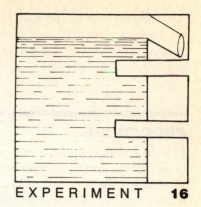

BUOYANCY OF LIQUIDS AND SPECIFIC GRAVITY

SPECIAL APPARATUS:

One small iron (or brass) cylinder, one larger aluminum cylinder, irregular solid, vernier caliper, rectangular block of wood, displacement vessel, set of weights, hydrometer, hydrometer jar, specific gravity bottle, overflow can.

GENERAL APPARATUS:

Laboratory balance, supply of unknown liquid.

THE PURPOSE OF THIS EXPERIMENT

is to study the buoyancy of liquids and to apply it to methods for finding specific gravity.

INTRODUCTION

You have no doubt heard the question, ''Which is heavier, iron or wood?'' What is the answer? If no particular volume is specified, the answer may be either iron or wood. But if one specifies that the volumes must be equal, then iron is heavier. Iron is heavier (more dense) because the atoms of iron have more mass than those in wood. The characteristic by which comparisons of weight should be made is called *density*, which is defined as the mass per unit volume. Since density is related to the kind of atoms in a substance, the density of materials is a property which characterizes that particular material. However, its numerical value depends on the units used for the mass and volume.

A term the numerical value of which does not depend on the units and also is characteristic of the kind of material is *specific gravity*. It is defined as the ratio of the weight of a given volume of a body (or substance) to the weight of an equal volume of water. Water is used as the comparison standard for the specific gravity of about all materials in the solid or liquid state.

The specific gravity of a liquid is defined in the same manner as the specific gravity of a solid; it is the ratio of the weight of a given volume of the substance to the weight of an equal volume of water. It is one of the most important

properties by which we characterize a liquid. For example, we determine the condition of charge of a storage battery by

Figure 16–1 Commercial Hydrometer.

measuring the specific gravity of the acid solution. We also determine the freezing point of the cooling agent in an automobile radiator and the cream content of milk by measuring the specific gravity. The instrument used in all these cases is some form of hydrometer. The commercial hydrometer (Figure 16–1) is a slender floating body with graduations from which specific gravity can be read directly.

PROCEDURE

1. Hold a small cylinder of some material, such as iron or brass, in one hand and a somewhat larger cylinder of aluminum in the other hand. Get a qualitative feel for their relative weights. Which is heavier and by how much, a small amount or a large amount?

2. Weigh each of the two cylinders on a laboratory balance and determine the volume of each from their dimensions as measured with a vernier caliper. Then compute and record the density of each. Make a mental note of the relative densities as compared to the relative weights.

3. Tie a string around one of the cylinders and lower it into a beaker (or can) of water and observe what happens to the pull on the string and also to the surface level of the water. Does this suggest another method for finding the volume of the cylinder? Fill an overflow can (Figure 16–2) until water slowly runs out of the spout and stops dripping. Then as the cylinder is lowered into the water, catch the overflow water in a can which has been weighed. Determine the weight of the overflow water and compute its volume from a knowledge of the density of water. If you do not know the density of water, consult your text.

4. Determine the density of the cylinder by using the volume of the displaced water and compare with that measured in Step 2. Also determine the specific gravity of the cylinder.

5. Now set a vessel of water on the circular platform attached to the laboratory balance and, by means of a string attached to the arm of the balance, suspend the cylinder in the water (totally submerged). Find its apparent weight while thus suspended and determine the difference between this weight and the original weight found in Step 2. How do you account for this difference? Does this difference (or loss) in weight correspond to any of the weighings you have already made? If so, which one?

6. Do your observations made in Step 3 reveal a possible method for finding the volume of an irregularly shaped solid? Repeat Steps 3, 4, and 5, using some irregularly shaped body and especially note the results from Step 5. Record the kind of material of the body used.

7. *Specific Gravity of a Liquid.* Weigh both the cylinder and the irregular solid in a liquid other than water, and determine the weight of the liquid displaced by each solid. Be sure the bodies and the container used are *dry* and clean. Remember that, for either solid, equal volumes of this liquid and of water have been displaced. Make use of the definition of specific gravity and calculate the specific gravity of the second liquid furnished you from the weights for each solid. Pour some of the liquid into a clean hydrometer jar and determine the correct value of the specific gravity by the reading of a calibrated hydrometer floating in it. Be sure to read the bottom of the meniscus. Compute the percent discrepancy of the average of the two values found by weighing.

8. *Specific Gravity by Pycnometer (Specific Gravity Bottle).* The specific gravity bottle (Figure 16–3) is provided with a perforated glass stopper so that when the bottle is filled with a liquid and the stopper inserted, the excess liquid will run out at the top of the stopper. This gives exactly equal volumes of different liquids when put in the bottle.

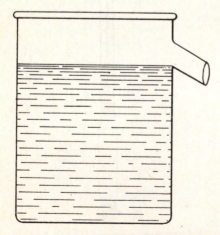

Figure 16–2 Overflow Can.

Figure 16–3 Specific Gravity Bottle.

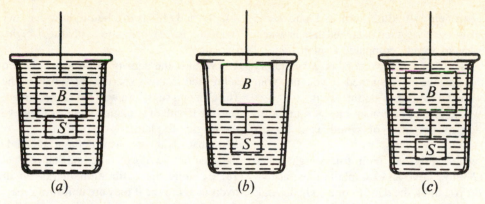

Figure 16–4 Wooden Block *B* Pulled Under Water by Sinker *S*.

Weigh an empty clean, dry pycnometer and record. Then fill it with the unknown liquid by pouring very carefully from a small beaker. See that the outside is dry, and weigh and record as weight of pycnometer + liquid *x*. Record the difference of the two weights as the weight of *V* cm³ of liquid *x*. Now empty the liquid back into the beaker and rinse the bottle with water. Next, fill with water, weigh, and record as before; also record the weight of *V* cm³ of water. From these weights of equal volumes of the two liquids, compute the specific gravity of liquid *x* and record.

9. *Specific Gravity of Solids Lighter Than Water.* Weigh a light body, such as a block of wood, in air. The block of wood must be coated with paraffin (or treated) to prevent absorption of water. In order to find the weight of water displaced by a floating body, it is necessary to tie a sinker to it to submerge it while suspended from the balance arm. If either the metal cylinder or the irregular body is heavy enough to serve as a sinker, it may be attached directly to the bottom of the block and suspended in water, as in Figure 16–4a. The weight of either of these metals in water is already known. If it seems more convenient, you may use the method indicated by Figure 16–4b and Figure 16–4c. By this method first weigh with block *B* in air and sinker *S* in water, as in *b*, then with both in water, as in *c*. For either method, let the weight of block *B* in air plus the weight of the sinker in water be W_1, and let the weight of both in water be W_2. Then $W_1 - W_2$ is equal to the weight of water displaced by body *B*, which is also the weight of a volume of water equal to the volume of the body. Inspection of Figures 16–4b and 16–4c will make this clear. Since the weight of the body in air is known, one may find the specific gravity of the block *B* by applying the definition of specific gravity. Show in your report the method of calculating the specific gravity.

QUESTIONS

1. From the observations and computations in Steps 1 and 2, which more accurately describes the nature of a particular kind of material, the density or the weight? Explain, *based on your observations*.

2. You determined the volume of a cylinder by two different methods. Which do you think is more accurate and why? Can you suggest still another method for finding the volume?

3. Can you suggest a method of measuring the specific gravity of a solid simpler than that using the displacement can? If so, describe it.

4. From the *observations made in this experiment*, write a statement of the relation of the apparent loss of weight of a body immersed in water to other weights of the body you have determined. Why does an immersed body appear to lose weight? Would this same relation exist for liquids other than water? Explain.

5. What relation did you find between the density and the specific gravity of the cylinder? Why does this relation exist? If the density were measured in pounds per cubic foot, would the relation be any different? Explain.

6. Suppose you have just weighed a vessel of water on the balance and it is still on the pan. If a block of wood is placed in the water and floats, what holds the block up? Will the vessel of water plus the floating block weigh any more than without the block? If so, how much? Explain.

7. Assume the overflow can is filled and preparations are ready for catching the overflow water. A block which will float is now carefully placed in the can, and the overflow water is caught and measured. The experiment is then repeated with gasoline in the can. Discuss the relative volumes and weights of the two overflow liquids.

8. One pound of iron and one pound of aluminum are each submerged in water and their apparent weights recorded. How do the apparent weights compare (qualitatively)? Explain.

9. One cubic centimeter of aluminum and one cubic centimeter of lead are each weighed in air and then in water. How do their losses of weight compare? Explain.

10. Suppose a vessel of water is weighed on a laboratory balance. Will the balance be disturbed if you put your finger into the water? Why? If in doubt, try it.

11. How much would a piece of timber 4 in. $\times$ 4 in. $\times$ 18 ft weigh if made of the kind of wood used in this experiment?

12. Do bodies weigh less than their true weight when immersed in air? Why? Give an example to illustrate.

13. In general, do you think the density of a body depends on its temperature? Why?

14. If a body floats in a liquid, how does the loss of weight (buoyant force) compare with the weight of the body? Explain.

15. What is the density in the cgs system of the irregular solid used in this experiment? What, in the mks system?

16. Can you suggest a way to use Archimedes' principle to find the length of a tangled bundle of wire without undoing the tangle? Explain.

17. What advantage has water as a standard in determining the specific gravity of other substances?

18. Examine the calibration marks on the commercial hydrometer to see if they are uniformly spaced. Would you expect them to be uniformly spaced? Why?

NAME *(Observer)* _____ Date _____

   (Partner) _____ Course _____

RECORD OF DATA AND RESULTS
Experiment 16—Buoyancy of Liquids and Specific Gravity

DENSITY OF A SOLID MATERIAL

Observation made	Cylinder no. 1	Cylinder no. 2
Kind of material		
Weight (grams)		
Length (cm)		
Diameter (cm)		
Volume (cm^3)		
Density (gm/cm^3)		

DISPLACEMENT DATA

Observation made	Cylinder	Irregular solid
Kind of material of solid used		
Weight of solid body used		
Weight of empty catch vessel		
Weight of vessel + overflow water		
Weight of overflow water		
Volume of overflow water		
Specific gravity of the solid		
Apparent weight of solid in water		
Loss of weight of suspended body		

Density of cylinder	Step 2 =	Step 4 =	Percent difference =

SPECIFIC GRAVITY OF A LIQUID

Kind of liquid used other than water   _____

Weight of metal cylinder in air   _____gm

Weight of irregular solid in air   _____gm

Weight of metal cylinder in other liquid   _____gm

Weight of irregular solid in other liquid   _____gm

Specific gravity of the liquid

    1. By loss of weight of metal cylinder   _____

    2. By loss of weight of irregular solid   _____

    3. By use of commercial hydrometer   _____

Percent discrepancy of the average of weighing methods   _____

SPECIFIC GRAVITY OF A LIQUID BY THE PYCNOMETER METHOD

Weight of empty pycnometer   _____gm

Weight of pycnometer full of liquid x _____gm

Weight of V cm^3 of liquid x _____gm

Weight of pycnometer full of water   _____gm

Weight of V cm^3 of water   _____gm

Specific gravity of liquid x _____gm

SPECIFIC GRAVITY OF A SOLID LIGHTER THAN WATER

Weight of wooden block in air   _____gm

Weight of wood in air + sinker in water   _____gm

Weight of wood and sinker in water   _____gm

Weight of water displaced by wooden block   _____gm

Specific gravity of wooden block   _____

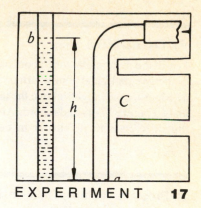

PRESSURE AND VOLUME RELATIONS FOR A GAS

SPECIAL APPARATUS:

U-tube manometer, rubber tube, two rulers, towel, graph paper.

GENERAL APPARATUS:

Boyle's law apparatus, table clamp, barometer, thermometer.

THE PURPOSE OF THIS EXPERIMENT

is to investigate the pressure—volume relations of a gas and to learn the use of a manometer as one form of a pressure gauge.

A. Description and Operation of Manometer

INTRODUCTION

The U-tube manometer shown in Figure 17–1 consists of a glass tube about 50 cm high on the open-end side and 30 cm on the other. Water is poured in at the open end until tube C is about half full. When tube C is connected to a gas cock by a rubber tube and gas is allowed to enter, the water level in C is forced down to some point a on the gas side, and up to some point b on the air side. The difference h in the heights of the water levels a and b is a measure of the effective (or gauge) pressure of the gas. Its value is given by

$$P = hdg \qquad [1]$$

where d is the density of water. The total (or absolute) gas pressure is equal to the sum of the pressure of the water column h and the atmospheric pressure in tube O. However, the gauge pressure (excess above atmospheric) is more commonly used.

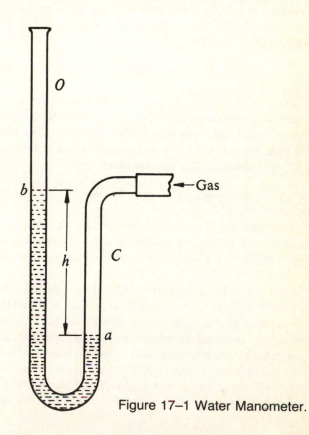

Figure 17–1 Water Manometer.

PROCEDURE

1. Fill the manometer tube about half full of water, connect to the gas cock, and gently turn on the gas to full pressure.
2. Measure the heights above the table of the water levels a and b (Figure 17–1) and record. Also determine and record the difference in heights h in centimeters. This height which you have just measured is often called the gauge pressure of the gas, since the pressure is proportional to it.
3. Compute the gauge pressure of the gas in centimeters of mercury (Hg), in inches of water, and in pounds per square inch, and record.

B. Pressure–Volume Relationships

INTRODUCTION

<u>NOTE TO INSTRUCTOR:</u>

Where students are unfamiliar with the logarithm approach, the paragraph immediately below Equation [2], including Equations [3] and [4], may be omitted.

Everyday experiences with toy balloons and other types of equipment readily reveal to us that the volume of an enclosed gas is related to both its temperature and the pressure applied to it. If we wish to consider the pressure P as depending on the volume V and the temperature T, we can think of P as a function of V and T, and write the relation

$$P = f(V, T). \qquad [2]$$

If we wish to investigate the relation of pressure to volume, it will be necessary to keep the temperature constant. In this case the above expression can be simplified to the form

$$P = kV^n \qquad [3]$$

where k is a constant the value of which depends on the nature of the container as well as the constant temperature. Hence, both k and n may be unknown quantities. If we take the logarithm of both sides of the equation, the V^n relation will be simplified. This process gives

$$\log P = n \log V + \log k \qquad [4]$$

which has the form $y = mx + b$, the familiar slope-intercept form of a straight line. This allows us to plot $\log P$ versus $\log V$ and compute n from the slope of the straight line.

One form of apparatus for investigating the pressure–volume relationship of a gas at constant temperature is shown in Figure 17–2. Two heavy glass tubes, one open and the other closed at the upper end, are connected by a heavy rubber tube T, and the system is filled with mercury until it rises to about the midpoint of the glass tubes. The enclosed gas (air) is in the closed tube between b and c. In one form of apparatus, a stopcock is used to close the tube at c.

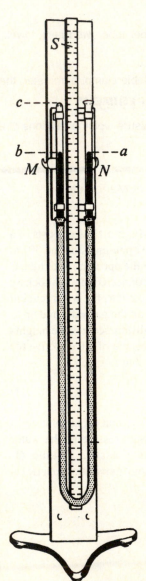

Figure 17–2 Boyle's Law Apparatus.

If the mercury is at the same height in both tubes, the pressure of the enclosed air is the same as in the open tube, namely, the existing atmospheric pressure. Both tubes are held in position by adjustable clamps M and N so that the mercury levels can be changed as desired. The levels are read on a meter stick S fastened between the tubes. Precautions should be taken to avoid errors in reading due to parallax. If the level at a is above the level at b, the pressure in centimeters of mercury in the closed arm (tube) exceeds the barometer reading (atmospheric pressure) by an amount equal to the difference in levels $a - b$. If a is below b, the pressure in the closed arm is less than the barometer reading by the amount $b - a$. Hence, we may obtain the total pressure of the enclosed gas in centimeters of mercury by adding to or subtracting from the barometer reading the difference in levels of the mercury columns in the tubes.

The volume of gas cannot be found in this apparatus unless the inside diameter of the tube is known. This is not usually the case, but because of the fact that the inside diamter is uniform throughout the length of the tube, the ratio of the heights of the air columns may be substituted for the ratio of the volumes.

PROCEDURE

NOTE TO INSTRUCTOR:

If the logarithm approach is omitted (see Note to Instructor under preceding Introduction), omit Steps 6 and 7 below (proceeding with Steps 8 and 9). If the logarithm approach is used, then use steps 6 and 7 (omitting Steps 8 and 9).

1. Since both pressure and temperature affect the volume of a gas, we must keep the temperature constant to get a true picture of the pressure–volume relationship. Hence, the temperature and ventilation of the laboratory room should be adjusted to suit our comfort before we begin the experiment and hope they can be maintained constant during the course of the investigation. When these adjustments have stabilized, read and record the temperature of the room and the barometer reading in centimeters of mercury.

2. Since the tubes on the equipment can be readily moved up and down, these adjustments provide for a wide range of pressures and volumes. If the closed end of the tube c (Figure 17–2) is closed by means of a stopcock, it is very important that the stopcock be kept tight to prevent leaks while the tubes are moved up and down.

Suppose we start with the high pressures first by lowering the closed tube so that it may be clamped near the bottom of the supporting stand. Then clamp the open tube near the top of the stand to provide the maximum difference. Watch the mercury levels for a few minutes for leaks that develop under the high pressure. Then read and record the positions of the two mercury levels and the end of the closed tube at c.

3. Obtain readings for about three other positions of the open arm as it is lowered to the bottom of the stand. Then, with the open end remaining fixed at the bottom, make three or four settings of the closed arm as it is raised to the top. This will give you seven or eight sets of readings for pressures ranging from the highest to the lowest possible values.

4. Record the temperature of the room again.

5. Compute the absolute pressure in centimeters of mercury and the height of the enclosed air column for each set of readings. These heights are to be used instead of volumes. Compute and record the products of pressure and volume. Keep in mind the lessons previously learned on significant figures.

6. Plot a graph of the logarithm of absolute pressure (gauge pressure + barometer reading) as ordinate versus the logarithm of volume as abscissa. Our proposed theory, Equation [4], suggests that we should expect a straight line. If your plotted points indicate this relation, sketch in a straight line and compute its slope.

7. From the (P, V) coordinates of some point on the line and the slope n, substitute into either Equation [3] or Equation [4], and solve for k. You should now be able to write Equation [3] with all constants as known quantities. Study the results and formulate some conclusions regarding the relation of pressure to volume for a gas.

8. Plot a graph of volume versus pressure with volume on the X axis. You need not begin your scales at zero in this case. Examine the graph carefully and note its shape.

9. Plot a second graph of volume versus $1/P$. Compute values for $1/P$ and list them in the log P column on the data record. Also examine the nature of this graph.

QUESTIONS

1. Would it be more accurate to use mercury instead of water in the manometer tube to measure the gas pressure? Why?

2. The manometer contained air between the gas and the water. Why was it not necessary to remove it?

3. What would have been the result if the gas pressure had been less than atmospheric when the gas cock was opened?

4. What do the data in the columns for pressure and volume indicate about their relation? Does your graph indicate the same relation?

5. Use the barometer reading as one atmosphere of pressure, and determine from the graph the volume (height) of the air column at a pressure of 0.75 atm.

6. Did the temperature change appreciably during the course of the experiment? How would a sizable increase have affected the products in the PV column? Why?

7. If the cross-sectional area of the air column in your experiment were 0.2 cm^2, what volume would the air occupy at atmospheric pressure? Show the steps in your computation.

8. What do the figures in the PV column of your data table indicate. Do these experimental results confirm the predictions of Equation [3]? Justify your answer.

9. In your equipment, heights of air columns were recorded and were used as volumes in plotting the graph. Assuming A to be the cross-sectional area of the inside of the glass tube, then $V = Ah$ could be used to express the volume. Make this substitution in Equation [3] and determine if it is valid to use h instead of V to establish the pressure–volume relationship for a gas.

10. If a slow air leak had developed in your system during the course of the experiment, how do you think it might have been revealed by your results?

11. Assuming the temperature in all parts of the room to be the same and all setups of apparatus in the room to have the same diameters, will all students in the class obtain the same equation for Boyle's law (see Question 9)? If not, explain.

12. Examine the values of the PV column in the data record and write an equation to express Boyle's law for your setup. Do you think your equation will be the same as that of other students working in the same laboratory? Explain.

13. If you performed Steps 8 and 9, which of the two graphs might be more useful in confirming Boyle's law? Why?

NAME *(Observer)* _____ Date _____

(Partner) _____ Course _____

RECORD OF DATA AND RESULTS

Experiment 17—Pressure and Volume Relations for a Gas

A. GAS PRESSURE

Height of water level *a* (Figure 17-1) above table _____ cm

Height of water level *b* (Figure 17-1) above table _____ cm

Gauge pressure (height *h*)_____ cm of water

Gauge pressure: (1) _____ cm of mercury

(2) _____ in of water

(3) _____ lb/in²

B. PRESSURE–VOLUME RELATIONS

Barometer reading _____ cm   **Temperature at beginning** _____   **Temperature at end** _____

Position of Hg in open tube	Position of Hg in closed tube	Difference in Hg levels, *h*	Position of upper end of air column	Pressure, *P* (cm of Hg)	Volume, *V* (height in cm)	Product, *PV*	$\frac{1}{P}$ or Log *P*	Log *V*

PART TWO

HEAT

2

LINEAR COEFFICIENT OF EXPANSION OF METALS

SPECIAL APPARATUS:

Burner and hose, meter stick, can (or beaker) to catch condensed steam, towel, two metal rods (different materials), thermometer.

GENERAL APPARATUS:

Expansion apparatus, boiler and hose, boiler stand.

THE PURPOSE OF THIS EXPERIMENT

is to determine the coefficient of linear expansion of certain metals.

INTRODUCTION

The fact that most objects expand when heated is common knowledge. The change in the linear dimensions of a solid is very nearly proportional to the temperature change over a considerable range of temperature. The increase in length per unit of length at 0°C per degree change in temperature is called the *coefficient of linear expansion*.

If L_0 is the length at 0°C, L_1 the length at temperature t_1, L_2 the length at temperature t_2, and α the coefficient of linear expansion, then the above definition may be expressed in symbols by

$$\alpha = \frac{L_2 - L_1}{L_0(t_2 - t_1)} \,. \tag{1}$$

In most experiments, L_1 will not differ greatly from L_0 and may be substituted for it in the denominator of the above relation. By making this substitution, the above relation may be written as

$$\alpha = \frac{L_2 - L_1}{L_1(t_2 - t_1)} \,. \tag{2}$$

By letting $t_2 - t_1 = \Delta t$, the change in temperature, and solving for L_2, the above becomes

$$L_2 = L_1(1 + \alpha \Delta t). \tag{3}$$

Why does one want to know the coefficient of expansion of a material? Design engineers and manufacturers of many types of equipment must know the coefficient of expansion of the material being used in order to know what will happen to the product when it undergoes a change of temperature. The results of the changes that occur may cause rivets or a bearing to loosen, a piston to tighten, or some object to bend and thereby change shape. There are dozens of such effects that require a knowledge of the coefficient of expansion on the part of the user as well as the manufacturer.

Any time one becomes interested in the uses of some new alloy, plastic, or other material, he must know the coefficient of expansion before he can predict how it will act when heated or cooled. One method of measuring this property of materials is to use a sample in the form of a long rod to measure its change in length as the temperature is changed, and then compute the coefficient of expansion. This is the method to be used in this experiment.

Figure 18–1 Coefficient of Expansion Apparatus—Micrometer Form (Courtesy of Central Scientific Co.).

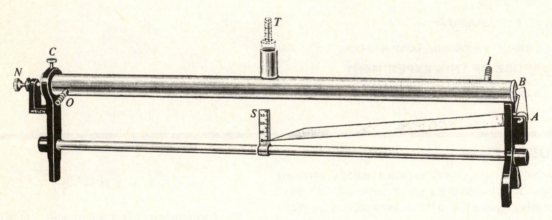

Figure 18–2 Linear Expansion Apparatus—Lever Form (Courtesy of Sargent–Welch Scientific Co.).

DESCRIPTION OF APPARATUS

The type of apparatus to be used in this experiment may be one of the forms illustrated in Figures 18–1 or 18–2. In either form, the metal rod to be studied is placed in the steam jacket and held in place by a stopper in each end. The steam for heating the rod is supplied by a boiler which is connected by a rubber hose to the inlet I. One end of the rod, protruding a little beyond the stopper, makes contact with an adjustable stop N, while the other end makes contact with a device designed to indicate the change in length as the rod expands.

In the micrometer form (Figure 18–1), the increase in length is determined by the difference in micrometer readings before and after heating. If desired, a battery and a buzzer (or voltmeter) may be electrically connected at P and R to make the rod a part of the electric circuit. The circuit is closed only when screws M and N both make contact.

In the lever form (Figure 18–2), the small change in length is magnified by the lever attachment. A small movement of the short end of the lever at B produces a much larger movement of the long end on the scale S as both rotate about the axis A. The magnification factor is the ratio of the lever arms AS/AB, both of which can be easily measured.

PROCEDURE

1. Fill the boiler about one-half full of water and start to heat it while other adjustments are being made. Do not connect the hose to the steam jacket yet. Record the kind of material of which each rod is made.

2. Measure the length of each rod, and record the length L_1.

3. Insert one of the rods in the jacket and tighten the screw at the upper left (C in Figure 18–1) to hold the jacket in place; adjust screw N so that good contact is made. Carefully insert the thermometer T through the stopper and adjust it in the central opening of the jacket until you feel the thermometer barely touch the rod. Read and record the temperature after the thermometer has had time to adjust to the temperature of the water jacket. This is temperature t_1.

••••••••••••••••••••••••••••••••

For *micrometer form* (Figure 18–1), follow instructions in Steps 4 and 5.

For *lever form* (Figure 18–2), follow instructions in Steps 6 and 7.

••••••••••••••••••••••••••••••••

4. Adjust the micrometer screw M until you feel it barely touch the end of the rod. Do not force the screw. If the electrical contact is used, the buzzer (or voltmeter) will indicate when the screw is touching. Obtain a reading on both the linear scale and the circular scale on the screw head. Any initial reading desired can be obtained by changing the adjustment of the screw N. Back the screw up a few turns and make four more trials of the initial reading and record.

5. After backing the micrometer screw off by at least 2 mm, admit steam to the jacket at inlet I and catch the condensed steam in a can (or beaker) at the outlet O. After the temperature has reached equilibrium conditions (remained steady for 3 or 4 min), read the thermometer and

record as temperature t_2; record five trials of the micrometer reading.

••••••••••••••••••••••••••••••••

6. While the equipment is cold (room temperature) adjust the rod position until the short lever is touching the end of the rod at B. Now adjust the screw N until the pointer on the long lever is near the bottom of the scale S and record the reading. Also measure both lever arms and record.

7. You are now ready to admit steam to the jacket at the inlet I. Arrange to catch the condensed steam in a can (or beaker) at the outlet O. After the temperature has reached equilibrium conditions (remained steady for 3 or 4 min), read and record the final temperature, and the reading of the pointer on the scale S. Note that the expansion at B has been magnified on the scale S by a factor equal to the lever arm ratio.

••••••••••••••••••••••••••••••••

8. Disconnect the steam supply, remove the thermometer, and pour cold water in the opening at T until the system has again come to room temperature. Replace the rod just measured with a second one of a different kind of material, and repeat the preceding manipulations, recording the data as before.

9. From the readings which you have obtained, determine the average value of the change in length for each material measured and compute the coefficient of linear expansion. Compare your values with the accepted values listed in Appendix B by finding the percent discrepancy. Show steps in your calculations in the report.

10. Make a second trial on one of the rods and compute the percent difference in the two trials. How does this compare with the percent discrepancy computed in Step 9?

QUESTIONS

1. To which measurement or reading do you attribute most of your error? Explain.

2. If you had measured the original length of the rod in inches, and the micrometer screw had given readings in inches, what value for α would you have obtained? Explain.

3. If lengths had been in centimeters but the temperatures used had been in degrees Fahrenheit, what value of α per degree Fahrenheit would you have observed?

4. What unit is given in the tables for the values of α? Why?

5. Which is the most serious, an error of 1 mm in measuring the original length or an error of 0.01 mm in

measuring the amount of expansion? Explain.

6. Do you think something more precise than a meter stick for measuring the length should be used? Why?

7. What would happen if the two specimens used in this experiment were clamped together, side by side, and heated?

8. In the case where two trials were made on the same material, how did your percent difference compare with the percent discrepancy of the average of the two from the accepted value in Appendix B? Discuss the significance of this comparison. What do your results reveal about the accuracy to be expected in this experimental arrangement?

9. Use the length of one of the rods at room temperature as L_2 and the value of α obtained from the tables to calculate, by Equation [3], the value for L_1 at 0°C.

10. Substitute the value of the length at 0°C, found from Question 9, for L_0 in Equation [1] and solve for the coefficient of linear expansion, using other data the same as before. Do you think that substituting L_1 for L_0 makes a serious error in Equation [2]?

RECORD OF DATA AND RESULTS

Experiment 18—Linear Coefficient of Expansion of Metals

Kind of Material						
Original length (cm)	$L_1 =$		$L_1 =$		$L_1 =$	
Temperature (C°)	$t_1 =$	$t_2 =$	$t_1 =$	$t_2 =$	$t_1 =$	$t_2 =$
For Figure 18-1 Micrometer settings (mm)						
Sum of settings (mm)						
Average of settings						
For Figure 18-2 Reading of Scale *S*						
Lever arm *AS*						
Lever arm *AB*						
Lever arm ratio *AS/AB*						
Change in scale rdg *S*						
Elongation of rod (cm)						
Coefficient of linear expansion, α						
Accepted value of α						
Percent discrepancy						
Step 10: Percent diff.						

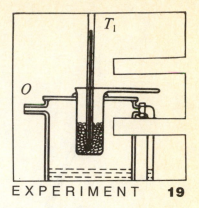

SPECIFIC HEAT AND TEMPERATURE OF A HOT BODY

SPECIAL APPARATUS:

Burner, calorimeter (complete with stirrer), metal cup containing about 400 gm of shot,* piece of metal with copper wire hook (Figure 19–2), support rod for the piece of metal, thermometer (in degrees Celsius), thermometer (in tenths of degrees Celsius), set of weights, towel.

GENERAL APPARATUS:

Boiler, boiler stand, laboratory balances.

THE PURPOSE OF THIS EXPERIMENT

is to determine the specific heat of lead shot (a solid piece of metal may be substituted) by the method of mixtures, and also to determine the temperature of a red-hot piece of metal whose specific heat is known.

INTRODUCTION

Heat, being a form of energy, may be measured in the same units as other forms of energy. The unit of energy most commonly used in the mks system is the *joule*, and it is quite appropriate to measure heat energy in joules. However, the calorie, which is equivalent to 4.18 J, is perhaps a more commonly used unit. The *calorie* is arbitrarily defined as the amount of heat energy required to increase the temperature of 1 gm of water 1°C.

The calorie is related to water as its reference material. A term characteristic of the material of which a body is composed is the *specific heat capacity* often abbreviated as specific heat. The specific heat, S, of a substance is the amount of heat energy which must be supplied to a unit mass of the material to increase its temperature 1°C.

Manufacturers often specify the heat capacity of some particular body in terms of its water equivalent. The *water equivalent* of a body or vessel is the mass of water that requires the same amount of heat energy to raise its temperature some specified amount as is required by the said body. If the thermal behavior of a body is the same as that of 10 gm water, its water equivalent is said to be 10 gm. The product of the mass and the specific heat of a body gives the water equivalent.

When a body undergoes a change in temperature, the heat energy lost or gained during the change is given by the product of its mass M, its specific heat S, and the change in temperature $t_2 - t_1$, as indicated by

$$H = MS(t_2 - t_1). \qquad [1]$$

The most common method used in the determination of heat energy exchanges is the *method of mixtures*, in which two or more systems with different temperatures are placed in contact in such a way that they interchange heat energy

---

*A little less accurate, but much simpler method is to suspend a solid piece of metal in the boiler instead of using shot. The piece of metal may be suspended in the boiling water by means of a string. In transferring it to the calorimeter, one unavoidably transfers a few drops of water but the error introduced is quite small. This eliminates the problem of drying the shot at the end of the experiment.

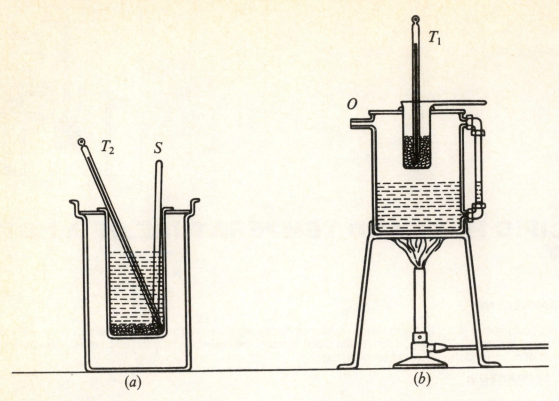

Figure 19/1 (*a*) Calorimeter. (*b*) Boiler.

until all of them acquire the same temperature, at which time the interchange stops. As a result of the interchange, the bodies at higher temperatures give out the same amount of heat energy in cooling as the bodies at lowest temperature absorb in being heated. Stated briefly,

$$\text{energy lost} = \text{energy gained.} \quad [2]$$

The operation of the method of mixtures is usually carried out in a double-walled *calorimeter* (Figure 19–1*a*). The specimen is placed in the inner cup, which is separated from the outer one by an insulating ring, leaving an air space between. Both cups are polished to reflect radiation. These features of the calorimeter reduce the energy lost to or gained from the room but do not eliminate this factor entirely. It is the usual practice to start the operation with the temperature in the calorimeter about as much below room temperature as it is anticipated the final temperature will be above that of the room, or vice versa. By this procedure the heat energy gained from the room during one part of the experiment will approximately cancel the heat energy lost during the other part.

PROCEDURE

A. Specific Heat of a Metal

1. Fill the boiler about half full of water and set the cup, containing about 400 gm of dry shot, in the opening at the top of the boiler. Place a thermometer T_1 (Figure 19–1*b*), which reads to at least 100°C, in the shot and start heating the water while the following manipulations are made. The shot need not be weighed accurately at this time.

2. Weigh the inner cup of the calorimeter plus stirrer when empty, and again when about two-thirds full of water at a temperature about 5°C below room temperature. Record.

3. When the water in the boiler has been boiling for about 5 min and the temperature of the shot has become constant, record the reading of thermometer T_1.

4. Now stir the water in the calorimeter and read thermometer T_2 to the nearest tenth of a degree. Record the temperature and quickly, *but carefully*, pour the hot shot (or other metal) into the calorimeter and stir the mixture immediately; then record the highest temperature reached. It may take several seconds for equilibrium conditions to be attained. Record this temperature as T_m.

5. Weigh the calorimeter cup and its contents (water +

shot) and record. From this weighing and the previous one, the weight of the shot can be determined.

6. Strain the water off the shot through a towel and spread the shot out to dry while performing the remainder of the experiment, unless some other method of drying is designated by your instructor. The shot must be dry when checked in at the storeroom.

7. Make the following solution a part of your report. Equate the heat energy lost by the hot shot to the heat energy absorbed by the cold water and its container, showing clearly the position of each item of data required in the equation. The energy lost or gained by each item will have the form of the right member of Equation [1]. Consider the metal stirrer as part of the calorimeter. Solve for the specific heat of the shot and compare with the accepted value given in Appendix B by calculating the percent discrepancy.

B. Temperature of a Hot Body

We shall now use the heat exchange procedure to determine the temperature of a hot body that cannot be measured with a thermometer.

8. Select a piece of metal of such size that all parts of it can be heated when suspended in a bunsen burner flame as shown in Figure 19–2. The metal should also have a high melting point, 800°C or higher (see Table 4, Appendix B).

9. Weigh the piece of metal to be used and record both the weight and the kind of material. Also record its specific heat as given in Appendix B.

10. Place a metal rod (bent as shown in Figure 19–2), in the steam outlet *O* of the boiler and suspend the piece of metal from the end so that it will hang in the hottest part of the flame (just above the inner cone). Allow it to heat while preparing the calorimeter, or while computing results for Part A.

11. Weigh the calorimeter when about two-thirds full of

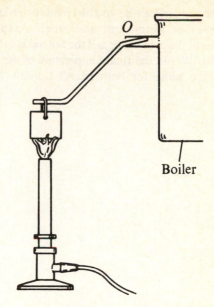

Figure 19–2 Arrangement for Measuring the Temperature of a Flame.

water which is 10–15°C below room temperature. Avoid cooling enough for dew to collect on it. Use a thermometer which reads to 100°C for this experiment.

12. When the metal becomes as hot as possible, read and record the temperature of the cold water, remove the thermometer, and quickly drop the hot piece of metal into the calorimeter. After the metal has cooled to a safe temperature, record the equilibrium temperature of the mixture.

13. By using the specific heat of the hot piece of metal (see tables in Appendix B) equate the heat energy lost to the heat energy gained, and compute the temperature of the hot metal. This should be the temperature of the flame.

QUESTIONS

1. How much is the water equivalent of your calorimeter?

2. Set up the equation used for the calculation of specific heat of shot, and show how it may be simplified by the use of the water equivalent of the calorimeter.

3. Name some possible sources of error in your experiment.

4. What precautions might be taken to improve the experiment?

5. If wet shot had been poured into the calorimeter of cold water, how would the value obtained for the specific heat have been affected? Why?

6. If dew had collected on the outside walls of the calorimeter containing the cold water in Part B, how would

the calculated temperature have been affected?

7. What evidence do the results of this experiment give to prove that water has a much higher specific heat than the metal shot?

8. Does the experiment suggest a method by which the temperature of a furnace might be obtained approximately? Explain.

9. If the temperature of the heated metal had been 1000°C, what would have been the resulting temperature of the mixture in Part B?

10. If you had started with water at room temperature in either part of this experiment, explain how your results would have been affected. Why is this the case?

11. Suppose that, in Part B, you had made two trials, with equal masses of aluminum and steel. Assume the temperatures of the cold water and the hot metal to be equal in both trials. Would the final temperature of the mixture have been the same for both trials? Consult Table 4, Appendix B, and explain your answer.

12. If, in Part B, the metal had been hot enough to make the water boil how would the validity of the results have been affected? Explain.

RECORD OF DATA AND RESULTS
Experiment 19—Specific Heat and Temperature of Hot Body

	Procedure A	Procedure B
Measurement being made	Specific heat of shot	Temperature of hot body
Kind of metal used		
Weight of calorimeter + water + metal		
Weight of calorimeter + water		
Weight of metal used		
Weight of empty calorimeter		
Weight of cold water		
Temperature of hot metal, T_1		(Calculated)
Temperature of cold water, T_2		
Temperature of mixture, T_m		
Specific heat of calorimeter + stirrer		
Specific heat of metal being heated	(Calculated)	(Table)
Specific heat of metal being heated	(Table)	
Percent discrepancy		

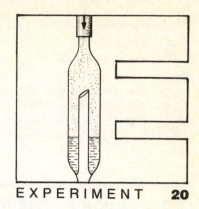

CHANGE OF PHASE—HEAT OF FUSION AND HEAT OF VAPORIZATION

SPECIAL APPARATUS:

Procedure A or B—Calorimeter (complete), one 1 to 51°C thermometer, Bunsen burner, extra can for water, towel, set of weights. Extra apparatus for Procedure B—Water trap, extra rubber hose, one −20 to 110°C thermometer.

GENERAL APPARATUS:

Supply of ice, steam generator and stand, laboratory balance.

THE PURPOSE OF THIS EXPERIMENT

is to observe the amount of heat associated with a change of the phase of water by measuring (a) the heat of fusion and (b) the heat of vaporization.

INTRODUCTION

In the experiment on specific heat, the relationship between the amount of heat energy supplied to a body and the corresponding temperature change was studied. However, none of the substances used in that experiment underwent a change of phase during the energy exchange.

If we fill a glass with ice at 0°C and allow it to stand in a warm room until the last bit of ice is melted, a thermometer would indicate the temperature to still be 0°C. The temperature of the contents has not changed, yet we know that the glass of ice has absorbed heat energy. Where did it go? What was its effect? We can easily observe that the solid crystalline ice has been changed into liquid water during the period under consideration. Our experience in breaking up a block of ice tells us that energy is required to change the arrangement of the crystalline structure. Heat energy, supplied to ice, will be utilized to break down the bonding forces of the solid and to convert the material into a liquid without a change in temperature.

The amount of heat energy L_f called the *heat of fusion* is defined as the amount of heat energy required to change a unit mass of a substance from the solid to the liquid phase (without a change of temperature). The accepted value of L_f for ice recorded in handbooks is 79.7 cal/gm. If warm water

is used to melt the ice, we should be able to determine the heat of fusion by measuring the amount of heat energy given up by the water in melting the ice. The heat energy exchanges can be measured in a calorimeter. By applying the energy gained = energy lost principle to the melting of M gm of ice at 0°C in a calorimeter containing m gm of water at a known temperature, we would have

heat energy to melt ice $(L_f M)$ + heat energy gained by ice water = heat energy lost by warm water + heat energy lost by calorimeter and stirrer.

This statement assumes that the final temperature of the mixture will be above 0°C.

All of us have observed the phenomenon of boiling and are no doubt aware that the escaping vapor and the boiling liquid are at the same temperature. Again, we have a change of phase without a change of temperature as the result of the absorption of heat energy. In this case the energy is used to do work against cohesive forces in separating the molecules and to make room for the escaping molecules by pushing back the air surrounding the vessel of liquid. When molecules in the vapor condense back into the liquid phase, this extra heat energy is given up to the surroundings.

The amount of heat energy L_v, called *the heat of vaporization*, is defined as the amount of heat energy required to change a unit mass of a substance from the liquid to the vapor phase (without a change of temperature). The heat of condensation (numerically equal to the heat of vaporization) can be measured by allowing a mass, of steam M_v at a known temperature to condense into a calorimeter containing a known mass of water and noting the rise in temperature. Again, applying the energy lost = energy gained principle, we would have

heat energy lost by vapor ($L_v M_v$) + heat energy lost by hot water formed from condensed steam = heat energy gained by cold water and calorimeter + stirrer.

PROCEDURE A—THE HEAT OF FUSION

1. Record the temperature of the room near your table, and warm some water to a temperature about 15°C above room temperature.

2. Note and record the kinds of material and the specific heats of the stirrer and the inner cup of the calorimeter. If the stirrer and cup are of the same kind of material, they may be weighed together. If of different materials, their specific heats will be different, and they must be weighed separately. After the cup has been weighed empty, without the ring, weigh again when a little more than half full of the warm water. This last weighing must be performed with care, because it cannot be checked later.

3. After placing the inner cup in the calorimeter, prepare several (8 or 10) pieces of ice about the size of a pecan. Stir the water and carefully read the temperature in preparation for the heat exchange process. As you gently stir the water, dry a piece of ice with a towel and slide it into the calorimeter, being careful not to splash water out. Continue to add the ice, one piece at a time, while stirring, until the temperature is about 5°C below room temperature. If the temperature goes much below this, condensation on the outside of the container will affect the heat energy exchange inside. Do not add too much ice toward the end so you can control the final temperature of the mixture. Read and record the lowest temperature reached as the last bit of ice has melted.

4. As you remove the inner cup from the calorimeter, examine it for the presence of condensed moisture on the outside wall. If moisture is present, discard your data and get a new supply of water at a somewhat lower beginning temperature. If no moisture is present weigh the calorimeter cup and contents and compute the mass of the ice.

5. Compute the heat of fusion and compare it with the accepted value, 79.7 cal/gm, by computing the percent discrepancy.

PROCEDURE B—HEAT OF VAPORIZATION

6. A boiler, such as shown in Figure 19–1, with the cup of shot replaced by a screw lid, may be used for heating the water. Fill the boiler about two-thirds full of water and heat it while preparing the calorimeter. Also record the barometer reading.

7. Because we are attempting to measure the heat energy given up by the condensation of steam as it enters the water in the calorimeter, we must take precautions in transporting the steam to prevent hot water from entering as a result of condensation along the connecting tube. A water trap (Figure 20–1)* inserted in the hose line between the boiler and calorimeter near the exit end will collect the condensed water permitting the steam to continue.

8. Measure the weight of the inner calorimeter cup when empty and again when about two-thirds full of water at 15°C below room temperature. Record to the nearest

Figure 20–1 Water Trap.

*If a conventional water trap is not available, a test tube or small bottle with a two-hole stopper may be substituted.

one-tenth gm. Also measure the temperature of the steam (or slowly boiling water).

9. Carefully read the temperature of the cold water and, while stirring, introduce steam into the calorimeter until the temperature is about 15°C above room temperature. Carefully read its maximum value. A final, but careful, weighing now will permit you to determine the mass of steam condensed in the water.

10. Compute the heat of vaporization and compare with the accepted value, 539.6 cal/gm, by computing the percent discrepancy. A second trial may be desirable to better evaluate the small mass of steam.

QUESTIONS

1. What becomes of the heat energy used in melting ice?

2. Is it likely that the results of this experiment would have been affected if the ice had not been dried before dropping it into the calorimeter? Why?

3. If enough ice had been added to cool the water below the dew point, how would the value obtained for the heat of fusion have been affected? Explain.

4. By using the accepted value (79.7 cal/gm) for the heat of fusion of ice, calculate the value in BTUs per pound. Refer to textbook or handbook.

5. Compute the value of the heat of fusion that you would have obtained if your value for the weight of the calorimeter plus water had been 1 gm too high. Why is it that such a small error in the weight causes such a large error in your final result?

6. In which case would ice be most effective in cooling a refrigerator, when wrapped in a blanket or when placed in the refrigerator with no wrapping? Explain.

7. How does the percent difference in your two trials for the heat of vaporization of water compare with your variation from the accepted value? Discuss the significance of this comparison.

8. In both parts of this experiment the suggested temperature change for the water in the calorimeter was about 30°C. How does the mass of steam needed compare with the ice required? Explain why.

9. If condensed water from the steam line had entered the calorimeter, how would your computed value of the heat of vaporization have been affected? Explain carefully.

10. If an error of 1 gm were made in obtaining the mass of the steam, what percent error would this introduce into your final result? Make the computation and then comment on the weighing precautions needed.

11. If the boiling took place quite vigorously, how might your data have been affected? Why?

12. If a mass of boiling hot water equal to the mass of the steam were poured into your calorimeter of cold water, how much would the temperature have increased? Which do you think would produce the most severe burn on the hand, 10 gm of steam at 100°C or 10 gm of water at 100°C? Explain.

13. Why does good insulation around the steam pipes leading from a steam heating plant to the radiators improve the efficiency of the heat output of the radiators?

14. As you observe water converted into steam without a change in temperature, what evidence do you have that heat energy is being absorbed? Give evidence based on observation, not on something read from a book.

15. How does perspiration give the body a means of cooling itself?

16. Why does the body feel cooler in moving air than in still air? Does a thermometer register a lower temperature in moving air? Discuss both questions.

NAME *(Observer)* _____ Date _____

(Partner) _____ Course _____

RECORD OF DATA AND RESULTS
Experiment 20—Change of Phase—Heat of Fusion and Heat of Vaporization

GENERAL INFORMATION ON EQUIPMENT AND ROOM CONDITIONS

Room temperature _____ Barometer reading _____

Material of calorimeter _____ Specific heat of calorimeter _____

Material of stirrer _____ Specific heat of stirrer _____

OBSERVED DATA AND CALCULATED RESULTS

Data Obtained	Heat of Fusion	Heat of Vaporization
Weight of empty calorimeter + stirrer		
Weight of calorimeter + original water		
Weight of water originally in calorimeter		
Weight of calorimeter + final contents		
Weight of ice (or condensed steam)		
Temperature of original water		
Final temperature of the mixture		
Temperature of ice (or steam)		
Heat of fusion (or vaporization) measured		
Accepted value of above heat constant		
Percent discrepancy from accepted value		

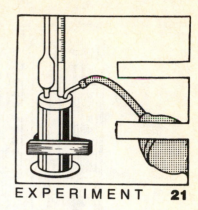

RELATIVE HUMIDITY

SPECIAL APPARATUS:

Shiny metal cup (or dew-point hygrometer), extra can (or beaker), thermometer (preferably graduated to 0.1°), towel.

GENERAL APPARATUS:

Supply of ice, some form of wet-and-dry bulb hygrometer (one or two for entire class), supply of ether (or acetone) if dew-point hygrometer is used.

THE PURPOSE OF THIS EXPERIMENT

is to determine the relative humidity of the atmosphere.

INTRODUCTION

A knowledge of the relation of relative humidity and dew point to the weather is very important to truckers, fruit growers, pilots, and others. It aids them in predicting fogs, frost, and icing conditions so that they may prepare for bad weather or avoid areas where bad weather prevails.

The *dew point* of the atmosphere is defined as that temperature to which the atmosphere must be cooled in order that it may become saturated with water vapor. When the dew point is below freezing, the water vapor freezes as it condenses and frost is formed.

The *absolute humidity* of the atmosphere is defined as the mass of water vapor present in a unit volume. It is usually expressed in grams per cubic meter or in grains per cubic foot. From this it will be noticed that absolute humidity is simply the density of the water vapor in the air.

We are more interested in what we sometimes call the dampness in the air. This is a relative term depending upon the degree of saturation and is called the relative humidity. *Relative humidity* is defined as the ratio of the existing absolute humidity to that required to produce saturation at the same temperature. Thus, if M represents the mass of water vapor per unit volume actually present and M_s the mass of water vapor per unit volume when saturated at the same temperature, the relative humidity is expressed as

$$R = M/M_s. \qquad [1]$$

Since the fractional part of the atmospheric pressure exerted by the water vapor is nearly proportional* to the absolute humidity (density of water vapor), the definition of relative humidity R may be expressed as

$$R = P/P_s \qquad [2]$$

where P and P_s are the pressures of the existing vapor and saturated vapor, respectively, at the same temperature. The pressure of saturated vapor can be measured rather easily, and these measured values are in Table 6 of Appendix B.

If the air with absolute humidity, M, is cooled to the dew point, the mass of water vapor M per unit volume will still be the same, but M will now represent also the mass of saturated vapor at the dew-point temperature. Hence, the mass of water vapor per unit volume at the dew point is the absolute humidity of the air at the existing temperature. The mass of saturated water vapor at any temperature is given in

*Not exactly proportional, because vapors near saturation do not obey Boyle's law accurately.

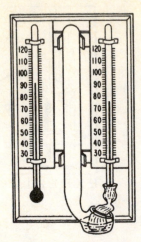

Figure 21–1 Wet-and-Dry-Bulb Hygrometer.

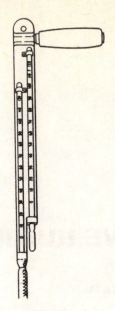

Figure 21–2 Sling Psychrometer.

Appendix B. With this information and the application of Equation [1], the relative humidity can easily be determined.

Any instrument which gives the information for making humidity determinations is called a hygrometer. One of the simplest and most commonly used hygrometers is the wet-and-dry-bulb hygrometer (Figure 21–1). It consists of two thermometers (usually with Fahrenheit scales), one of which gives the true temperature of the air. The other thermometer is covered with a wet cloth and gives a temperature that varies with the rate of evaporation of water vapor. The drier the air, the greater the evaporation from the wet bulb, and the lower its reading. When the air is saturated with moisture there is no evaporation, and the two thermometers indicate the same temperature. If the air around the hygrometer is to maintain room conditions, it must be circulated by some fanning process. This may be done by direct fanning, or it may be done by slinging the hygrometer through the air. Hygrometers constructed to be used by the sling method are called *sling psychrometers* (Figure 21–2). A chart showing the relation between the

relative humidity and the difference between the readings of the two thermometers must be used with the wet-and-dry-bulb hygrometer.

Another type of hygrometer, called the *dew-point hygrometer*, is illustrated in Figure 21–3. It consists of a highly polished nickel-plated cylinder provided with an aspirator arrangement, whereby air may be pumped through a volatile liquid, such as ether or acetone. The air escapes through the outlet tube. As the liquid is agitated by the air stream, it evaporates and thus cools the cylinder and contents to a temperature that causes water vapor to condense on the polished outside wall. A thermometer is inserted to give the temperature at all times, and thus indicates the temperature at the moment that dew begins to form, or disappear. From this dew-point temperature and the tables in Appendix B, the relative humidity of the surrounding air can be determined.

PROCEDURE

NOTE TO INSTRUCTOR:

If you are furnished with a dew-point hygrometer (Figure 21–3), omit Steps 1–3 and begin with Step 4. Otherwise, begin with Step 1 and omit Steps 4–6.

1. Read and record the temperature of the room in the area where you are working. Fill the shiny cup about one-third full of tap water and set it on a printed sheet of paper. You should be able to see a clear image of the print in the walls of the cup.

2. Add small quantities of ice water, while stirring thoroughly, until moisture appears on the cup. Its appearance may best be noted by the blurring of the image of the print. Estimate the thermometer reading to tenths of a degree, and record the temperature at which the moisture

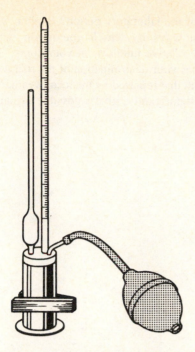

Figure 21–3 Dew-Point Hygrometer.

film appears. *Avoid breathing directly on the cup*.
3. Now allow the temperature to rise slowly, and record the temperature at which the moisture film evaporates. This may best be accomplished by pouring out a little of the water and slowly adding tap water. The more slowly the temperature is changed in the neighborhood of the dew point, the more accurately can the dew point be determined. Record the average of these two temperatures as the dew point. Repeat several times, endeavoring to get the appearance and disappearance temperatures as close together as possible.

4. Fill the dew-point hygrometer cylinder about half full of ether (or acetone) to serve as the cooling agent; then assemble the unit as shown in Figure 21–3. *Caution!* Avoid direct breathing of the vapor.
5. While slowly pumping air through the liquid, watch for the appearance of moisture on the polished surface. The method of detecting its appearance described in Step 2 may be useful here also. Read the temperature to the nearest 0.1° and record.
6. Stop the aspirator action and, as the liquid warms, read and record the temperature at which the moisture film evaporates. Repeat the above steps for five trials, and record the average dew point, along with the temperature of the room.

7. From the tables in Appendix B and the definitions given in the preceding paragraphs, compute and record both the absolute humidity and the relative humidity from your value of the dew point.
8. Use some form of a wet-and-dry-bulb hygrometer and record the temperatures indicated by the two thermometers after their readings become constant under the proper fanning conditions. In using the sling psychrometer, best results will be obtained if water a few degrees below room temperature is used to moisten the wet bulb. Find the difference of the two readings, and read the relative humidity from a psychrometric chart (Appendix B).
9. Determine the percent difference between the average value of the relative humidity found by the dew-point method and that found by the wet-and-dry-bulb hygrometer.
10. If it seems feasible to do so (ask instructor), determine the relative humidity at other places by the wet-and-dry-bulb method. If information is available on your campus, or at the local weather bureau, obtain and record the relative humidity of the atmosphere in your city.

QUESTIONS

1. Would the dew-point depression be greater on a dry day or on a damp day? Why?
2. How is the relative humidity of air affected when it is heated? Why? How is the absolute humidity affected? Why?
3. Compare the rate of evaporation of perspiration from the body on a damp day and on a dry day, provided all other conditions are the same.
4. If, when the air temperature is 68°F, the relative humidity is 60%, estimate the dew point (see table in Appendix B).
5. Would a cooling system which blows air into the room

through a water spray be more efficient in the Gulf Coast area or in Arizona? Explain why.
6. If both had polished surfaces, which would serve best for the dew-point determination, a metal cup or a glass cup? Explain why.
7. You were cautioned not to breathe on the cup. Explain the reason for this precaution and the possible effect on your results if it were not heeded. If you should rub your finger on the cup to detect the presence of moisture, could this affect your results? Explain.
8. Why is it desirable to have moving air in the area of the hygrometer? If the water used is at room temperature,

explain how wetting one of the thermometer bulbs results in a cooling effect? Be explicit.

9. Was the reading of the wet-bulb thermometer as low as the dew point? Explain the reason for the relationship observed.

10. In a normal laboratory room containing 24 people, would you expect the relative humidity to change during the course of the laboratory period? Explain why.

11. Suggest ways by which you might increase and decrease the relative humidity in your home.

12. How does your determination of the relative humidity compare with that reported by the local weather bureau? If they differ significantly, give a possible explanation.

NAME *(Observer)* _____ Date _____

   (Partner) _____ Course _____

RECORD OF DATA AND RESULTS
Experiment 21 — Relative Humidity

DETERMINATION OF DEW POINT

Trial	Temperature at which dew appears	Temperature at which dew disappears	Mean-temperature dew point
1			
2			
3			
4			
5			
Room temperature =		Average dew point	

RELATIVE HUMIDITY RESULTS

DEW POINT METHOD		WET–DRY BULB HYGROMETER		
Vapor pressure, P, at dew point		Position at which data were obtained	Lab Table	Other
Vapor pressure, P_s, at room temperature		Dry-bulb reading		
M, in gm/m^3, at dew point		Wet-bulb reading		
M_s, in gm/m^3, at room temperature		Wet-bulb depression		
Relative humidity, P/P_s		Relative humidity		
Relative humidity, M/M_s		Percent difference of two methods		
Absolute humidity		Relative humidity (Weather Bureau)		

PART THREE
WAVE MOTION AND SOUND

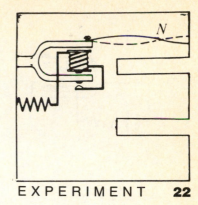

EXPERIMENT　22

A STUDY OF VIBRATING STRINGS

SPECIAL APPARATUS:

Two strings (different sizes), set of slotted weights, weight holder, pulley and clamp, 2-meter stick, two rulers, graph paper (furnished by student).

GENERAL APPARATUS:

Electrically driven vibrator (or tuning fork), source of current (120 V ac, for vibrator, 8 V dc, for tuning fork), connecting wires, switch, analytical balance.*

THE PURPOSE OF THIS EXPERIMENT

is to set up stationary waves in a stretched string and then to study the relation of wavelength to stretching force and mass of string, and to determine the frequency of an electrically driven vibrator.

INTRODUCTION

When two equal wave trains traveling in opposite directions act upon a series of particles, the resulting phenomenon is called *stationary waves*. This type of wave motion may be produced by either longitudinal or transverse waves. If one end of a light, flexible string is attached to a vibrator (Figure 22–1) and the other end passes over a fixed pulley to a weight holder, the waves travel down the string to the pulley and are then reflected, thereby producing a reflected wave moving in the opposite direction. If the tension and length of the string are properly adjusted, these two oppositely directed wave trains are superimposed on the string in such a way as to give alternate regions of no vibration N (Figure 22–1), and regions of maximum vibration A. These regions N and A are called *nodes* and *antinodes*, respectively, and the segment between two nodes is called a loop.

If one changes the tension in a vibrating string, it will be noted that the changes in tension cause a change in the number of segments (loops) between the ends of the string. A change in the number of loops means a change in the

length of a loop thus producing a change in wavelength. We must also consider the possibility of the wavelength being affected by a change in the size (or mass) of the string, or by a change in the length of the string. Still other factors that must be considered are the frequency and the velocity of the wave which may have a bearing on the wavelength. Hence, it seems possible that the wavelength of a standing wave in a vibrating string may be a function of the *tension*, *mass of string*, *length of string*, *frequency of the source*, and *velocity of the wave*.

It is a well-established fact that the wavelength λ in any type of wave motion is related to the frequency f and the velocity v, by the relation

$$v = f\lambda. \qquad [1]$$

In this experiment the wave motion will be maintained by a mechanical vibrator (Figure 22–1) with a frequency that will remain constant. Whatever variables affect the velocity will influence the wavelength as indicated by

---

*If the instructor prefers to furnish the student with the values of the mass per unit length of the string, the students will not need the analytical balance.

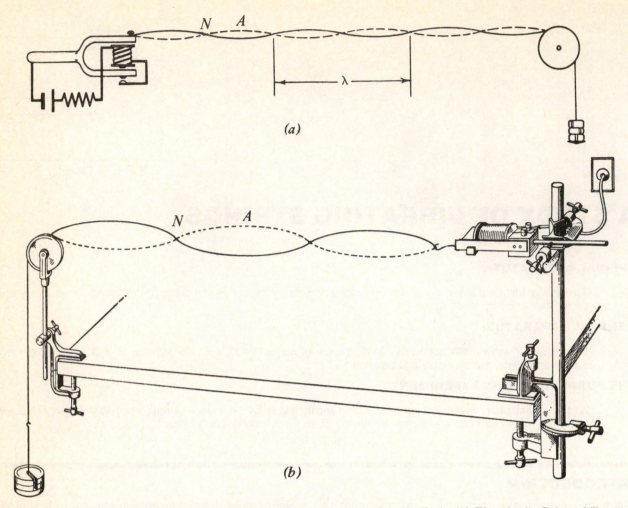

Figure 22–1 Standing Waves in a Stretched String. (*a*) Vibrating Tuning Fork. (*b*) Electrically Driven Vibrator.

Equation [1]. Hence, if we limit our study to the effect of other variables, we can always refer to Equation [1] for the effect on the velocity.

Of the variables left, we can vary any one of them while we keep the others constant. If we use one particular length of a given type of string, both the length and the mass are fixed, and we are left free to investigate the relation between wavelength and tension. Observation will indicate that as the tension F increases, λ also increases, which would indicate the possibility of some form of direct proportion. Suppose we try the relation

$$\lambda = kF^n \qquad [2]$$

where k is a proportionality constant and n is an exponent of

unknown value, because we do not know the exact nature of the proportion.

Since the frequency f remains constant, we can assume that it has some fixed relationship to k in Equation [2].

The mass of a given type of string would obviously be proportional to the lengths of the string used, and we could never separate these variables to study the independent effect of each on the wave pattern. However, mass per unit length, m, depends only on size and density, and we do have independency. For a given string, m would be constant and thus tied to k in Equation [2].

All of the above considerations will furnish guidance in our experimental approach to investigate the relationship and effect of the variables involved in the wave pattern in a vibrating string.

PROCEDURE

NOTE TO INSTRUCTOR:

The latter part of the procedure, which deals with analysis of the data, is designed to provide the choice of method which best fits the background of the student. The two methods, A and B, begin with Step 8 below.

1. If the mass per unit length of the strings is not known, untie all knots, measure the full length of each string carefully, and then weigh each on an analytical balance to the nearest milligram. If you are not familiar with an analytical balance, ask the instructor to explain the procedure before attempting to weigh on it.

2. Mount a pulley 120–150 cm from the vibrator (or tuning fork) so that a string, when attached to the vibrator at one end and passing over the pulley, will have its length perpendicular to the direction of vibration (see Figure 22–1).

3. Attach the lighter of the two strings to the vibrator and add weights to the weight holder until a distinct standing wave with about seven or eight loops is produced. If the weights available are not small enough for a fine adjustment, a small movement of the pulley toward or away from the vibrator may accomplish the desired result.

4. If we examine the vibrating string near the point of attachment to the vibrator, it will be noted that this point cannot be a true node (a stationary point). Hence, in our determination of the wavelength, the loop nearest the vibrator should not be used. Count the number of remaining loops and measure the total length of the group. From this measurement the average value of λ can be computed. Also record the amount of the stretching force F, expressed in terms of the mass added to the end of the string.

5. Keep the distance between the pulley and the vibrator the same, except for fine adjustments, and increase the stretching force until a distinct standing wave is produced with the number of loops reduced by one between pulley and vibrator. Again measure the total length of the group of distinct loops, and record all information as in Step 4.

6. Continue the process of reducing the number of loops by one until you have added enough force to produce two or three loops; record all data as before. Stop the vibrator but leave your setup as is for later use.

7. Compute the wavelength λ for each of the trials with the lighter string, and convert the unit of the stretching forces to newtons.

Method A

8A. Examine your recorded values of F and λ obtained from Step 6. When λ is doubled, how much is F increased? Since F increases faster than does λ, we know that the n in Equation [1] is greater than 1. Compute λ^2 and record in the data table.

Suppose we try for a straight-line relation by plotting F versus λ^2, with F in newtons on the Y axis and λ^2 in square meters on the X axis. Determine the slope and record in the lower data table with the proper units.

9A. If a plot of F versus λ^2 gives a straight line, then λ^2 is proportional to F, or λ is proportional to $\sqrt{F}$. This says that the value of n in Equation [2] is one-half, and the equation takes the form $\lambda = k\sqrt{F}$, or $\lambda = kF^{1/2}$.

10A. Now let us keep F constant and investigate the relation of λ to the mass per unit length m. With a heavier string in place, use a stretching force exactly equal to one of the larger forces in the preceding observations. Adjust the position of the pulley until a distinct standing wave is produced; then measure and record λ.

11A. For the case of two different strings with the same stretching force, compute the ratio of λ for string 1 to λ for string 2. With the mass per unit length for the two strings designated m_1 and m_2, respectively, compare the ratio λ_1/λ_2 to the ratio m_1/m_2. Do you find a direct relation or inverse relation? Try comparing with m_2/m_1. Does this more nearly approach a direct proportion? In Step 8A, what procedure was used to get a direct proportion between λ and F? Try the same scheme with λ and m.

12A. With a fixed string, m remains constant and thus becomes a part of k in Equation [2]. The frequency of the vibrator remains constant and is also a part of k.

13A. It can be shown that the velocity of a wave in a stretched string is given by the relation $v = \sqrt{F/m}$. Use Equation [1] with this velocity relation to eliminate v from the above equation and then, using some particular set of your measured values of λ, F, and m, compute the frequency of the vibrator. Show all of these computations as a part of your report. Check with the instructor on the manufacturer's value of the vibrator frequency and compute your percent discrepancy.

Method B

Looking back at Equation [2], we note that both k and n are unknown quantities. The relationship of the variables in this equation can best be examined by taking logarithms of both sides of the equation and writing it in the form

$$\log \lambda = \log k + n \log F. \qquad [3]$$

Because k is a constant, this equation has the form $y = mx + b$, where $y = \log \lambda$, $m = n$, $x = \log F$, and $b = \log k$.

Hence, if log λ is plotted versus log F, we should expect a straight line of slope n, and log k will be the intercept on the Y axis. Since the mass and length of the string cannot be eliminated, we might expect that they, being constant, are incorporated into the constant k. Hence, the effect produced by using a different length, or different mass, would be reflected in different values of k without affecting the slope of the line. The lines would be parallel but have different intercepts.

8B. Plot a graph using log λ as ordinate and log F as abscissa, and determine the slope of the line for the value of n in Equation [3].

9B. Extrapolate the line to the Y axis and determine the intercept which should be log k. A more direct method of finding k is to pick an arbitrary point on the graph for a value of F and λ, substitute in Equation [3], and solve for k. Show the calculations required in Steps 8B and 9B as a part of your report and write Equation [2] with your values of k and n. You now have an empirical equation of the wavelength as a function of the tension of one particular string as it vibrates with a standing wave.

10B. Now let us look at the effect of a change in length of the string on the wavelength. Select some force you have already used, and when a distinct standing wave has been reestablished, move the pulley until you have changed the number of loops to either one more or one less than in the original arrangement. Now measure the wavelength as before and compare it with the value obtained in Step 7. Discuss the results in your report.

11B. In the Introduction we concluded that the mass per unit length of the string was incorporated into the constant k for a particular string. If we should now change the mass, we should expect a change in k. Make one trial with a heavier string, and when a distinct standing wave has been produced, record the wavelength λ and the tension F.

12B. Use your value of n from Step 8B, and by substituting your values of λ and F (Step 11B) into either Equation [2]

or Equation [3], compute the value of k for the heavier string.

13B. It is quite probable that you have found the wavelength to be smaller and the value of k to be larger for the heavier string than for the lighter one. If so, does it seem possible that k is inversely proportional to the mass per unit length? Let the mass per unit length, M/L, be designated by m and make a comparison of the ratios k_1/k_2 and m_2/m_1. If you get a poor comparison, try taking the square root of the larger of the two ratios and compare again.

Optional. If time permits, and the background of the students is sufficient, the following steps will prove to be an interesting extension of the study.

14B. Since λ is directly proportional to k, and k is inversely proportional to the square root of m, then λ is inversely proportional to $\sqrt{m}$. If we now let $k = K/\sqrt{m}$, then Equation [2] can be written as

$$\lambda = K(F^n/\sqrt{m}) \qquad [4]$$

where K is a new constant containing the remaining quantities which were held constant.

15B. The only quantity not yet considered is the frequency of the vibrator. This quantity is not easily changed, but perhaps we can find its value for the vibrator being used. Recall your experiences of shaking a rope with your hand. How does the wavelength change with a change in the rate of vibration of the hand? Hence, how does wavelength vary with frequency? If the new K in Equation [4] contains only the frequency, then replace K with $1/f$, and with some set of known values for all other quantities substituted into the equation, solve for f and check with the instructor on the manufacturer's value of the vibrator frequency.

Make all of the calculations associated with the preceding steps an important part of your report.

QUESTIONS

1. In general what effect does increasing the tension in a vibrating string have on the wavelength?

2. In which cases did you find a small change in the force to have the greatest effect on the standing wave adjustment? Was the percent difference smaller than in other cases? (Illustrate with calculations.)

3. What effect does increasing the tension (stretching force) have on the frequency of the wave? Give evidence from your data in support of your answer. What determines the frequency of the wave?

4. What kind of graph did you obtain? What does this show about the relation of the quantities plotted? Examine all the quantities in Equation [3], and explain why this relation should exist.

5. With the same stretching force but different-sized strings, what relations do the computations of Step 13B show to exist between the wavelengths and masses per unit length of the string? By using λ_1 and m_1 for the wavelength and mass per unit length, respectively, for string 1, and λ_2 and m_2 for string 2, substitute in Equation [3] for each string and show that the above-mentioned relation should exist.

6. Assume the distance between the pulley and the vibrator to be 150 cm for the first string used, and compute the stretching force in newtons required to cause this whole length to vibrate in one loop.

7. Why should one not use the loop nearest the vibrator in the length measurement?

8. If the pulley, over which the masses are suspended, were moved slowly toward the vibrator for a considerable distance, what effects would you likely observe? If not sure, try it and see.

9. All strings on a violin are the same length. What characteristic differences do they have that gives them different frequencies (different pitch)? What other way can the frequency be changed?

10. Revise Equation [4] by putting in your value of *n* and replacing *K* with its equivalent in terms of the frequency *f*.

11. Examine Equation [4] and then consider the effect produced when a violinist pushes one of the strings against the neck of the violin. Which quantities in Equation [4] remain constant and which are changed? Does the pitch get higher or lower?

12. A piano is capable of producing many frequencies. Which factors in Equation [4] are utilized in producing the variety of tones? Explain how it is done.

13. Should the length of the string be incorporated into Equation [4]? Justify your answer with experimental evidence?

14. On the basis of Newton's second law of motion and the inertia of the heavier string, make a qualitative statement about the comparative velocities of the waves in the two strings under equal tensions.

15. Using one of your values for the wavelength and the computed frequency of the vibrator, compute the corresponding velocity of the wave in the stretched string.

NAME *(Observer)* _____ Date _____

(Partner) _____ Course _____

RECORD OF DATA AND RESULTS

Experiment 22—A Study of Vibrating Strings

String 1. Mass _____ gm Length _____ cm $\dfrac{\text{Mass}}{\text{Length}}$ _____ gm/cm _____ kg/m

String 2. Mass _____ gm Length _____ cm $\dfrac{\text{Mass}}{\text{Length}}$ _____ gm/cm _____ kg/m

OBSERVED DATA

String	Load mass (kg)	Force, F (N)	Length measured L	Number of loops N	Wavelength λ	Method A λ^2 / Method B $\log \lambda$	Log F
1							
2							

CALCULATED RESULTS AND ANALYSIS OF DATA

Method A			Method B		
Slope of graph F *versus* λ^2 =			Slope of graph $\log \lambda$ *versus* $\log F$ =		
Comparison of strings — below			Log k = k (graph) = k (calc) =		
λ_1/λ_2 =	m_1/m_2 =		String 2 λ_2 = $\sqrt{F}$ = k_2 =		
m_2/m_1 =	$\sqrt{m_2/m_1}$ =		Comparison of strings	k_1/k_2 = m_2/m_1 = $\sqrt{m_2/m_1}$ =	
Frequency of vibrator	f (calc) =	Percent discrepancy	Frequency of vibrator	f (calc) =	Percent discrepancy
	f (mfg) =			f (mfg) =	

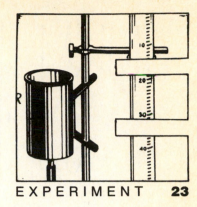

VELOCITY OF SOUND IN AIR—RESONANCE-TUBE METHOD

NOTE TO INSTRUCTOR:

Two alternate procedures are given for this experiment in order to adapt the method to either of two different designs of apparatus. Procedure A is for the conventional resonance tube containing water. Procedure B is for a resonance tube using two telescoping metal tubes to provide a variable length arrangement. Aluminum tubes with outside diameters of 1.5 in. and 1 3/8 in., respectively, and a wall thickness of 0.058 in. work quite well. Plastic tubes would also serve the purpose.

SPECIAL APPARATUS:

Procedure A—Container for water, thermometer, two tuning forks (different frequencies), rubber mallet, meter sticks, pinch clamp. Procedure B—Two telescoping metal tubes, end plug (cork or wood) for tube of smaller diameter, two tuning forks (different frequencies), rubber mallet, two meter sticks, thermometer.

GENERAL APPARATUS:

Procedure A only—Resonance tube, supply of string.

THE PURPOSE OF THIS EXPERIMENT

is to determine the wavelength in a resonating air column, and from this to determine the velocity of sound in air.

INTRODUCTION

If a tuning fork is set in vibration and held over an air column, the loudness of its note will be greatly increased if the air column has a length which will vibrate in sympathy with the fork. Such an air column is said to be in resonance with the fork. The waves set up in the air column are called standing waves. The shortest tube (closed at one end) that will give resonance is one-quarter of a wavelength, $\lambda/4$, but if the tube is made longer, resonance will occur also at odd fourths, namely, $3\lambda/4$, $5\lambda/4$, and so on.

If f is the frequency of the source and λ the wavelength of the standing wave, then the velocity of the sound is given by

$$v = f\lambda. \qquad [1]$$

A closed pipe (air column) has a node N at the closed end and an antinode A at the open end (Figure 23–1). Unfortunately, the antinode is not located exactly at the open end but a little beyond it. A short distance is required for the equalization of pressure to take place. This distance of the antinode above the end of the tube is called the *end correction*, and is equal to about 0.6 of the radius of the pipe.

Because of the end correction, the pipe length in Figure 23–1a will be slightly less than $\lambda/4$. However, the distance between two nodes as shown in b and c will give the exact value of $\lambda/2$. Since the distance between two nodes is $\lambda/2$, we can obtain the wavelength λ, and, if the frequency of the source is known, the velocity of sound at room temperature can be obtained by Equation [1]. From this correct value of $\lambda/2$ the correct value of $\lambda/4$ is known, and by subtracting the

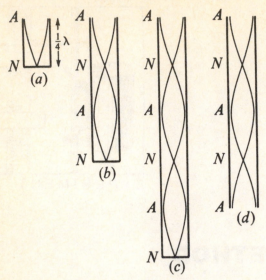

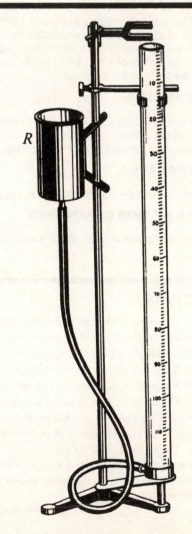

Figure 23–1 Resonance Diagrams for Different Lengths of Air Columns.

length of the pipe in *a* from λ/4 the end correction is obtained.

If the resonating tube is open at both ends as in Figure 23–1*d*, it may be noted that antinodes will appear at both ends. The tube length in this case is one wavelength. The shortest open tube that would resonate has a length of λ/2, a node at the center, and an antinode at each end.

The velocity of sound in air at 0°C is 331.5 m/sec, and, as the temperature rises, it increases at the rate of about 60 cm/sec per degree Celsius. Hence, the velocity v_t at temperature t is obtained from the velocity v_0 at 0°C by the relation

$$v_t = v_0 + 0.6t \text{ m/sec.} \qquad [2]$$

PROCEDURE A

1. Record the frequency of each tuning fork, the temperature of the room in the region near your apparatus, and the inside diameter of the tube.

2. Fill the tube (Figure 23–2) nearly full of water. The level can be adjusted by raising and lowering the reservoir *R*. *Caution. Do not at any time let the vibrating fork strike the top of the glass tube. It will break it.*

3. Start one of the tuning forks vibrating by striking it with the rubber mallet and, while holding it above the tube as shown, adjust the water level for the shortest length of the air column which gives a maximum resonance (increased loudness). Mark the point by tying a string around the tube, and move the water level past the string a few times to check the correctness of the position selected.

4. Now lower the water level until the next lower resonance position is found and mark it. Continue in this way as far as the tube permits. Estimate your uncertainty in centimeters in locating the position of a resonance level, and record in the lower section of data table. With this uncertainty at each resonance level, compute the percent uncertainty in λ.

5. Measure and record the positions of all the resonance levels located.

6. Use the other tuning fork to repeat all the preceding operations and record as before.

7. Subtract the reading for the first resonance position from each of the others in turn, and record as $L - L_1$, where L_1 is the distance to the first resonance position. The first difference will be λ/2, the next one will be λ, the next will be 3λ/2, and so on. From each of these differences determine λ, and then compute the average.

8. Use the average value of the wavelengths and compute by Equation [1] the velocity of sound in air at room temperature for each source used.

Figure 23–2 Resonance Tube (Courtesy of Sargent–Welch Scientific Co.).

9. By using Equation [2] compute the value of v_t at room temperature, based on the accepted value of 0°C, and determine the percent discrepancy of the average value obtained in your experiment.

10. From the average of the wavelengths for the fork of lower frequency, compute the value of $\lambda/4$, and then determine the end correction and compare with the value $0.6r$. The end correction $= \lambda/4 - L_1$, as indicated in the Introduction. Compute the percent discrepancy in λ if calculated from $\lambda = 4L$, and record.

PROCEDURE B

Caution:

Always use the rubber mallet to set the tuning fork into vibration. Do not strike either the metal tube or the table top with the tuning fork. Be careful not to drop or bump the metal tubes against any solid object. Dents will hinder sliding of the tubes (Figure 23–3).

1. Record the frequency of each tuning fork, the temperature of the room in the region near your apparatus, and the inside diameter of the larger of the two telescoping tubes.

The following procedure steps are best performed with the tubes lying flat on the laboratory table with one end projecting over the table edge.

2. While one partner holds a vibrating tuning fork (the one of higher frequency) at the mouth of the larger size tube, the other partner varies the length of the tube by sliding the smaller tube (stoppered at one end) back and forth as illustrated in Figure 23–3a. Begin with zero length (plunger all the way in), and slowly pull the small tube out; listen for the increased loudness of the resonating air column at various positions. Be sure the pitch of the resonating sound is the same as that of the fork alone.

3. First, locate the resonance position nearest the open end of the tube. Move the plunger slowly back and forth across this resonance point until you locate the position for maximum intensity. Then measure the length of the tube by inserting a meter stick inside. Record this length as the position of resonance from the open end of the tube. Estimate your uncertainty in centimeters in locating the position of a resonance point and record in the lower section of data table. With this uncertainty at each resonance level compute the percent uncertainty in λ.

4. Now pull the plunger out until the next resonance point is located and measure the new length of the resonance column and record. Continue this procedure as far as the length of the tube will permit.

5. Remove the plunger tube and reverse its position as shown in Figure 23–3b. This arrangement allows a longer resonating column. Continue the procedure of Step 4 by locating all the additional resonance points provided by the extension.

6. Use the other tuning fork and repeat all the preceding operations; record as before as data for fork 2.

7. Remove the end plug from the smaller tube to achieve a tube of variable length open at both ends. Use the original fork of higher frequency and determine and record at least three successive lengths for three successive resonance positions in the open tube. Record the data.

8. Make all calculations for the closed tube data in accordance with the directions in Steps 7–10 of Procedure A.

9. Determine the wavelength of the sound produced in the open tube (Step 7), and compare it with the wavelength in the closed tube by finding the percent difference.

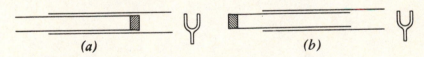

(a)   *(b)*

Figure 23–3 Telescoping-Tube Resonance Arrangement.

QUESTIONS

1. Explain how you would proceed in finding the frequency of an unmarked fork, using this apparatus and the information you now have.

2. From your results what value would be obtained for the velocity of sound at 0°C.

3. Neglect the end correction and assume the tube length to be for the first resonance position found for fork 1. What is the next highest fork frequency that would give resonance without changing the tube length?

4. Does it appear from your results in only two trials that the velocity of sound in air depends upon either the frequency or the wavelength of the waves produced by the tuning forks? Why?

5. If the end correction is neglected and the wavelength taken to be four times the length of the pipe, in which would the greater error be likely to occur, a long pipe or a short pipe, assuming the diameter to be the same for both? Explain.

6. If the temperature of the air in the resonance tube is 60°C, what fork frequency is required to produce resonance at the same positions as found in this experiment for fork 1?

7. What is the shortest open tube (open at both ends) that will resonate with your lower frequency fork? How does this length compare with your shortest closed tube? (If you used Procedure A for the experiment, Figure 23–1*d* will serve as a hint to the answer.)

8. Suppose we compare the percent discrepancy in the experimental value of v_t with the percent uncertainty in the measurement of λ as shown by your percent uncertainty in locating the resonance level. Does this comparison reflect a need for careful measurement? Explain.

NAME *(Observer)* _____ Date _____

(Partner) _____ Course _____

RECORD OF DATA AND RESULTS

Experiment 23 — Velocity of Sound in Air — Resonance-Tube Method

Temperature of room _____   Diameter of tube _____

Frequency of fork 1 $f_1 =$			Frequency of fork 2 $f_2 =$		
Position of resonance	$L - L_1$	Wavelength, λ_1	Position of resonance	$L - L_1$	Wavelength, λ_2
Average value of λ_1			Average value of λ_2		
Velocity of sound (Equation 1) =			Velocity of sound (Equation 1) =		
Avg. V (Equation 1) = m/sec		V_t (Equation 2) = m/sec		Percent discrepancy	

RECORD OF CORRECTIONS AND UNCERTAINTIES

End correction results		Uncertainty in resonance level	
¼λ −L₁ = cm	$0.6r =$ cm	Estimated uncertainty = cm	
Percent difference =		Percent uncertainty in λ (calc.) = %	
Avg. $\lambda_1 =$ cm	$4L_1 =$ cm	Percent uncertainty in λ (calc.) = %	
Percent discrepancy by using λ = $4L_1$ =			

DATA FOR OPEN TUBE — PROCEDURE B

Frequency of fork used =		
Position of resonance	$L - L_1$	Wavelength, λ
Average value of λ =		Percentage difference
Avg. λ closed tube =		

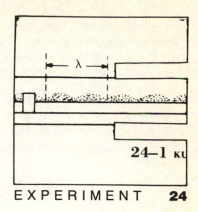

VELOCITY OF SOUND IN A METAL—KUNDT'S-TUBE METHOD

SPECIAL APPARATUS:

Meter stick, thermometer, box of resin, cloth.

GENERAL APPARATUS:

Kundt's-tube apparatus, cork dust, table clamp.

THE PURPOSE OF THIS EXPERIMENT

is to measure the velocity of sound in a metal rod by the Kundt's-tube method and to determine Young's modulus of the metal.

INTRODUCTION

The Kundt's-tube apparatus (Figure 24–1) consists of a hollow glass tube closed at one end. The tube is mounted on a frame in such a way that it can be moved longitudinally with respect to the frame. A metal rod m is clamped to the support frame, exactly at its center. This metal rod carries a disk at one end which is inserted into the glass tube but does not touch the tube. The tube contains cork dust which is distributed along the length of the tube.

When the rod is stroked lengthwise with a resined cloth, longitudinal standing waves are set up in it with minimum vibration (node) at the clamp and maximum vibration (antinode) at each end. Since the distance between successive antinodes (or nodes) in a standing wave is one-half wavelength, the wavelength of the tone in the rod is twice the length of the rod.

The sound waves are produced in the rod as the layers of molecules are successively displaced and released by the resin cloth and the frequency of the vibrations in a given metal rod depends on the length of the rod and the position of the clamp. The vibrations are transmitted to the disk, which in turn transmits them into the air column at the same frequency. The wavelength and velocity change as the wave train goes from one medium to another but the frequency remains the same.

As longitudinal waves leave the rod at the end containing the disk, they proceed down the tube, which acts as a closed pipe (closed at one end). Hence, if the distance between the disk and the closed end is such as to produce resonance, the cork dust will be agitated at the antinode positions and remain relatively still at the node positions (Figure 24–1).

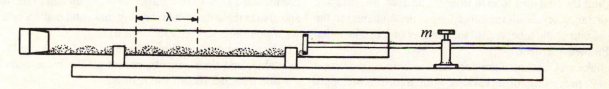

Figure 24–1 Kundt's-Tube Apparatus.

The length of one dust loop (the distance between successive nodes) is one-half of the wavelength in air.

If V_m is the velocity of sound in the metal and λ_m the wavelength, the frequency f is given by

$$f = V_m/\lambda_m. \qquad [1]$$

If V_a and λ_a are the velocity and wavelength, respectively, in the air column at room temperature, the frequency is given by

$$f = V_a/\lambda_a. \qquad [2]$$

The frequency of the sound depends only on the source and is therefore the same in both the metal and the air.

It can be shown that the velocities of sound in air, V_1 and V_2, at two different temperatures, T_1 and T_2, respectively, are related to the temperatures as shown in the equation

$$V_1/V_2 = \sqrt{T_1/T_2} \qquad [3]$$

where T_1 and T_2 are expressed as absolute temperature. If one of these temperatures is 0°C (273°K) and the other temperature is not too far from 0°C, then Equation [3] can be expanded and the result approximated by

$$V_a = V_0 + 0.61\,t \qquad [4]$$

where V_0 is the velocity of sound in meters per second at 0°C (273°K), and V_a is the velocity in meters per second at t°C. The factor, 0.61, is the increase in velocity in meters per second per degree rise in temperature in degrees Celsius.

V_0 in Equation [4] has been determined as 331.5 m/sec. Thus, we have all the information needed to obtain V_m, the velocity of sound in the metal rod.

The velocity of a compressional wave in a metal depends on the elastic properties and the density of the metal, and its value in centimeters per second (or meters per second) is given by

$$V_m = \sqrt{E/d} \qquad [5]$$

where E is Young's modulus (coefficient of elasticity) and d is the density. E must be expressed in absolute force units per unit area, dynes per square centimeter or newtons per square meter. See your text for a discussion of Young's modulus.

PROCEDURE

1. Remove the glass tube from the support and distribute the cork dust uniformly throughout most of the length of the tube by carefully shaking it back and forth, and then replace it in the support.

2. Record the kind of material of which the rod is made, and measure and record its length. Adjust it so that it may be clamped at its midpoint, being careful that the disk does not press against the walls of the glass tube (Figure 24–1). Leave it free to vibrate. Also note and record the temperature of the room near your apparatus.

3. Rub just a little resin on a cloth and stroke the rod lengthwise to get a high-pitched tone, but do not let your hand slip off the end of the rod. This will cause both ends to vibrate transversely, and the vibrating disk may break the glass tube. If the cork dust is not agitated, change the length of the air column by moving the glass tube a short distance. Continue the procedure after each stroking until the best resonance condition (maximum agitation) is obtained. If the rod gets too warm, cease stroking it until it cools.

4. Omit the first dust loop in front of the disk, and measure the distance across all the other loops (or segments) to the closed end of the tube; record the number of loops included in the measurement.

5. Shake and distribute the cork dust as before, and repeat all the preceding observations with a different length between the disk and the closed end of the tube.

6. From the measured distance the number of dust loops included, determine the average wavelength (see Figure 24–1) of sound in the air column for each trial and record as λ_a. Then calculate the velocity of sound in air from Equation [4].

7. By using the average wavelength from the two trials, compute the frequency of the sound by Equation [2].

8. From the wavelength in the rod and, using Equation [1], compute the velocity of sound in the metal and compare it with the accepted value (see tables in Appendix B).

9. From the tables in Appendix B determine the density of the metal used and, using Equation [5], compute its coefficient of elasticity. Make a *careful check of the units*, and then compare your result with table values for the coefficient of elasticity (Young's modulus).

10. If the stopper closing the glass tube at the end is removable, adjust the distribution of the cork dust and remove the stopper. Now stroke the rod and adjust the position of the tube until maximum agitation is obtained. Examine the cork dust agitation at the open end of the tube and also in the area of the vibrating disk, and make a note of what you observe (see Question 6).

11. If a metal rod of a different kind of material is available, and time permits, repeat all the measurements for one trial and record the results.

QUESTIONS

1. Would you expect to find a node in the air column next to the disk? Explain.

2. How would you suggest using equipment of this type to measure the velocity of sound in carbon dioxide? Explain how the necessary information would be obtained.

3. What would be the effect on the distance between nodes in the cork dust if the experiment had been done on a hot summer day?

4. What are your percent discrepancies in the velocity and coefficient of elasticity? (Use table values as correct.)

5. If the rod were clamped in two places, each at one-fourth the distance from the end, what difference would you find in the cork dust pattern? Why?

6. What was your observation concerning the cork dust agitation at the end of the tube when it was opened by removing the stopper?

7. Draw two diagrams, one of a closed tube and one of an open tube, and illustrate the character of the wave form of a standing wave in each.

8. If the rod in your apparatus had been half as long and the glass tube had been filled with hydrogen at 0°C, what would have been the distance between nodes in the cork dust pattern? (Refer to Appendix B for information on hydrogen.)

9. As stroking of the rod continues, cork dust gradually accumulates at the nodes. Explain why this happens.

10. Why does the disk on the end of the rod have to be at some particular distance before the cork dust is agitated by the sound waves?

11. If sand grains were used in the glass tube to show the wave pattern, would you expect the general nature of the wave pattern to be different? Explain. What advantage, if any, may cork dust have in comparison to sand?

12. In Step 9 of the procedure you computed the coefficient of elasticity. Show the steps of the method used to arrive at the proper units for the coefficient.

NAME *(Observer)* _____ Date _____

(Partner) _____ Course _____

RECORD OF DATA AND RESULTS

Experiment 24 — Velocity of Sound in a Metal — Kundt's-Tube Method

OBSERVED DATA

Temperature of the room _____

Material of rod	Density of rod material	Length of rod	Wavelength in the rod λ_m	Length of tube measured	Number of dust loops measured	Wavelength in air λ_a
1.						
1.						
2.						

RECORD OF CALCULATED RESULTS

Velocity in air V_a	Frequency of sound f	Velocity of sound in metal V_m	Velocity of sound in metal (Table)	Percent discrepancy	Coef. of elasticity calculated	Coef. of elasticity (Table)	Percent discrepancy
1.							
2.							

Show calculations as part of report.

PART FOUR
ELECTRICITY AND MAGNETISM

4

MAPPING OF ELECTRIC FIELDS

SPECIAL APPARATUS:

Simple switch, bundle of connecting wires, detector (galvanometer, earphones,* or high-impedance voltmeter), graph paper.

GENERAL APPARATUS:

Field mapping board with probes, conducting sheet with electrode configuration,† source of potential (dry cell, audio oscillator,* or dc power supply).

THE PURPOSE OF THIS EXPERIMENT

is to examine the nature of electric fields by mapping the equipotential lines and then sketching in the lines of force.

INTRODUCTION

An *electric field* is a region in which forces of electrical origin are exerted on any electric charges that may be present. If a force F acts on a charge q at some particular point in the field, the electric field strength E at that point is defined as the force divided by the charge, and the *magnitude* is given by

$$E = F/q. \qquad [1]$$

Since E is a vector quantity, it also has direction, and we arbitrarily define the *direction of an electric field* as the direction of the force on a positive test charge placed at the point in the field.

Faraday introduced the concept of lines of force as an aid in visualizing the magnitude and direction of an electric field. A *line of force* is defined as the path traversed by a free test charge as it moves from one point to another in the field. Figure 25–1 shows several possible paths that a test charge might take in going from the positively charged body A to the negatively charged body B. The relative

magnitude of the field intensity is indicated by the spacing of the lines of force and the arrows indicate the direction.

Electric Potential. Since a free test charge would move in an electric field under the action of the forces present, work is done by the field in moving charges from one point to another. If external forces act to move a charge against the electric field, then work is done on the charge by the external force. If the charge q (Figure 25–2) is placed at a point very far from the charge Q, where the repulsion is essentially zero, the work per unit charge to move it from this point to point A is called the *absolute potential* at A. If the charge is now moved from A to B, against the electric field, work is again done, and the ratio of work W to charge q is called the *potential difference* between the two end points of the path traversed. Hence we may write $V = W/q$, where V is the difference of potential.

However, if the charge is moved along a path at all times perpendicular to the lines of force, such as from A to C, there is no force component along the path, and hence,

---

*The earphones require an audio oscillator.
† The electrode configurations furnished with the equipment may be used, or, with the aid of silver paint, the instructor can make other configurations on blank conducting sheets.

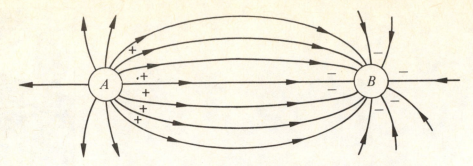

Figure 25–1 Lines of Force in an Electric Field.

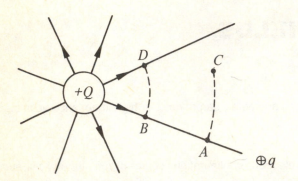

Figure 25–2 Potential Difference in an Electric Field.

there is no work done. Then points *A* and *C* are said to be at the same potential, and the path traveled is called an equipotential line. The path from *B* to *D* might be another equipotential line, but with a different value of absolute potential than the line *AC*.

One can readily see that many equipotential lines, or surfaces, are possible in an electric field. Experimentally, it is much easier to trace the path of equipotential lines than to trace the lines of force. When a network of equipotential lines has been mapped out, the lines of force, being everywhere normal to the equipotential lines, can be readily plotted.

DESCRIPTION OF APPARATUS

Two types of equipment used for mapping electric fields are shown in Figures 25–3 and 25–4. A high-resistance, slightly conductive sheet is used as the medium for mapping the field. On the sheet is an electrode configuration of conducting silver paint which provides the electric field when connected to a source of potential. The equipment in Figure 25–3 uses a galvanometer connected between two probes to locate points having the same potential. When there is no current through the galvanometer the two probes are at the same potential. Thus, a series of points can be found at the same potential which then defines an equipotential line in the electric field.

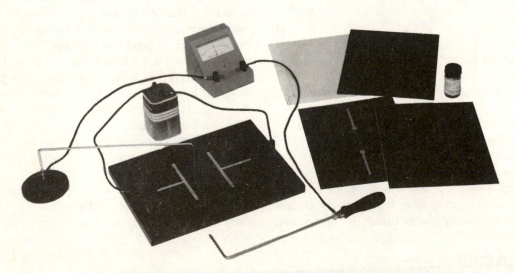

Figure 25–3 Electric Field Mapping Equipment (Courtesy of Sargent–Welch Scientific Co.).

The equipment in Figure 25–4 uses a high-impedance voltmeter connected between a terminal fixed to one of the electrodes and a free probe which can be moved around the conducting sheet to locate points having the same potential. A line drawn between points of the same potential defines an equipotential line in the electric field.

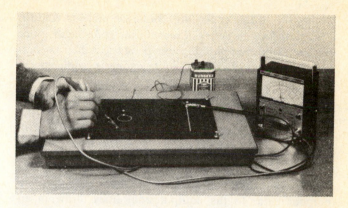

Figure 25–4 Electric Field Mapping Equipment (Courtesy of Pasco Scientific Co.).

PROCEDURE

If using the equipment in Figure 25–3, follow Steps 1 through 6 and Steps 11 through 14, omitting Steps 7 through 10. If using the equipment in Figure 25–4, omit Steps 1 through 6 and follow Steps 7 through 14.

Your instructors will provide you with one or more sheets, each with an electrode configuration on it, or they will provide material and instruction on how to create your own electrode configuration. Your principal data sheet for this experiment will be your mapped electric field. Provide an additional organized sheet containing recorded information and the answers to assigned questions.

1. Carefully place the sheet on the board with the contact terminals firmly on the painted outlet from the electrode. Then connect the assembly as shown in Figure 25–3, leaving the switch open until you are ready to operate the equipment.

2. Place the stationary probe on the sheet some place in the general region between the electrodes, preferably near the edge of the paper. The potential at this point will serve as your reference potential in the location of a series of other points with the same potential.

3. The galvanometer is a very delicate instrument and is easily damaged unless the current is limited to very small values. The amount of current sent through it will be proportional to the difference in potential of the two probes as they make contact with the sheet. Since we are now ready to begin, where, with respect to the stationary probe, should the movable probe be placed so as to protect the galvanometer? If you have any doubts check with the instructor. (If earphones are used as a detector, then no sound indicates zero potential difference.)

4. You will note a cross-section arrangement on the sheet which will aid you in locating the coordinates of the probe positions. Arrange a sheet of graph paper with the electrode configuration in the proper position on a coordinate system

that will correspond with the one surrounding the electric field to be mapped. As the points are located their position can be plotted on the graph paper.

5. With the switch closed and the movable probe at a safe position, move it until the galvanometer shows zero deflection, thus indicating that the two points located in the electric field are at the same potential. Now locate a series of other points with this same potential until you have some eight or ten points reaching across the general region of the field. If a smooth curve is drawn through these points, what could you say about other points on this curve?

6. Choose a new location for your reference probe, say 1 to 2 cm from the previous position, and locate another series of equipotential points. Continue this process until you have mapped the entire electric field region. Contine with Step 11.

•••••••••••••••••••••••••••••••

7. Mount the electrode sheet to the corkboard using the pins provided. Connect the silver electrodes to the power source.

8. Using a pin, connect the fixed terminal to one of the silver electrodes. (The voltmeter polarity connected to this lead should be the same as that of the power source connected to the electrode.) Be sure all contacts are clean and a good connection is made. Touch the free probe to the other electrode and adjust the range of your voltmeter to some convenient value close to the full range of the meter. Record this value. Now touch the free probe to the same electrode to which the fixed lead is connected, and record the voltage. If this value is greater than 1 percent of your previous reading across the electrodes, you should check your connections.

9. Pick a point between the electrodes on the conducting sheet to start your mapping and mark it with a + using a soft lead pencil. Touch the free probe to this point and record the voltmeter reading. Move the probe around the sheet and locate a series of points which provide the same voltmeter reading. Mark these points and then connect them with a smooth line. This line defines an equipotential line.

10. Repeat Step 9 for another equipotential line. The new starting point should be between 1 and 2 cm from the previous line. Continue this process until the entire conducting sheet is mapped.

•••••••••••••••••••••••••••••••

11. Remembering that the lines of force are everywhere perpendicular to the equipotential lines, draw in an arbitrary number of smooth uninterrupted lines to represent the lines of force, and place arrows on them to indicate direction. Consider the positive electrode as a positively charged body. This system of lines of force gives a graphical representation of the general nature of the electric field for this one configuration of electrodes.

12. You will note that the lines of force are closer together in some regions than in others. Is it also true for the equipotential lines?

Place both probes at a point on the coordinate grid that corresponds to a point on a line of force where the spacing of the lines of force is wide. Now examine the change in potential along the line of force, in both directions, as the hand probe is slowly moved away from the reference probe. Repeat this procedure for other points on this same line of force, some of which are in regions where the lines are closely spaced. What is the nature of your observation?

13. In analyzing your observations try to correlate them with the definition of absolute potential V_a at a distance S from charge Q, and recall that the value is given by $V_a = Q/S$. It will be helpful to also correlate your observation with the definition of electric field intensity in terms of the potential gradient, given by the relation

$$E = - \frac{\Delta V}{\Delta S}$$

14. *Optional Experiment.** If time permits, the instructor may wish to assign you a second mapping problem with a different electrode configuration, such as some nonsymmetrical arrangement.

QUESTIONS

Answer these questions in discussion form as if the answers are conclusions which you draw from an analysis of your observations.

1. If the space on the sheets surrounding the electrode configuration were completely nonconducting, explain how your observations with the charged probes would be affected.

2. If the surface on the sheet where highly conductive, how would the difference of potential between two points on the grid region be affected? Explain.

3. If the conductivity of the sheet were not uniform, what effect would this have on the nature of your plotted pattern?

4. What is the nature of the path by which the electric charge flows from one electrode to the other on the sheet? How does your pattern support your answer?

5. Examine your electric field pattern and state what relationship (if any) is observed between the regions of close spacing of the lines of force and the regions of close spacing of the equipotential lines. Explain this relationship in terms of the definition of potential gradient given in Step 13.

6. State the results of your observations in Step 12 and explain their significance in terms of the hints given in Step 13. Do your observations verify the negative sign in the potential gradient relation? If so, explain how.

7. If a more careful and a more complete mapping job were done, do you think you might encounter situations where two equipotential lines would intersect each other? If two such lines should intersect, and the point of intersection is on a line of force, explain what this would mean in terms of the concept of electric potential.

8. How much work, in symbolic form, would be required to move a charge q across the center portion of your electric field from one side of the sheet to the other if the charge follows an equipotential line? Explain.

9. Suppose a line of force is not perpendicular to an equipotential line at some point P but makes an angle θ with it. Make a sketch of this intersection, using a vector for the field intensity E at point P, and explain the significance of the situation with respect to the motion of charge q along the equipotential line.

*Note to instructor. It may be desirable to either omit this optional experiment or use a second laboratory period to thoroughly analyze some particular configuration which fits the needs of the class. Other configurations are easily made by the use of silver paint.

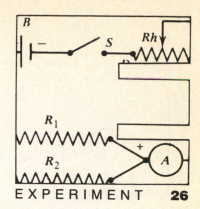

A STUDY OF SERIES AND PARALLEL ELECTRIC CIRCUITS

NOTE TO INSTRUCTOR:

This experiment can be done in the laboratory without any previous knowledge of circuits on the part of the students. It will be more interesting and challenging to students if they have not studied the theory of circuits. These are experiments designed to familiarize students with electrical hookups and the use of simple electrical equipment. A class discussion of the items in Appendix C might well serve as a laboratory lecture to introduce the students to their first experiment in electricity.

SPECIAL APPARATUS:

Ammeter (range 0–0.5 A, preferable), voltmeter, variable resistance box beginning with 0.1 ohm, fixed resistor between 10 and 20 ohms, slide-wire rheostat of about 22 ohms, single-pole switch, bundle of connecting wires.

GENERAL APPARATUS:

Source of direct current (2–4 V).

THE PURPOSE OF THIS EXPERIMENT

is to learn how to use electrical equipment and to study the distribution of current and energy in an electric circuit.

INTRODUCTION

In a gaseous conductor, such as a neon sign tube, and in a liquid conductor, such as an electrolytic cell, positively charged particles flow in one direction and negatively charged particles flow in the other. In order that we may always be understood, some convention must be adopted for designating the direction of the current, and the direction positive charges flow has been chosen. In a metallic conductor, such as a wire, the only mobile particles are negatively charged electrons, which move in a direction opposite to that chosen for the conventional current. Throughout this laboratory manual we shall speak of the direction of the electric current as the direction positive charges would move if they were free to move.

The current in a conductor is related to the velocity of moving charges and may be designated as the quantity of charge passing a point per unit time. The usual unit of current is the *ampere* (A) and, although it is defined by

other relations, a current of one ampere exists in a wire if approximately 6.21×10^{18} electrons flow through a given cross section of wire in one second. The quantity of charge carried by this number of electrons is called one *coulomb* (C); therefore, one ampere of current is equal to a flow of charge of one coulomb per second. We frequently hear statements such as "This toaster sure does consume a lot of current." A part of the purpose of this experiment will be to test the validity of this statement.

Circuit elements which offer resistance, or "opposition," to current are sometimes called resistors and their opposing effect is called *resistance*. The unit of resistance is called the *ohm*. In this experiment, the current in a resistor or a combination of resistors will be measured with an ammeter (a contraction of the word amperemeter). If an ammeter is to measure the current in a resistor, it must be connected into the circuit so that all of the charge that

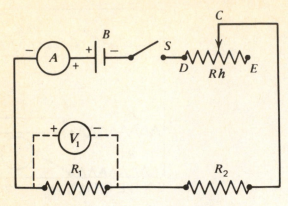

Figure 26–1 A Series Circuit.

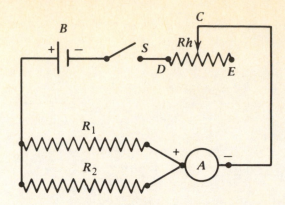

Figure 26–2 A Parallel Circuit.

flows through the resistor also flows through the ammeter. This type of connection, as illustrated in Figure 26–1, is called a series connection. If the connection is such that the two resistors are connected between the same two points, in order to divide the current, the elements are said to be connected in parallel (Figure 26–2).

Since there is resistance to the flow of charge there must be some force causing the charge to flow through the resistors. This further implies that energy is involved and we would expect the energy of the charge leaving a resistor to be less than the energy of the charge entering the resistor.

It is actually the potential energy of the flowing charge that changes so we speak of the potential energy difference or potential difference across a resistor. The electrical unit of potential difference is the *volt* (V). If a volt exists between two points it requires one *joule* (J) of energy to move one Coulomb of charge between the two points. A voltmeter is used to measure the potential difference (PD) between two points in an electric circuit. Since it measures the change in potential between two points it must be connected to the two points or in parallel with the circuit components. See Figure 26–1.

PROCEDURE

Before beginning any part of this experiment, read all of Appendix C, and examine the pieces of equipment which you are to use as you read about them. Record the ranges of the voltmeter, the ranges of resistors, the type of current source used, and the supply voltage if known.

The following projects should be *neatly* reported and discussed, one at a time, as they are carried out. All statements about your circuits and results should be self-explanatory so that the one who reads your report can follow it clearly without referring to the manual. Record all meter readings and other information below the appropriate circuit diagram illustrated on the data form. Use an additional sheet of paper if necessary. Also, note that the circuit diagrams on the data form show the correct meter connections.

A. Series Circuit
Preliminary circuit arrangements. *Without connecting to the current source* (battery or power supply), connect the ammeter A, the resistors R_1 (resistance box) and R_2, the rheostat Rh, and switch S in a series arrangement as shown in Figure 26–1. The purpose of the rheostat is to provide a variable current control as needed. The point C on the rheostat is the binding post at the end of the rod across the

top which connects to the slide knob. Ask the instructor to suggest a safe range of use for the ammeter, and be sure to arrange the polarity as indicated in the circuit diagram. The current should be no more than the maximum allowed for your resistance box. The needle of the ammeter can move across the scale only if the conventional (positive) charge enters the meter at the + terminal.

Voltmeter readings can be made simply by touching the appropriate contact points with the voltmeter leads. Be sure good contacts are made before reading the meter. The positive (+) side of the voltmeter should always be closest to the positive (+) side of the electrical source. Lightly touch the contacts first and insure that the meter deflects in the right direction. Avoid parallax errors when reading meters.

The switch should always be opened immediately after readings are made. The resistors have no polarity preference, since they are designed to carry current in either direction. Let the resistance box be R_1 and the fixed resistor R_2. Set R_1 at about 20 ohms.

Ask the instructor or laboratory assistant to check and approve your connections before closing the switch.

1. If the needle on the ammeter or voltmeter is not exactly on the zero mark, record the zero reading and be prepared to make a correction on all readings made from the meter, or ask the instructor to adjust it for zero reading. First record the direct reading from the meter and then the corrected reading after adding, or subtracting, the zero correction.

Close the switch and note the changes in the current while adjusting the rheostat. Adjust the resistance in the rheostat until the current is about 0.2 A. If your current source is a dc power supply, you may also control the current by adjusting the input voltage. Read the current and record as I_a beside the diagram designated (a) on the data form. Also record the values of R_1 and R_2.

2. With the switch S closed, measure and record the voltage across R_1, R_2, $R_1 + R_2$, a connecting wire, and the electric source. Open the switch between readings while recording data or proceed without hesitation to measure all values while your partner records the data. Compare the voltage across $R_1 + R_2$ with the sum of the voltages across each resistor. Calculate and record the percent difference.

3. Without changing any other adjustments of your circuit, connect the ammeter between R_1 and R_2 as shown in circuit diagram (b), making sure that the + terminal of the meter is connected to R_1. With this arrangement the current passes through R_1 before reaching the ammeter. Carefully read and record this current as I_b.

4. Again, without changing any other adjustments, move the ammeter to the position shown in circuit diagram (c) on the data form. Read the meter and record the current as I_c. Now look at the current value recorded for each of these three arrangements, and then relate the results to the statement in the Introduction to this experiment about the toaster. Write your comment below the recorded value of I_c.

5. Now remove the resistor R_2 from the circuit. Arrange your connections as shown in circuit (d) on the data form, *but do not change the settings on the power supply or rheostat.* Before closing the switch, readjust the setting on the resistance box to a higher value to compensate for the absence of R_2. On closing the switch, readjust the resistance box setting until the current is the same as it was when both R_1 and R_2 were in the circuit. We shall now call this new value of the resistance R_s. Record this value of R_s and record the current as I_d. Compare R_s to $R_1 + R_2$ and compute the percent difference.

6. Now using one or two sentences for each, comment on the following:

   a. Current in all parts of a series circuit.
   b. Voltage across resistors connected in series.
   c. Total effective resistance of resistors connected in series.

B. Parallel Circuit

Preliminary circuit arrangements. Connect the circuit as shown in Figure 26–2, again using the resistance box as R_1 and the fixed resistor as R_2. Rh is the slide-wire rheostat to

be used in controlling the size of the current, and it should be set for maximum resistance until the current needs are established. Adjust the value of R_1 to some multiple of 10 between 20 and 50 ohms. Be sure the + terminal of the ammeter is connected as shown in Figure 26–2. *Before closing the switch ask the instructor or his assistant to check your circuit connections.*

7. Close the switch cautiously while watching the ammeter to be sure that you are connected to a safe range. Adjust the rheostat until the ammeter reads 0.5 A or less. Record the current value along with the resistance values R_1 and R_2 beside diagram (a) on the data form. In this position, the ammeter reads the total current in the circuit, the combined amount through both R_1 and R_2. Designate this current as I.

8. With switch S closed, measure and record the voltage across R_1, R_2, and $R_1 + R_2$. Open the switch between readings while recording data or proceed without hesitation to measure all values while your partner records the data. What conclusions can you draw about voltages across parallel resistors?

9. Without changing any adjustments, connect the ammeter in the circuit branch of R_1 as shown in diagram (b) on the data form. The ammeter is now in series with R_1 alone and will register only the current in R_1. Close the switch, carefully read the current, and record as I_1 beside diagram (b), thus designating it as the current through R_1.

10. Now, keeping all adjustments the same, connect the ammeter in series with R_2 as shown in diagram (c). Record the current as I_2, thus designating it as the current through R_2.

Examine the three current values you have recorded and determine the relationship between them. Record your findings below the recorded value of I_2.

11. With the rheostat adjustment the same as in Step 8, remove the resistor R_2 from the circuit, thus leaving a circuit with R_1 and R_h in series as shown in diagram (d). After increasing the setting of the resistance box to a safe value, close the switch and adjust the resistance until the ammeter reads the same as recorded for circuit (a). Since these two currents are the same, we must conclude that we have replaced the parallel arrangement of R_1 and R_2 of circuit (a) with an equivalent resistance of R_e as shown in circuit (d). Record the current as I' and the resistance as R_e.

Is R_e equal to the sum of R_1 and R_2? If not, is it some value in between? Or, is it some value less than either of the other two? If your answer to the last question is yes, use the values of R_1 and R_2 from circuit (a) and try a calculated value of R_e from the relation $1/R_e = 1/R_1 + 1/R_2$, and record the computed value as R_e'. Compute the percent difference between R_e and R_e'.

12. Using one or two sentences for each, comment on the following:

   a. Total current in parallel resistors compared to individual currents.

b. Total voltage across parallel resistors compared to individual voltages.

c. Total effective resistance of resistors in parallel.

QUESTIONS

1. Is it correct to say that a toaster consumes (or uses up) a lot of current? Justify your answer with data obtained in this experiment. Write a statement about the current in the toaster which you think might be more appropriate.

2. When a group of elements are connected in series as shown in Figure 26–1, which element does the current get to first when the switch is closed? Justify your answer from observations made in this experiment.

3. Can two resistors in series be replaced by a single resistor and give an equivalent resistance to the circuit? If so, how must the single resistance be related to the two in series?

4. Does your experiment present evidence that the charge flow divides in a parallel circuit? Explain.

5. At a junction in a circuit where the three wires meet, such as at the negative terminal of the ammeter in Figure 26–2, how would you relate the amount of charge flowing into a junction to the charge flowing out of the junction? This relation is a very important law of electric circuits.

6. When resistors are connected in parallel, which one, according to resistance value, carries the larger portion of current? Does the ratio of the two currents appear to be directly, or inversely, proportional to the resistance? Justify your answer with calculations from your data.

7. When you had a parallel circuit, which of the resistances was the smallest, R_1, R_2, or the equivalent resistance R_e, found in Step 11? Explain.

8. What do your results show concerning the relative values of the current in the different parts of a series circuit? What do these results indicate about choosing a position for the ammeter in a series circuit?

9. When the voltmeter was connected across the switch near the battery, how did the reading compare with the assumed known voltage of the battery? If the resistors R_1 and R_2 were elements in an electric toaster connected to an ordinary house outlet, what would you expect a voltmeter to read when connected across the outlet contacts? (Disregard any differences between ac and dc current.)

10. What was the reading of the voltmeter when connected across the ends of a lead (or connecting) wire? Work is sometimes defined as the accomplishment of motion against an opposing force. Tie this definition in with the definition of difference of potential and the definition of resistance in terms of opposition offered to the current, and explain the significance of this voltmeter reading.

11. When the resistors are connected in series what relation do you find between the individual voltage and the individual resistance? Justify your answer with figures from your data.

12. In a series circuit, what relation do you find between the voltages across the individual resistors and the voltage across the entire series group? Support your answer with experimental data.

13. Does the insertion of additional resistance in a series circuit materially affect the difference of potential across the entire group? Some types of Christmas tree lights consist of about eight lights connected in series. If one bulb burns out, and a connection is completed around the empty socket so that the remaining seven lights will "burn," what happens to the voltage (PD) across the individual lights? Explain.

14. Suppose you have a string of the series-type Christmas tree lights with two or three of the bulbs burned out. You need to test your bulbs to see which ones are the good ones. Would you suggest connecting each bulb individually to the 120-V outlet? Why? Do you have a better suggestion based on what you have learned from this experiment? If so, how would you do it? (Assume the equipment used in this experiment is available to you.)

RECORD OF DATA AND RESULTS

Experiment 26—A Study of Series and Parallel Circuits

INFORMATION ABOUT THE CIRCUIT ELEMENTS

Type of power source _____ Voltage _____ V
Ammeter: Ranges _____ A Zero reading _____ A
Voltmeter: Ranges _____ V Zero reading _____ V
Resistors: Fixed _____ ohms Rheostat _____ ohms
Ranges of the resistance box _____ ohms

WIRING DIAGRAMS AND OBSERVATIONS IN THE EXPERIMENT

A. Series circuits

(a)

R_1 _____ ohms R_2 _____ ohms

	Meter reading	Corrected reading
I_a		
V_1		
V_2		
V_3		
V_4		
V_5		

$V_1 + V_2 =$ _____ V

Percent diff. (vs V_3) = _____

(b)

(c)

I_b meter _____ A I_c meter _____ A

I_b corr. _____ A I_c corr. _____ A

(d)

I_d _____ A I_d corr. _____ A

R_s _____ ohms $R_1 + R_2$ _____ ohms

Percent diff. (R_s vs $R_1 + R_2$) _____

Step 6

(a)

(b)

(c)

B. Parallel circuits

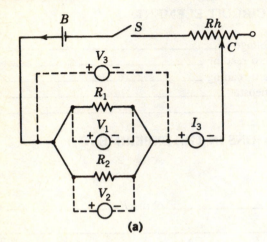

(a)

R_1 _____ ohms R_2 _____ ohms

	Meter reading	Corrected reading
I_a		
V_1		
V_2		
V_3		

Relation between V_1, V_2, and V_3

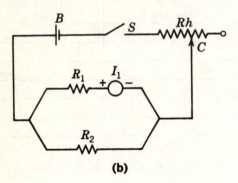

(b)

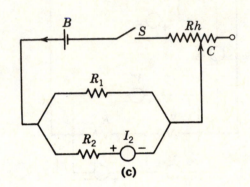

(c)

I_1 meter _____ A I_2 meter _____ A

I_1 corr. _____ A I_2 corr. _____ A

Relationship between I_1, I_2, and I

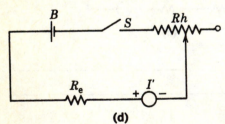

(d)

I' meter _____ A I' corr. _____ A

R_e _____ ohms R_e' com _____ ohms

Percent diff. (R_e vs R_e') _____

Step 12

(a)

(b)

(c)

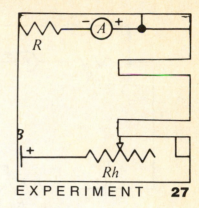

METHODS OF MEASURING RESISTANCE

SPECIAL APPARATUS:

Bundle of connecting wires, voltmeter (0–30 V), ammeter (0–1.5 A), resistance box, set of coils for unknown resistances, galvanometer, tap key, switch with protective resistance (K_2 in Figure 27–3).

GENERAL APPARATUS:

Rheostat (20–25 ohms), two dry cells (or other power supply), slide-wire Wheatstone bridge.

THE PURPOSE OF THIS EXPERIMENT

is to become familiar with two methods of measuring resistance: (a) the voltmeter–ammeter method and (b) the slide-wire Wheatstone bridge.

A. The Voltmeter–Ammeter Method

INTRODUCTION

When a current I' exists in a conductor whose ends have a difference of potential of V', the experiment shows that

$$I' = V'/R \qquad [1]$$

where R is the resistance of the conductor. This relationship is known as Ohm's law, and may be used to calculate the resistance of a conductor if the potential difference and the current are known. When V' and I' are expressed in volts and amperes, respectively, the resistance R is given in ohms. The resistance is a characteristic of a conductor and depends only on its dimensions, the kind of materials of which it is made, and its temperature. It does not depend on either V' or I', but its value is given by the equation

$$R = V'/I'. \qquad [2]$$

A straightforward technique for measuring the resistance of a resistor would be to measure V' and I' with a voltmeter and ammeter and find the ratio. However, the introduction of meters into the circuit alters the circuit so that the measured values may not be the same as the actual values.

Figures 27–1 and 27–2 show two methods of connecting a voltmeter and ammeter to determine the resistance R.

The first method of connecting the voltmeter and ammeter for measuring R will be called circuit 1 (Figure 27–1). The other items in the circuit are the tap key K, the battery B, and the rheostat Rh. The ammeter A measures the

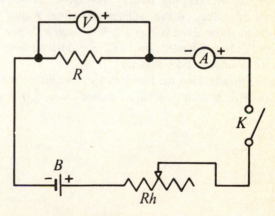

Figure 27–1 Circuit 1.

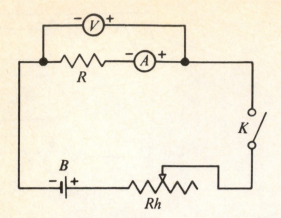

Figure 27–2 Circuit 2.

$$I_\text{v} = V/R_\text{v}.$$

Since the current I' through R is $I - I_\text{v}$ we have

$$R = \frac{V'}{I'} = \frac{V}{I - I_\text{v}} = \frac{V}{I - V/R_\text{v}} \qquad [3]$$

for the corrected value of R.

The second method of connecting the meters is shown in circuit 2 (Figure 27–2). In this arrangement, the ammeter reads the correct value of the current I' desired, but the voltmeter V reads the potential drop across both the resistor R and the ammeter A. If R_a is the resistance of the ammeter, then

$$V = I(R + R_\text{a}),$$

$$V/I = R + R_\text{a},$$

and

$$R = V/I - R_\text{a}. \qquad [4]$$

Note in Equation [3] that the larger the resistance of the voltmeter the less the effect on the measurement. Also, from Equation [4], we see that the smaller the resistance of the ammeter the less the effect on the measurement.

total current I in the main circuit, but a portion of this current, I_v, is shunted around R through the voltmeter. Hence, the ammeter reading does not give the exact value of the current I' in R, and, consequently, V/I does not give the exact value of R. The voltmeter reads the potential difference V between its own terminals as well as across R, so that here the voltmeter reading (V) is equal to V'.

If the resistance of the voltmeter R_v is known, the value of I_v is given by

PROCEDURE

1. Use a battery (or other power supply) arranged to supply about 3 V to the circuit and connect as shown in Figure 27–1. The resistance R should be between 1 and 5 ohms and the rheostat Rh should be 20–25 ohms. Be sure to connect to safe ranges on the meters, say, 3 V on the voltmeter and at least a 1-A range on the ammeter. Higher ranges might be best until you know what the values will be, then if safe, change to lower ranges.

2. In order to determine how resistance is related to current and potential difference, we might vary the current through the resistor R (keeping battery voltage constant) and observe the effect on the voltmeter reading. Adjust the rheostat to allow about 0.1 to 0.2 A through R. When the current has been adjusted, read and record the readings on both the ammeter and voltmeter.

3. Now, by adjusting the rheostat, increase the current in steps of 0.1 A until you have reached about 1.0 A and

record the readings of both meters for each adjustment.

4. The relationships can best be observed by plotting a graph with voltmeter readings on the Y axis and ammeter readings on the X axis. What relationship does your graph indicate? Find the slope of the curve and incorporate it as a part of your report with a discussion of its relation to Ohm's law. What quantity have you measured?

5. If the resistance of your voltmeter is known, use Equation [3] to correct for the effect of the voltmeter. What is the percent difference between the original measured value and the corrected value?

6. If the resistance of your ammeter is known, change your circuit to circuit 2 shown in Figure 27–2. Make a single measurement and record your values of V' and I'. Compute R using Equations [2] and [4] and compute the percent difference.

B. The Wheatstone Bridge Method

INTRODUCTION

The Wheatstone bridge offers a convenient and very precise method of measuring electrical resistance. The apparatus (Figure 27–3) consists of four resistances R, X, M, and N connected together as shown and then connected to a battery E. The resistances M and N are two parts of the wire AB. A galvanometer G is connected between C and some other point D on the wire. Current from the battery, upon reaching point A, divides, so that part I_1 is in branch ACB and part I_2 is in branch ADB. If the contact point D is moved along the wire until there is no current in the galvanometer, then C and D will be at the same potential, which means that all of the current I_2 is in DB. Because Ohm's law states that the voltage drop (potential difference) across a resistor is given by $V = IR$, then the voltage drop (IR drop) from A to C is equal to the IR drop from A to D, and we have

$$I_1 R = I_2 M \qquad [5]$$

and likewise

$$I_1 X = I_2 N. \qquad [6]$$

By dividing Equation [5] by Equation [6], we get

$$R/X = M/N. \qquad [7]$$

If wire AB is uniform in size, resistances M and N are directly proportional to lengths AD and DB, which we shall call L_1 and L_2, respectively. Therefore,

$$M = kL_1$$

$$N = kL_2$$

where k is the resistance per unit length of wire AB. Then Equation [7] reduces to

$$\frac{R}{X} = \frac{kL_1}{kL_2} = \frac{L_1}{L_2}. \qquad [8]$$

From this fundamental equation of the Wheatstone bridge, we can calculate the value of X if R is a known resistance and the two lengths L_1 and L_2 are measured. The unknown resistances furnished for this experiment will probably be in one of the forms shown in Figure 27–4 or 27–5. The resistances consist of spools of wire mounted on a small block, with binding posts attached for convenience in connecting them in the circuit. Other types of resistors can also be used.

PROCEDURE

Before attempting to perform this experiment, read Section 4 in Appendix C on resistance boxes and galvanometers.

7. Connect your circuit as shown in Figure 27–3, using a dry cell for the battery E, a resistance box as R, and one of the designated resistors as X. Use the shortest wires you have to make the connections for R and X. The contact at D is made by a key which slides along wire AB. Be sure that all connections are tight. One loose connection may introduce more resistance than the resistance you are attempting to measure.

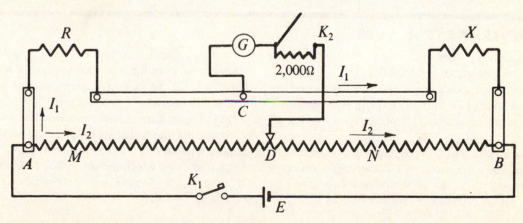

Figure 27–3 Slide-Wire Wheatstone Bridge.

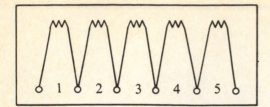

Figure 27–4 Set of Five Resistance Coils.

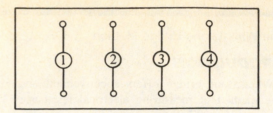

Figure 27–5 Set of Four Resistance Coils.

8. With switch K_2 open, set the contact switch at the center point of the wire as a starting point. With switch K_2 open, the circuit is still completed through the resistor connected across the switch, but the high resistance (2000 ohms or more) reduces the current to a safe value for protection of the galvanometer.

Since the resistances to be measured are unknown, the trial-and-error method will have to be used, so one may start by giving R a value of, say, 0.1 ohm. Now with battery key K_1 closed, lightly tap the contact key at D and note the direction of the deflection of the galvanometer. Next give R a value of, say, 10 ohms, and again note the direction and approximate amount of the deflection when the keys are closed. *Note: Always close battery key K_1 first; then tap the galvanometer key D.* If the two deflections are in opposite directions, the value of R that will give zero deflection must lie between 0.1 and 10. If both deflections are in the same direction, their relative magnitudes tell you which one is the nearest to the value of R required for zero deflection. Vary R until the smallest deflection possible is obtained. If a value is found which gives zero deflection with the contact D still at the center of the wire, the two lengths L_1 and L_2 are equal, and X is equal to R. However, if such an adjustment is not possible, slide the contact D until a position is found which does give zero deflection but do not hold the switches closed while making changes in adjustment. When the bridge appears balanced close switch K_2 and check the balance point under more sensitive conditions.

9. When the final adjustment is made for zero deflection, record wire lengths L_1 and L_2 and the value of R.

10. Repeat Observation 8 with at least one other unknown resistor furnished for the experiment.

11. Now connect the two resistors you have used in series and measure their total resistance as though it were a single resistor. Then connect them in parallel and measure the combined resistance.

12. Interchange the positions of R and X in the bridge circuit and repeat Observations 8 through 11.

13. By the appropriate use of Equation [8], calculate the resistance of the unknown for each set of readings and find the average for the original and interchanged positions of each resistance measured.

14. It can be shown (see your textbook) that the total resistance R_1 of a group of resistances connected in series is given by the relation

$$R_s = R_1 + R_2 + \ldots + R_n \qquad [9]$$

and the value R_p of a parallel-connected group is given by

$$1/R_p = 1/R_1 + 1/R_2 + \ldots + 1/R_n \qquad [10]$$

Do your experimental results confirm the predictions of the theory indicated by Equations [9] and [10]? Compare by computing the percent difference.

QUESTIONS · PART A

1. What relationship does your graph show between the potential V and the current I? Explain.

2. Does your graph indicate that the potential depends on anything else besides the current? Explain your reasoning. Do you think the resistance R depends upon I and/or V? Why?

3. Ohm's law, relating the resistance to the current and potential, may be written as $R = V/I$. Explain the results of your experiment in terms of this law which gives the appearance that R might be a function of both V and I.

4. In which cases did you note the smallest percent difference between the corrected value of R and the value found by using the meter readings directly? Explain why these are smaller.

5. Do you think there are any cases whereby one is justified in computing the value of R directly from the meter readings without making the correction? If so, explain the circumstances by which the simplified computation, $R = V/I$, is justifiable. Assume a general situation rather than a specific set of data.

6. The resistance of the ammeter leads and their contacts will be of the order of 0.015 ohm and is variable. In view of this fact, what must be concluded concerning the accuracy of measuring any low resistance by the ammeter–voltmeter method?

7. Some voltmeters have a total resistance of 500,000 ohms or more. If the resistance you are measuring is of the order of 100 ohms, which circuit would you recommend when using a very high resistance voltmeter? Explain your answer.

QUESTIONS - PART B

8. Which of the two methods (A or B) used in this experiment do you think provides the most precise experimental measurement of a resistance? Explain your answer.

9. By changing the value of R, the bridge may be made to balance at any desired point along the length of wire AB (Figure 27–3). If it were balanced at a point 2 cm from one end, would the accuracy of the results be as good as when balanced near the center? Why?

10. What is the advantage of interchanging the positions of R and X and taking a second set of measurements? *Hint: Suppose the diameter of the slide wire was not quite uniform.*

11. What is the purpose of using short wires for connecting R and X in the bridge?

12. Suppose the resistances of the leads for R and X are not negligible, say some value r for each pair. Add this resistance to both R and X in Equation [8] and determine what condition of balance will make the lead resistance cancel out.

13. How does the protective resistance in series with the galvanometer affect the balance position? Explain.

14. If the connections at binding post B are loose, what effect might this have on the value obtained for X? Explain.

15. Would there be any advantage in using a slide wire ten times as long as you used? Explain.

NAME *(Observer)* _____ Date _____

(Partner) _____ Course _____

RECORD OF DATA AND RESULTS
Experiment 27—Method of Measuring Resistance

A. THE VOLTMETER-AMMETER METHOD

Current (Amperes)									Slope of line $\Delta V/\Delta I$
Potential Difference (Volts)									– – – – – – –

Step 5	$R_v =$		$R_{corr.} =$		Percent difference =
Step 6	$V' =$		$I' =$		$R =$
	$R_a =$		$R_{corr.} =$		Percent difference =

B. THE WHEATSTONE-BRIDGE METHOD

Resistor number	Known resistance R	Length of wire L_1	Length of wire L_2	Computed value of X	Average X Exper.	X, Calculated from Equation [9] or [10]	Percent difference
1							
	*						
2							
	*						
Optional 3							
	*						
Series	*						
Parallel	*						

*For record of data and results when *R* and *X* are interchanged (see observation 12).

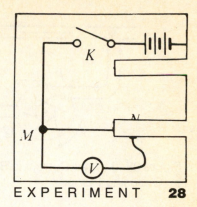

MEASUREMENTS OF POTENTIAL DIFFERENCE WITH A POTENTIOMETER

SPECIAL APPARATUS:

Rheostat (15–25 ohms), portable galvanometer, resistance box (10,000 ohms), bundle of connecting wires, voltmeter, two or three cells for unknown electromotive forces (one being an old dry cell), standard cell, simple switch, double-pole double-throw switch, sliding-contact switch.

GENERAL APPARATUS:

Slide-wire potentiometer (1–2 m long), 2–6 V dc source.

THE PURPOSE OF THIS EXPERIMENT

is to study the principle of the potentiometer, to measure electromotive forces of cells with the potentiometer, and to determine the terminal voltage and internal resistance of cells when furnishing current.

INTRODUCTION

When the potential differences of cell terminals are measured with a voltmeter, some current is required to actuate the instrument. Since this current is furnished by the cell, the voltmeter reading V will be less than the electromotive force (emf) E as indicated by

$$V = E - Ir \qquad [1]$$

where I is the current and r the internal resistance of the cell. Hence, the terminal voltage of a cell will always be less than the emf if current is drawn from the cell.

The *potentiometer* is a device for measuring potential difference without requiring current through the cell. This is done by balancing the unknown emf against a variable potential difference, the value of which is known by balancing it against a standard emf.

The principle of the potentiometer will be explained by the use of analogy. In Figure 28–1, let AB represent a section of a pipe carrying a current of water. Between points M and N a pressure difference exists, which, other things remaining the same, will increase with the distance between the points, the higher pressure being at M. If the

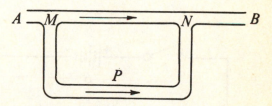

Figure 28–1 Water Current in Branched Pipe Maintained by Difference in Pressure.

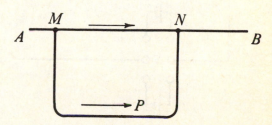

Figure 28–2 Electric Current in a Branched Conductor Maintained by a Difference in Potential.

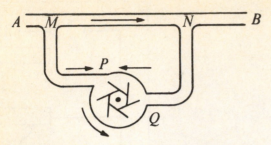

Figure 28–3 Water Current in Branched Pipe Stopped by Opposing Pump.

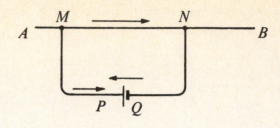

Figure 28–4 Electric Current in Branched Conductor Stopped by Opposing EMF.

pipe is tapped at M and N and a branch pipe P attached, there will be a current in P due to the difference in pressure. In an analogous manner, there will be an electric current in branch conductor P (Figure 28–2) due to the difference in potential between M and N.

Refer next to Figure 28–3. A rotary pump may be inserted into branch MPN, rotating in the direction indicated. By the proper action of the pump, a pressure difference can be maintained between P and Q just strong enough to balance the pressure difference between M and N. In this case, there will be no current in branch MPN, and this absence of current could be shown by a meter placed in branch MPN. In an analogous manner, the battery PQ in the branch conductor (Figure 28–4) could have an emf of the proper size to just balance the potential difference MN, and there would be no current in conductor $MPQN$. Hence, there is no current in the cell PQ, and its emf is exactly equal to the potential difference between M and N.

The arrangement and connections of electrical equipment necessary to accomplish this balancing of electromotive forces are shown in Figure 28–5. The main battery B is connected to a long wire MD through a rheostat R_1. The branched circuit contains a double-pole double-throw switch S, a resistance box R_2, and a portable galvanometer G. A standard cell E_s and an unknown cell E_x are connected to the switch to act as the balancing cells.

This circuit connects to the wire at N_1 by means of a sliding tap key. If, when the cell E_x is connected in the branch circuit, the sliding-contact key is moved along the wire until point N_1 is reached, at which the potential difference between M and N is equal to E_x, there will be no current in the galvanometer G, as indicated by the zero deflection. When E_s is connected, the balance point is at some other position, N_2, since E_s is different from E_x. Now, since the potential drop along a uniform wire is proportional to the length, we have

$$E_x/E_s = MN_1/MN_2. \qquad [2]$$

This equation makes it possible to compute E_x, if E_s is known.

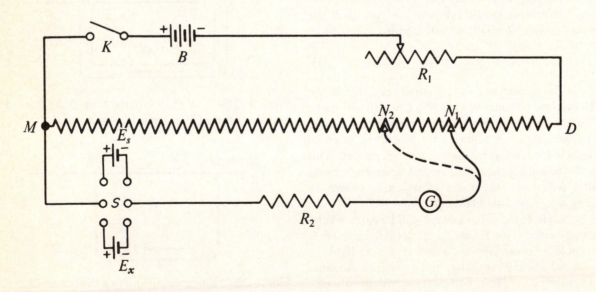

Figure 28–5 Potentiometer Circuit.

PROCEDURE

*Caution: Handle the standard cell with care, always keep it in an upright position, never short its terminals, and **Do Not** connect it to a voltmeter. The cell should never be connected into any circuit except where the emf will be practically balanced and it will draw no appreciable current.*

1. The Principle of the Potentiometer.

Connect your circuit as shown in Figure 28–5 using a dc power source (or two or three dry cells in series) for B, a standard cell for E_s, and a dry cell for E_x. Be sure that the positive (or negative) terminal of each battery (or cell) is connected to the same side of switch S, but leave one terminal of each battery free until your circuit has been checked by the instructor. Make the point M the zero end of the scale under the wire. Have switches closed only while using the equipment.

Set the resistance R_2 at about 10,000 ohms and close the switch K in the main circuit. Now with the dry cell connected by switch S tap the sliding key at some point N_1 on the wire and note the direction of the galvanometer deflection. Do not hold the contact down and slide along the wire while balancing; just tap it, and if the galvanometer deflects, make a new adjustment and tap again. Zero deflection means zero current, irrespective of the original galvanometer reading. Adjust the rheostat R_1 until the dry cell shows a balance at a point about three-quarters of the wire length from M. When the balance point has been located, the sensitivity may be increased by reducing the resistance R_2.

Readjust the value of R_2 to about 10,000 ohms and, without changing the value of R_1, locate the balance point N_2 for the standard cell, which should be about two-thirds the wire length found for the dry cell. Always use a high value of R_2 for each new trial.

The emf of the usual standard cell is 1.0183 V. Use this value for E_s to compute the emf of the dry cell from Equation [2], and record as a simple calculation below the data form.

2. Direct-Reading Potentiometer and the Measurement of Potential Difference.

When measurements of several unknowns are to be made, it is quite convenient to adjust the equipment so that the potential difference (emf) can be determined directly from the balance point on the wire. If the length of your wire is 160 cm or more, set the contact sliding switch at 1.0183 m (101.8 cm) and, with switch K closed and the standard cell in the branch circuit, adjust the rheostat R_1 until a balance is obtained. Merely tap the switch on the wire at each trial in making this adjustment. If the length of your wire is only 100 cm, set the sliding contact at 0.509 m (50.9 cm), which is one-half of 1.018.

Now find the balance point for each of the unknown cells furnished to you for the experiment, and record; check the standard cell adjustment after each unknown has been balanced. Change the rheostat setting only if the standard cell balance point shows the need for readjustment. The length of wire MN_1 in meters should be either equal to, or one-half of, the emf of the unknown, depending on the original length of the wire used. Determine the emf of each cell, and record. Now open the switches, but do not disturb the adjustment of the rheostat.

3. Terminal Voltage and Internal Resistance of a Cell.

When a voltmeter is connected directly to a battery (Figure 28–6), the only resistance in the circuit is that of the voltmeter R_v and the internal resistance of the battery r. Since the potential across the connector leads is zero, the potential across the voltmeter terminals AB is the same as that across the battery terminals CD. Hence, the current I in the circuit, which is also the current through the voltmeter, is given by $I = V/R_v$ (Ohm's law). This is also the current drawn from the cell E. R_v is usually given on the face of the voltmeter in ohms per volt. If the resistance label gives 1000 ohms/volt for a meter with a 15-V scale, then the meter resistance is 15,000 ohms.

Note very carefully the zero reading of the voltmeter. If it does not read zero, either make the proper correction for all readings or ask the instructor to adjust the zero setting. Connect each *unknown* cell in turn to the proper range of the

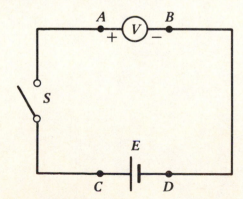

Figure 28–6 Voltmeter Across a Battery.

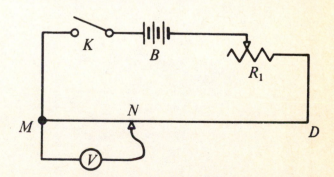

Figure 28–7 Voltage Divider.

voltmeter, and very carefully determine and record the correct terminal voltage.

From the voltmeter reading and the resistance of the voltmeter (usually recorded on the meter) compute, by the use of Ohm's law, the current in each cell.

Now compute the internal resistance of each unknown cell by the use of Equation [1], and record.

4. *Principle of Voltage Divider*. Replace the branch circuit of Figure 28–5 with a voltmeter as shown in Figure 28–7. Place the sliding-contact switch at a small distance from *M*, and note the voltmeter reading when switch *K* is closed. Slowly move the contact point *N* along the wire toward *D*,

and note the change in the voltmeter reading. This arrangement is called a voltage divider, because it divides the voltage along *MD* into steps of any size desired, up to the maximum voltage drop over the entire wire.

Now remove the slide wire *MD* from the circuit and connect the rheostat to the battery circuit so as to use it as a voltage divider; repeat the above procedure with the voltmeter. Be sure to use a voltmeter range as large as the emf of the battery *B*. As a report on this procedure, draw two wiring diagrams, one showing a low-voltage reading and one showing a high-voltage reading.

QUESTIONS

1. With the adjustment for a direct-reading potentiometer, what was the potential drop over the entire length of the wire *MD*? How do you account for this being considerably less than the emf of your working battery *B*?

2. If the wire in your equipment were replaced with one ten times as long but with the same resistance, what would be the potential drop over its entire length? Explain.

3. Would the longer wire mentioned in Question 2 affect the sensitivity of the potentiometer? Explain.

4. Give two advantages of having a large resistance R_2 in the branch circuit. What disadvantage does a large resistance have here?

5. Explain why the terminal voltage of a cell as indicated by a voltmeter is less than the emf.

6. Under what two conditions will the potential difference between the terminals of a cell be numerically equal to the emf? Which of these conditions is realized in the potentiometer?

7. For which cell was the terminal voltage most nearly

equal to the emf? Explain.

8. Although an old dry cell has nearly the same emf as a new one, explain why the current capacity is so different.

9. If the slide wire in this experiment has a resistance of 2.5 ohms/m, compute the current in the main circuit when one of the cells is balanced against the potential difference in the wire.

10. Use the current obtained in Question 9 and the answer to the second part of Question 1 to compute the approximate value of the resistance R_1 in the main circuit.

11. If the battery *B* furnishing the current for the circuit had a smaller emf than the emf to be measured, what would be the result?

12. Would an ammeter connected to the main circuit with battery *B* be of any value? Explain.

13. What would be the shortest slide wire you could have used in this experiment and have had a direct-reading potentiometer? Explain.

NAME *(Observer)* _____ Date _____

(Partner) _____ Course _____

RECORD OF DATA AND RESULTS

Experiment 28—Measurements of Potential Difference with a Potentiometer

Electromotive force of standard cell _____ **V**

Cell designation				
Balance length, standard cell				
Balance length, unknown cell				
Emf of cell (potentiometer)				
Resistance of voltmeter, R_v				
Terminal voltage of cell (voltmeter)				
Current in cell, I				
Internal resistance of cell, r				

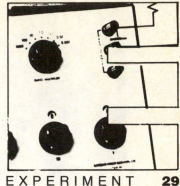

A STUDY OF THE FACTORS AFFECTING RESISTANCE

SPECIAL APPARATUS:

Set of coils, galvanometer, bundle of connecting wires, thermometer, burner, two rulers, towel, graph paper.

GENERAL APPARATUS:

Dial-type Wheatstone bridge (or slide wire plus resistance box), temperature coefficient apparatus consisting of coil, calorimeter, tripod support, and variable low-voltage dc power supply.

THE PURPOSE OF THIS EXPERIMENT

is to investigate the factors that affect the resistance of conductors and to analyze their relationships to resistance.

INTRODUCTION

Circuit elements which offer resistance, or "opposition" to the current, are called resistors, and their opposing effect is called resistance. In Experiment 27, observations showed that the resistance of a conductor does not depend on either the current in it or the potential applied across its terminals. In this experiment we shall examine some of the factors which we might suspect would determine the value of the resistance.

Since resistance opposes the flow of electric charge, we might suspect that this opposition is related to the number and kind of atoms in the conductor and to the random motion of the atoms and their electrons. Upon careful examination of these points one might conclude that the resistance of a conductor would probably depend on the length, the cross-sectional area, the kind of material, and the temperature.

As we have noted in other experiments, with several variables present, only one variable at a time should be changed, while all others are kept constant. Since these variables are completely independent of each other, this method can be readily applied. In functional notation, we might write the relationship of these variables in the form

$$R = f(l, A, k, t) \qquad [1]$$

where the resistance R of a conductor is said to be a function of (or depends on) the length l, the cross-sectional area A, the kind of material k, and the temperature t.

DESCRIPTION OF APPARATUS

The precision required for good results in this experiment can best be obtained by using some form of Wheatstone bridge to measure the resistance. The bridge usually available for such measurements may be one of the following: (1) the *slide-wire bridge* illustrated in Figure 27–3, or (2) one of the two *dial-type bridges* shown in Figure 29–1.

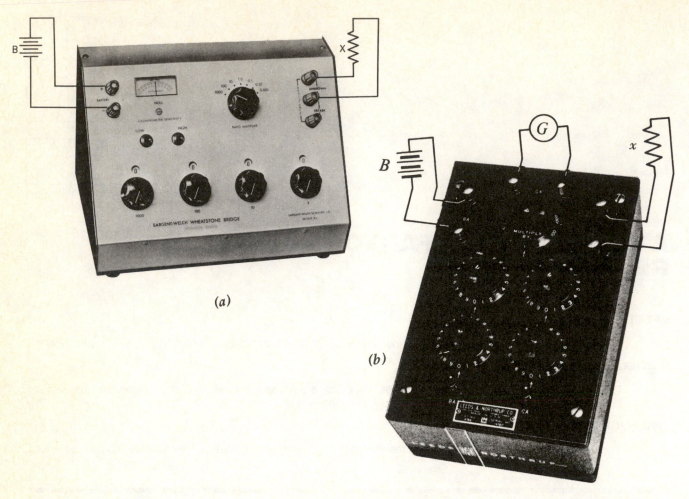

Figure 29–1 Dial-Type Wheatstone Bridge. (*a*) with Built-In Galvanometer (Courtesy of Sargent-Welch Scientific Co.). (*b*) with Externally Connected Galvanometer (Courtesy of Leeds and Northrup Co.).

The procedure for using the slide-wire-type bridge is described in Experiment 27 and can be followed for use in this experiment.

The dial-type bridge illustrated in Figure 29–1 provides a faster and more convenient method when measuring a large number of resistances. In this instrument the slide wire is replaced by a set of coils connected to a ratio dial which gives a direct reading of the ratio of the two arms of the bridge replacing the wire. The ratio can be made to have any value from 1000 to 0.001, in multiples of ten, by manipulating only one dial. The group of four dials marked 1000, 100, 10, and 1, respectively, serve as the known resistance *R*. The unknown resistance *x* may be determined by multiplying the total resistance *R* recorded on these four dials by the reading on the ratio dial (or multiplier), when the galvanometer shows the bridge as balanced. Or, by formula,

$$x = (R) \times (\text{dial ratio}). \qquad [2]$$

The two instruments pictured in Figures 29–1*a* and 29–1*b* use the same principle of operation, but they differ in the way the galvanometer is connected. In Figure 29–1*a*, the galvanometer is a built-in unit with all connections made internally. The switches for connecting to the battery and galvanometer, in both models, are closed and opened by pressing a button. By pushing either of the buttons labeled **Low** and **High**, the battery circuit is closed first and then the galvanometer circuit (automatically). The switch button marked **Low** has a high resistance in series to protect the galvanometer until an approximate balance is attained, after which the **High** sensitivity switch is used to get a precise balance. This model requires a battery potential of about 5 V.

For the bridge pictured in Figure 29–1*b* the galvanometer is manually connected to the terminals labeled *GA*. The switch buttons, marked *BA* and *GA*, at the other end of the box operate independently, one for the battery and one for the galvanometer. Some models of the Sargent–Welch bridge have external galvanometer connections and operate the same as the Leeds & Northrup model in Figure 29–1*b*. The Leeds & Northrup bridge is designed to operate with a 1.5-V dry cell as the potential source.

PROCEDURE

1. Connect the required potential source as described above, a galvanometer G (if required), and the unknown resistance x, as shown in either type of bridge. The lowest of the three terminals on the model pictured in Figure 29–1a will not be used. The unknown resistance used in this experiment may be similar to that shown in either Figure 27–4 or Figure 27–5. Data for size, length, and kind of wire are usually marked some place on the spool set.

2. Use the shortest wires in your bundle to connect coil 1 to the binding posts marked x, *being sure that all connections are tight*. Set the ratio dial at the position marked 1 and the value of R at the position for 1 ohm. Following the instructions given above, close the appropriate switches and note the direction of the galvanometer deflection. Now set R at 2 ohms and repeat the procedure. If the deflection changes direction, you know that the resistance x is between 1 and 2. If the deflection is in the same direction, but weaker, you know that x is greater than 2. However, if the deflection is greater than when set on 1, you know the value lies between 0 and 1. When an adjustment has been made which gives the smallest deflection, make R ten times as large and set the ratio dial on 0.1 and adjust until you get R to the nearest 0.1 ohm. Now increase R again by ten times and set the ratio dial on 0.01; adjust until you get the nearest possible balance. For Figure 29–1a you may now use the high sensitivity button. You could increase the sensitivity by another factor of ten if need be. The value of x is found by multiplying the total value of R by the ratio factor indicated on the ratio dial, as indicated by Equation [2]. Record your values to about three significant figures.

3. For our next coil, suppose we keep all factors in Equation [1] constant except the length. Hence, choose a coil of the same diameter and kind of material as in Step 2, but with a different length, and repeat that procedure for measuring its resistance. Record the measured resistance along with other data given on the coil set.

4. Now let us keep the length constant (same as in Step 2) and measure the resistance of a coil of different diameter (or cross section). Record wire data as before.

5. For the next measurement, let us choose a coil made of a different kind of material but with the same length and diameter as one previously measured. The measurement shows the effect of the kind of material on the resistance of a conductor.

6. It has been assumed that the temperature of the room, and consequently, the temperature of the coils being measured, has remained constant. In order to measure the temperature effect on resistance it will be necessary to use a coil designed for immersion in a liquid bath as shown in Figure 29–2. The coil may be immersed in water in the inner cup of a double-walled calorimeter. The temperature is best controlled by putting water in the outer cup also and carefully heating the combination with a Bunsen burner.

Select the longest wires (2–3 ft) in your bundle for connecting the coil to the box bridge, so the heat will not be too near the bridge.

7. Because the resistance of the connecting wires to the coil may be from 1 to 5 percent of the total resistance being measured, it will first be necessary to measure their resistance. This may be done by connecting both wires to the same binding post on the coil. Make this connection, and after the bridge has been balanced with the ratio dial on 0.001, record the resistance of the connecting wires.

8. Next connect the wires to the coil, as shown in Figure 29–2, and measure the resistance of the coil at the temperature indicated for tap water by your thermometer. This will be the resistance for a coil of a particular length, diameter, and kind of material at approximately room temperature. As the temperature is changed, all of the other factors remain fixed.

9. While constantly, but slowly, stirring the water around the coil, measure the resistance at about three other temperatures at about 20°C intervals. Remove the burner from the bath while balancing the bridge so as to hold the temperature constant during the measuring process. The water in the outer vessel helps to hold the temperature constant.

10. Since the connecting wires are not a part of the coil, their resistance must be subtracted from each of the above measured values to get the true resistance of the coil.

11. The temperature relationship to resistance can best be obtained by graphical representation with temperatures plotted on the X axis and corrected resistance plotted on the Y axis. The temperature scale should begin at 0°C, but the resistance scale need not do so. However, leave enough space below the first plotted point to allow for the

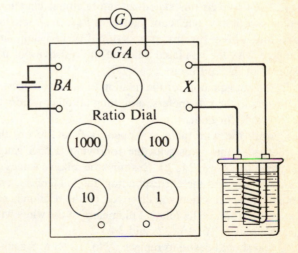

Figure 29–2 Circuit for Measuring the Temperature Coefficient of Resistance.

extrapolated value of the resistance at 0°C. If your points show a linear relation, draw a straight line through them and extrapolate (dotted line) to 0°C.

11a. Optional. If a thermistor is available, an investigation of the effect of temperature changes on its resistance provides a very interesting experiment. A curve of resistance versus temperature might be compared with that of copper.

Analysis of the Data

12. For the length analysis, compare the ratio of the resistances for two different lengths with the ratio of the corresponding lengths. Make this comparison of ratios a part of your report and state your conclusions about the effect of a change in length on the resistance of a conductor. Do the same for the effect of the cross-sectional area on the resistance. Is this relationship the same as for length? Record results in the last data table.

13. The resistivity of a material is defined as the resistance of a conductor of unit length and unit cross-sectional area. This relates the resistance to a specific length and size. This aspect of our data can be checked by comparing the ratio of the resistances of two coils of different materials, of equal lengths and areas, with the ratio of the resistivites listed in tables for the kinds of materials measured. Do your data confirm the tabulated information?

14. For metallic conductors the relation of the temperature effect is predicted by theory to be of the form $R_t = R_0 \alpha t + R_0$ where R_t is the resistance at $t°C$, R_0 is the resistance at 0°C, and α is a constant called the temperature coefficient of resistance. You will note that this equation has the form $y = mx + b$ which predicts a linear relation between R_t and t, with a Y intercept of R_0 and a slope of $R_0 \alpha$. The extrapolated intercept gives the value of R_0, and α can be found from the slope relation. Determine the slope of your graph and compare your computed value of α with tabulated values of the temperature coefficient of resistance of the material of your coil. Record all results from the graph in data table.

QUESTIONS

1. Explain why one might suspect that the resistance of a conductor is related to the cross-sectional area.

2. Based on the definition of work, explain how one would expect the resistance to be related to the length of a conductor.

3. An electric current is due to a stream of electrons moving along a conductor. Since all electrons are identical, how can you explain the dependence of resistance on the kind of material of a conductor.

4. Explain why one would expect the temperature to affect the resistance of a conductor.

5. Write a statement which reveals your findings about the relation of resistance to both length and cross-sectional area. Then convert this statement into a simple equation, and by substituting one set of values for R, l, and A into the equation, solve for the proportionality constant and compare with the tabulated value of the resistivity of the material of the coil used.

6. Derive an equation for the relation of the resistances of two wires to their diameters, assuming all other factors to be the same for both.

7. Consult the wire table in Appendix B and record the diameters of the wires with the following B & S gauge numbers: 6, 12, 18, 24, 30. Examine the relative values of the diameters and the corresponding gauge numbers, and state the relation which you discover. (If the relation is not apparent, examine the relative diameters of the wires with gauge numbers 10, 16, 22, and 34.)

8. If you should desire to replace a No. 12 (B & S gauge) copper wire with an aluminum wire having the same length and resistance, what diameter of aluminum wire will be required?

9. What kind of relation does your graph show to exist between the resistance and temperature of a conductor? If the temperature changes from 40 to 80°C, does the resistance become twice as much? Explain, and then state another possible way of expressing a proportionality relation of temperature to resistance.

10. To what temperature would the coil used in this experiment have to be heated to double the value of the resistance it has at 0°C?

11. If the temperature coefficient of resistance remains the same for all temperatures, compute the temperature at which the resistance of your coil is zero. What is the significance of zero resistance?

12. What unit does the temperature coefficient of resistance have? What other quantity do you know about that has similar units?

13. Do the results of this experiment suggest the possibility of a temperature-measuring device (resistance thermometer)? Explain. What advantages, if any, would such an instrument have over other thermometers?

14. What advantages do you think the dial-type Wheatstone bridge has over the slide-wire type?

15. If you were using a Wheatstone bridge for the first time and were interested in the "how" and "why" of the operation of the circuit, which type bridge would you prefer, the dial box or the slide wire? Why?

16. If you investigated the temperature versus resistance relation for a thermistor, discuss the results and explain their significance. You may want to do some reading on the properties of a thermistor.

RECORD OF DATA AND RESULTS

Experiment 29—A Study of the Factors Affecting Resistance

RESISTANCE VERSUS LENGTH, AREA, AND KIND OF MATERIAL

Coil No.	Kind of Material	Length (cm)	Diameter (cm)	Area (cm^2)	Temperature (°C)	Resistance (ohms)

RESISTANCE VERSUS TEMPERATURE

Resistance of connecting wires _ _ _ _ _ _ _ _ R_0 from graph _ _ _ _ _ _ _ _

Temperature						
Resistance as measured						
Corrected resistance						

ANALYSIS AND SUMMARY OF THE DATA

Step	Resistance Relationships Investigated		Percent discrepancy
12	R vs Length L	$\dfrac{R_1}{R_2} = \qquad \dfrac{L_1{}^*}{L_2} =$	
12	R vs Area A	$\dfrac{R_1}{R_2} = \qquad \dfrac{A_1}{A_2} = \qquad \dfrac{A_2{}^*}{A_1} =$	
13	R vs Resistivity, ρ	$\dfrac{R_1}{R_2} = \qquad \dfrac{\rho_1{}^*}{\rho_2} =$	
14	Quantities determined from graph relating resistance to temperature	R_0 from graph intercept $=$	
		Slope of graph $= R_0\alpha =$	
		Temp. Coef. of Resistance, $\alpha =$	
14	α, measured $=$	α, table value $=$	Percent discrepancy

*Use these ratios as correct values for computing percent discrepancy.

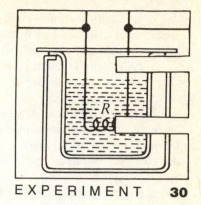

THE HEATING EFFECT OF AN ELECTRIC CURRENT

SPECIAL APPARATUS:

Bundle of connecting wires, voltmeter, ammeter, electric calorimeter, thermometer, rheostat (2–5 ohms), switch, set of weights, two rulers, graph paper, and a watch.

GENERAL APPARATUS:

Storage battery or other 6-V dc source, laboratory balance.

THE PURPOSE OF THIS EXPERIMENT

is to study the relation of electric energy to heat energy by means of an electrical calorimeter and to determine the ratio of two units of energy, the calorie and the joule.

INTRODUCTION

The energy from a seat of electromotive force (emf) may be converted into mechanical energy by means of an electric motor, into chemical energy in an electroplating device, or into heat as attested by the operation of such household appliances as the electric iron and toaster. These heating appliances represent pure resistance in a circuit, and, when the switch is opened, the only change resulting from the expenditure of electric energy is the production of heat. James Joule did many experiments (1843–1850) with various types of equipment in an attempt to determine the relation of heat to the energy which produced it. In the case of electrical energy being converted into heat, he discovered that the heat produced in a given resistor was directly proportional to the square of the current and the time it is on.

If we consider a simple electric circuit such as shown in Figure 30–1, containing a pure resistance R, the difference of potential V across the resistor is the work per unit charge required to maintain a current in the resistor. Then the total work W in joules to carry Q coulombs from A to B is given by

$$W = VQ. \qquad [1]$$

But since $V = IR$ (Ohm's law) and $Q = It$,

$$W = I^2Rt = VIt \qquad [2]$$

where W is the heat energy in joules developed, I the current in amperes, t the time in seconds, R the resistance in ohms, and V the potential difference in volts. This relation expresses what Joule first measured experimentally and is called *Joule's law*.

Through a completely different avenue of approach, a unit for measuring the quantity of heat has been designated as the kilocalorie in the mks system. The kilocalorie is that quantity of heat which must be added to, or removed from, 1 kg of water to change its temperature by 1°C. When the

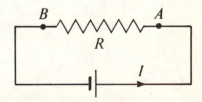

Figure 30–1 Simple Resistance Circuit.

gram is used as the mass unit, the heat unit is then called one calorie, where 1 kcal = 1000 cal. Since heat is a form of energy, the calorie is also a unit of energy, but it is arrived at by a completely different process than the unit joule.

The principle of the conservation of energy, or the first law of thermodynamics, predicts that there is a definite relation between mechanical energy and heat energy. Hence, the two independently determined units of energy, the joule and the calorie, bear a definite relation to each other. The purpose of this experiment is to determine how these two units are related.

The electric calorimeter will be used as the heating system for investigating the relation of the energy supplied by an electric circuit to that absorbed from the circuit by a known amount of water. The amount of heat may be measured experimentally by immersing the heating coil R (Figure 30–2) in a calorimeter cup containing some liquid

the specific heat of which is known. Let M_c and M be the masses of the calorimeter and liquid, respectively, and S_c and S their specific heats. Then, the heat H (in calories) absorbed by the vessel and contents when warmed from temperature T_1 to T_2 is given by

$$H = M_c S_c(T_2 - T_1) + MS(T_2 - T_1) + w(T_2 - T_1) \quad [3]$$

where w is the water equivalent (definition follows) of the resistance coil R. The *water equivalent* of a body (or vessel) is the amount of water which would require the same amount of heat as does the body to raise the temperature 1°C. If there is no heat lost to the surroundings, the amount of heat absorbed by the vessel and its contents (Equation [3]) is equal to the energy in calories furnished by the electric current (Equation [2]).

PROCEDURE

1. Examine the voltmeter and ammeter carefully, noting the + and − terminals and whether the meters have more than one range (scale). Also review Sections 4 and 5 in Appendix C on connections for voltmeters, ammeters, and rheostats.

2. Connect equipment as in Figure 30–2 but do not close switch K unless coil R is in the liquid in the calorimeter. Use the high ranges on the ammeter and voltmeter until you make a trial setting for your experiment. Ask instructor to check the circuit.

3. Weigh the inner cup of a calorimeter when empty and again when about two-thirds full of water (or other liquid specified by your instructor) at a temperature six or seven degrees below room temperature, and record the weights.

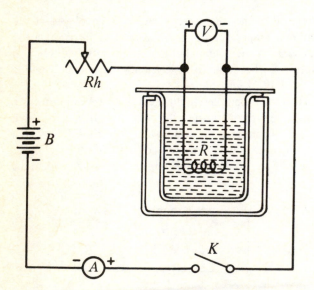

Figure 30–2 Heating-Coil Circuit

Caution:

Do not use liquid cool enough to cause condensation on the outside of the cup.

4. Assemble the complete calorimeter with the coil in the liquid and, after closing the switch for a trial setting, adjust the rheostat until the ammeter indicates some convenient value in the range of 2–3 A. Note the voltmeter reading. Now open the switch and connect to the lowest ranges of the meters that will take the current and voltage just noted.

5. Stir the liquid, using a thermometer as a stirrer if no other kind of stirrer is provided, and read and record the temperature and the time at the moment the switch is closed.

6. While keeping the current constant by adjusting the rheostat, record the temperature and voltage every 2 min until the water has reached a temperature about as much above room temperature as the beginning temperature (Observation 4) was below. Stir the liquid at frequent intervals and always stir it just before reading the temperature. Record the time of opening the switch, and while stirring the liquid read and record the highest temperature reached. Record these data in the second table of the data form.

7. In order to have more than one set of data for the investigation, it will be desirable to make a second run, using different values for the mass of water, the current, the time, and the temperature change. For one trial, plot a graph of T on the Y axis versus t on the X axis.

8. In Equation [3] set $T_2 - T_1 = \Delta T$ and reduce the right member to a single term. With $W = H$, equate the right members of Equations [2] and [3] and predict the type of graph to be expected by plotting ΔT versus t. If, in Step 7, the current were increased for the last half of the run, make a freehand sketch on your graph to indicate the effect.

9. Obtain the needed specific heats from Table 4 of Appendix B and the water equivalent of the coil from the instructor, and then compute, by means of Equation [3], the heat in calories absorbed by the calorimeter and its contents for each set of data.

10. For each of the two runs, compute the ratio of the heat energy (H) in calories absorbed by calorimeter and contents to the energy in joules supplied by the electric circuit. Compute and record the percent discrepancy of your average ratio (W/H) compared to the accepted value, 4.19 J/cal.

QUESTIONS

1. Which unit did you find to be the larger, the joule or the calorie? Write the value of one in terms of the other.

2. What effect did the use of different values of the current, the time, the mass of water, and the temperature change have upon the ratio of these two units? Explain. Do you think a change in heating coils would have resulted in a different ratio? Explain.

3. Do your experimental results verify the law of conservation of energy as expressed in the first law of thermodynamics? Explain. Give a special wording of the law to incorporate the results of your experiment.

4. In Step 3 you were cautioned not to use water cool enough to cause condensation on the outside of the cup. Explain the reason for this precaution and state the possible effects its oversight might have had on your results.

5. Did the voltmeter reading remain constant during the course of each run of the experiment? If not, explain some of the possible causes for its variation.

6. If, during the course of an experiment run, you should find it desirable to increase the current to speed up the operation, would this affect the calculation procedure? Explain the changes, if any.

7. If you had heated the water to a temperature a little below the boiling point before stopping the run, how might your results have been affected? Explain.

8. Examine your data relative to the relation of the temperature changes and the 2-min time intervals. If these two variables, temperature (T) and time (t), were plotted, what might be the general shape of the graph? If the current were changed during the run, how would the shape of the graph be affected? Illustrate with freehand sketches.

9. At the rate of 5 cents per kilowatt hour, how much would it cost to heat one liter of water from room temperature to the boiling point? Make use of the relations determined in this experiment.

10. Why connect to the high ranges of the meters for the trial adjustment of this experimental setup? What ranges, in general, should one use on the meters for the actual run? Explain.

RECORD OF DATA AND RESULTS

Experiment 30—The Heating Effect of an Electric Current

Water equivalent of coil _____ Specific heat of calorimeter _____

Liquid used _____ Specific heat of liquid, S _____

Observations	Trial 1	Trial 2
Weight of cup + liquid (gm)		
Weight of empty cup (gm)		
Weight of liquid (gm)		
Current (A)		
Total time of run (sec)		
Average Voltage (V)		
Heat energy, W (J)		
Heat energy, H (cal)		
Ratio of heat energy units, W/H		
Avg. Value of W/H =	Table Value =	Percent discrepancy

RELATION OF TEMPERATURE TO TIME

Observations, Trial 1			Observations, Trial 2		
Time	Temperature	Reading of voltmeter	Time	Temperature	Reading of voltmeter

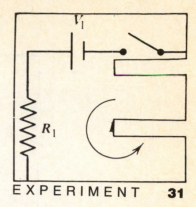

CIRCUITS CONTAINING MORE THAN ONE POTENTIAL SOURCE

SPECIAL APPARATUS:

Milliammeter (0–100 mA), voltmeter (0–10 V), three resistors (100–200 ohms), two single-throw switches, nickel–cadmium rechargeable battery (1.2–1.5 V), bundle of connecting wires.

GENERAL APPARATUS:

Power supply (6 V dc).

THE PURPOSE OF THIS EXPERIMENT

is to make measurements on a circuit containing two sources of electrical power and verify the usefulness of Kirchhoff's rules for such circuits.

INTRODUCTION

Many electrical and electronic circuits contain more than one source of potential difference such as that in Figure 31–1. These circuits cannot be solved by using combinations of series and parallel components, and the more general approach given by Kirchhoff's rules must be used. The first step in applying these rules is to arbitrarily select and label the direction of the currents through each part of the circuit. These selections must then be adhered to throughout the application of Kirchhoff's rules. If, after solving the equations, a negative value for one of the currents is found, it simply means that the direction is opposite to that selected; however, the magnitude obtained is still correct.

The first of Kirchhoff's two rules states that *the sum of all the currents coming into a junction must equal the sum of all the currents leaving that junction.* Using the currents as shown in Figure 31–1 and applying the rule to junction *J*, we get the equation

$$I_1 - I_2 - I_3 = 0. \qquad [1]$$

The second rule states that *the algebraic sum of the voltages around any closed loop must equal zero.* When moving from a point of high potential to a point of low potential (a potential drop), we assign a plus sign to the voltage; when moving through a potential increase, we assign a minus sign to the voltage. Recall that electric charge always moves from points of high potential to points of low potential except through sources of electric power, such as batteries or power supplies. Beginning with V_1 and applying this rule around loop L_1 as shown in Figure 31–1 we obtain the equation

$$-V_1 + I_1R_1 + I_2R_2 = 0. \qquad [2]$$

For loop L_2 we have

$$V_2 - I_2R_2 + I_3R_3 = 0. \qquad [3]$$

A third loop is possible around the outside of the circuit, which gives

$$V_2 - V_1 + I_1R_1 + I_3R_3 = 0. \qquad [4]$$

Note that this equation can also be obtained by adding Equations [2] and [3] so that it contains no new

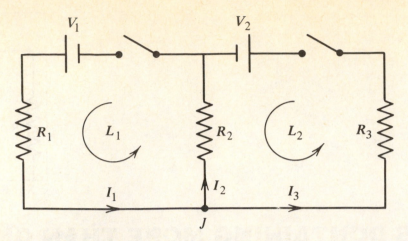

Figure 31–1 Circuit with Two Potential Sources.

mathematical information.

Solving Equations [1], [2], and [3] simultaneously, we obtain

$$I_1 = \frac{V_1R_3 + V_1R_2 - V_2R_2}{R_1R_2 + R_1R_3 + R_2R_3} \qquad [5]$$

$$I_2 = \frac{V_1R_3 + V_2R_1}{R_1R_2 + R_1R_3 + R_2R_3} \qquad [6]$$

$$I_3 = \frac{V_1R_2 - V_2R_1 - V_2R_2}{R_1R_2 + R_1R_3 + R_2R_3} \qquad [7]$$

PROCEDURE

1. With the power supply as a source of current, use the voltmeter–ammeter method to determine the resistance of each of the three resistors illustrated in Figure 31–1. Record the results in the data table. Designate the resistors as R_1, R_2, and R_3, and be sure to remember which is which for later reference.

2. Set up the circuit shown in Figure 31–1, using the 6-V power supply as V_1 and the battery as V_2. The values of the three resistors should be within the range 100–200 ohms, but with different values. Check the polarity of your power supply and battery to be certain they are connected as shown in the diagram. If you are not sure that your circuit is correct, ask your instructor to check it before closing the switches.

3. Close the switches and measure and record the voltage across each resistor and each power source *paying very careful attention* to the polarity of the terminals of each circuit element as you apply the voltmeter leads. Gently touch the voltmeter leads to the appropriate places and watch the meter to be sure it deflects in the right direction. Draw the circuit diagram on your data sheet and mark the polarity of the ends of each resistor as + or − as determined by your measurements. When not making measurements, open both switches but do not change the setting on the power supply.

Check the validity of Equation [2], [3], and [4] by substituting your measured voltage values in the appropriate places. In the data table where each equation is indicated as ΣV, write the equation; then show your substituted voltage values and the resulting sum. If your results do not check reasonably close, repeat your measurements, remembering that your voltmeter readings are the actual values of I_1R_1, and so on.

4. Calculate the values of I_1, I_2, *and* I_3 by substituting the measured values of V_1, V_2, R_1, R_2, and R_3 into Equations [5], [6], and [7]. Are the current directions as shown in Figure 31–1 correct for your circuit? Check that these calculated directions of current are consistent with the polarity diagram made in Step 3.

5. Insert the milliammeter in the circuit at an appropriate point to measure I_1, using the already determined directions of current to insure correct connections. Gently close the switches and watch the meter to be sure it deflects in the right direction. Read the meter as accurately as possible. Following the same procedure, measure I_2 and I_3. Substitute these measured values in the current junction rule given by Equation [1]. Show this calculation in the data table where the equation is indicated by ΣI. If your results do not check reasonably close, repeat your measurements.

6. Compare your measured values of I_1, I_2, and I_3 with those calculated in Step 4 by calculating your percent differences.

QUESTIONS

1. What is the direction of current in battery 2? Would the type of potential source used in this position make any difference? *Hint:* Note that the battery used in this position was a rechargeable battery.

2. It is frequently stated that Equation [1] is just the conservation of electric charge. Explain why this is so.

3. It is frequently stated that Equations [2] and [3] are statements of conservation of energy. Explain this carefully, being sure to note what energy you are considering.

4. Explain fully the physical interpretation given to a negative current resulting from using equations [5], [6], and [7]. Also explain what, if any, effect this has on applying the equations obtained from Kirchhoff's rules.

5. Suppose in writing the loop equations from Kirchhoff's second rule we change our signs so that we use a plus sign for potential increases and a minus sign for potential drops. How would this change Equations [2] and [3]? What effect would this have on Equations [5], [6], and [7]? Explain.

6. Would it be reasonable to assume the direction of I_3 is opposite that shown in Figure 31–1? Explain.

7. Redraw the circuit of Figure 31–1 with the polarity of V_1 reversed. Specify possible current directions in all branches of the circuit, and set up the junction and loop equations equivalent to Equations [1], [2], [3], and [4]. Be sure you have not made any sign errors.

8. Redraw the circuit of Figure 31–1 with the resistor R_2 replaced by a potential source V_3 having its positive end up in the diagram. Specify possible current directions in all branches of the circuit and set up the junction and loop equations equivalent to equations [1], [2], [3], and [4]. Be sure you have not made any sign errors.

9. If the resistors were reversed end for end in the circuit of Figure 31–1, how would the direction of the currents be affected? Explain.

10. Verify Equations [5], [6], [7] by solving Equations [1], [2], and [3] simultaneously.

NAME *(Observer)* _____   Date _____

(Partner) _____   Course _____

RECORD OF DATA AND RESULTS

Experiment 31—Circuits Containing More Than One Potential Source

DATA ON RESISTORS

Circuit Diagram—Step 2
with labeled polarity

Resistor measured	Voltmeter reading	Ammeter reading	$R = V/I$
R_1			
R_2			
R_3			

RESULTS OF VOLTAGE MEASUREMENTS IN THE CIRCUIT (FIGURE 31–1)

Circuit element measured	Power supply	Battery	R_1	R_2	R_3
Voltmeter reading					
Equation [2] $\Sigma V =$					
Equation [3] $\Sigma V =$					
Equation [4] $\Sigma V =$					

RESULTS OF CALCULATED AND MEASURED CURRENT VALUES

Current value determined	I_1	I_2	I_3
Calculated value Step 4			
Measured value Step 5			
Percent difference Step 4 versus Step 5			
Equation [1] $\Sigma I =$			

Show steps of calculations below

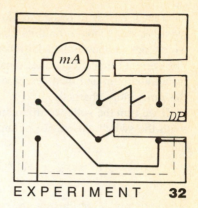

EXPERIMENT　32

A STUDY OF CAPACITANCE AND CAPACITOR TRANSIENTS

SPECIAL APPARATUS:

Milliammeter (0–10 mA), milliammeter (0–50 mA), high resistance voltmeter (0–10 V), capacitor (1000 μF), resistor (10,000 ohms), double-pole double-throw switch, bundle of connecting wires, graph paper.

GENERAL APPARATUS:

Power supply (6–10 V dc).

THE PURPOSE OF THIS EXPERIMENT

is to study the effects of capacitors in dc circuits and related transient phenomena.

INTRODUCTION

The combinations of two conductors separated by a nonconductor is called a *capacitor*. In contrast to a resistor which dissipates electrical energy in the form of heat, a capacitor stores electrical energy in the electric field between the conductors. This electric field is maintained by equal and opposite charges stored on the conducting plates of the capacitor. When the charge on a fixed capacitor is increased, the potential difference between its plates increases. The ratio of the total charge on one of the conductors divided by the voltage between the plates is always a constant called the *capacitance*. If Q is the charge in coulombs and V is the potential difference in volts, then the capacitance (C) is in farads and is given by

$$C = \frac{Q}{V}.$$ [1]

The capacitance of a capacitor is determined by the geometry of the conductors and the material separating them.

When a switch is closed on any dc circuit there is always some time required for the current to reach its steady state value. For circuits containing only resistance these time

periods are very small (approximately the time required for light to travel around the circuit). With many capacitance–resistance circuits this transient period is long enough to be easily studied in the laboratory.

When the switch is closed in a circuit like that in Figure 32–1, the capacitor will initially offer no resistance to the flow of charge in the circuit, because it is a charge reservoir which easily gives up charge from one conductor and collects charge on the other. As the capacitor accumulates

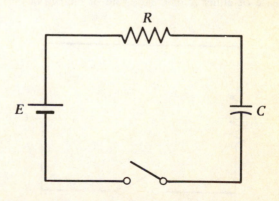

Figure 32–1 Capacitor Charging Circuit.

charge, it begins to resist this flow of charge because the positive side tends to hold its electrons, and the negative side tends to oppose the addition of more electrons. Thus, the current immediately after closing the switch will jump to a maximum and then decrease to zero as the capacitor becomes fully charged.

If we sum the voltages around the circuit of Figure 32–1 with the switch closed, we obtain

$$E = V_R + V_C = IR + \frac{Q}{C}. \qquad [2]$$

Immediately after closing the switch, $Q = 0$, so the potential across the resistor is E, and the current jumps to its maximum value which is E/R. E, R, and C are constants in Equation [2] so that as the charge Q increases on the capacitor, the current I in the circuit must decrease. This charge continues until the potential difference across the capacitor is E, and the current in the circuit is zero.

A similar phenomenon occurs when a charged capacitor is discharged through a resistor. The charged capacitor in Figure 32–2 behaves like a voltage source. Since the potential difference across the capacitor is a maximum when the switch is closed, the current through the resistor will also be a maximum. As the capacitor discharges, its potential difference decreases, and thus the current through the resistor also decreases.

The *rate* at which a capacitor charges or discharges is not uniform. This means that the capacitor accepts large amounts of charge in short intervals of time during its initial stages of charging, but during the latter stages of charging it accepts very small amounts of charge in the same intervals of time. The two factors which determine this rate are the resistance of the resistor and the capacitance of the capacitor. If the resistance is large, the maximum current is small, and thus the rate at which charge can flow on or off the capacitor is small. If the capacitance is large, it will take longer for the capacitor to accumulate enough charge to change its potential difference by a given amount.

The above discussion indicates that the rate of change of current with time is a function of both R and C. With an increase of either R or C, the rate of change of current decreases, thus indicating that the rate of charge is a function of the product RC. This product has the units of time (seconds), and, because R and C remain constant for a given circuit, there is a constant time element involved which is related to the rate of change in current. For a circuit containing capacitance and resistance, this constant time element is called the *capacitive time constant* τ, and its value is given by

$$\tau = RC. \qquad [3]$$

It can be shown that this is the time in seconds for a charging or discharging current to change by a factor of e^{-1} (or 0.368), where e is the base of natural logarithms. Thus, during a charging process, an RC circuit will change its current from I_1 to $0.368I_1$ A in the interval of *one time constant*.

The complete mathematical treatment of the charging and discharging of a capacitor requires the methods of calculus. The following discussion will contain only the results of such an analysis. A good qualitative understanding and an intuitive feel for the transient behavior of capacitors can be gained by doing this experiment, even if you are not familiar with the types of equations below.

The mathematical expression for the current, while charging or discharging a capacitor, is an exponential function given by

$$I = I_m e^{-(t/RC)}. \qquad [4]$$

I is the current at time t, I_m is the maximum current, R is the total series resistance, and C is the capacitance of the capacitor. Let I_1 be the current in Equation [3] at time t_1, and let I_2 be the current at time t_2. Then, the ratio of I_2 to I_1 is given by

$$\frac{I_2}{I_1} = e^{-[(t_2 - t_1)/\tau]} \qquad [5]$$

where τ is the time constant defined in Equation [3]. Note that when $t_2 - t_1 = \tau$, the current at time t_2 will be a factor of $e^{-1} = 0.368$ less than the current at time t_1.

One can also study the charge on a capacitor as a function of time and the terminal voltage as a function of time. Equations [6] and [8] give the charge on a capacitor as a function of time for charging and discharging, respectively. Equations [7] and [9] give the voltage across the capacitor when charging and discharging, respectively.

$$Q = Q_m(1 - e^{-[t/RC]}) \qquad [6]$$

$$V = V_m(1 - e^{-[t/RC]}) \qquad [7]$$

$$Q = Q_m e^{-(t/RC)} \qquad [8]$$

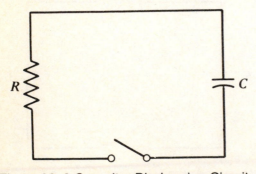

Figure 32–2 Capacitor Discharging Circuit.

$$V = V_m e^{-(t/RC)} \qquad [9]$$

Equations [8] and [9] show the same decay for charge and voltage as discussed above for current when discharging a capacitor. Equations [6] and [7] show that the charge and voltage build up rather than decay when charging a capacitor. Note that the decay constant plays the same role in Equations [6] through [9] as it did in Equation [4].

PROCEDURE

1. Study the circuit shown in Figure 32–3. Note that when the double-pole double-throw switch is closed to the right, the capacitor will be charged through the resistor and milliammeter. When the switch is to the left, it will be discharged through the resistor and the milliammeter. Trace the circuit for both positions of the switch and make sure you understand what is happening in each position.

Connect the circuit as shown in Figure 32–3. The power supply should be set between 6 and 10 V. Before closing any switches or turning on your power supply, *check to be certain* that the milliammeter and the capacitor are connected with correct polarities as indicated in the diagram and marked on them. If you are uncertain, ask your instructor to check your circuit.

2. Before beginning you and your partner should agree on who will read the milliammeter and who will call out the time. Be prepared for short time intervals. Then, close the switch in the direction which will charge the capacitor, record the maximum current, and continue to record the current every 10 sec for about 2 min. Now reset your time to zero, throw the switch to the discharge position, and again record the current every 10 sec.

Leave the switch in the discharge position for several minutes after completing your readings so that the capacitor can completely discharge. Collect data at least two more times while charging and discharging your capacitor so that you will have a good set of data points.

3. Plot a graph on linear graph paper of the average current versus time for the data obtained when charging the capacitor. Plot a second graph for the data obtained when discharging the capacitor. On each graph pick any current I_1 and determine the time required for the current to change to $0.368 I_1$. Recall that this time period is the time constant of your circuit, τ_1. Pick a second current I_2 on each graph and again determine the time required for the current to change to $0.368 I_2$. Theoretically, these four values for the time constant should all be the same. With the average of these four values and the known value of the resistance, determine the capacitance of the capacitor by using Equation [3]. Compare this computed value with that stamped on the capacitor by finding the percent discrepancy.

3a. *Optional (Check with Your Instructor.)* If we take natural logarithms of both sides of Equation [4], we obtain

$$\ln I = \ln I_m - \frac{t}{RC}$$

or, in terms of common logarithms, we have

$$2.3 \log I = 2.3 \log I_m - \frac{t}{RC}$$

which may be written as

$$\log I = \left(-\frac{1}{2.3 RC} \right) \times t + \log I_m. \qquad [10]$$

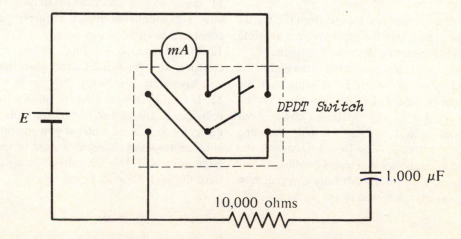

Figure 32–3 Circuit for Studying Capacitor Transients.

Since this has the form of a linear equation, a plot of log I versus t would result in a straight line with a slope of $-(1/2.3)RC$.

This linear plot may be made in either of two ways: (a) plot log I versus t on regular linear graph paper, or (b) use your original data for *the charging process* and plot I versus t on semilog graph paper with the average current values on the log axis and time t on the linear axis.

If you use plan (a), you will need to prepare a column on your data form for log I. If the column for trial 4 is not filled, use it; otherwise, just draw a column in the margin area. For plan (b), ask your instructor what cycle log paper to use.

Find the slope of your graph and then the time constant for this resistance–capacitance circuit. Note that the slope will be negative. Using the known value of the resistance and Equation [3], determine the capacitance of the capacitor, and compare it to the value written on the capacitor by computing the percent discrepancy.

4. With the circuit still connected, as in Figure 32–3, connect a high resistance voltmeter across the capacitor terminals. Make sure the positive lead of the voltmeter is connected to the positive side of the capacitor. Take data in the same fashion as done in Step 2, except this time record the voltage instead of the current as a function of time.

Also, keep time continuously for charging and discharging; i.e., start time at zero seconds when you begin charging, and at 80 sec switch to discharging and continue to record the voltage as a function of time until you get to 160 sec. Mark your data where the switchover is made. Leave the switch in the discharge position for about 2 min after you finish, and then repeat the entire measurement at least twice more.

5. Plot the average capacitor voltage versus time on linear graph paper. Determine the time constant from the discharged section of this graph by finding the time required for the voltage to change by a factor of 0.368. Compare this to the value found in Step 3.

6. With your data from Step 4 it is possible to determine the charge on the capacitor as a function of time. From Equation [1], if we know the capacitance, we can find Q for any V. Using the capacitance determined in Step 3 and the average voltage data from Step 4, calculate the charge on the capacitor as a function of time. Then plot a linear graph of charge versus time.

7. Compare your linear graphs of current versus time with your graph of charge versus time. Note when they are increasing or decreasing and how fast they are increasing or decreasing. Comment on these similarities and differences on your data sheet.

QUESTIONS

1. Derive Equation [5] using Equation [4].
2. Using Ohm's law, calculate the initial and final voltage across the 10,000-ohm resistor. How does this compare to the voltage across the capacitor?
3. How are the voltages across the resistor and capacitor related to the voltage supplied by the power supply?
4. When the switch was first closed to charge the capacitor, what was the effective resistance of the capacitor? What was the effective resistance of the capacitor when it was fully charged?
5. If the 10,000-ohm resistor is removed from the circuit and the switch closed to charge the capacitor, it is possible that the milliammeter might be destroyed. Explain.
6. If the 10,000-ohm resistor is connected in parallel to the capacitor instead of in series, it is possible that the milliammeter might be ruined. Explain.
7. Calculate the fraction of the maximum charge your capacitor will acquire in 1, 2, and 10 min, assuming charging begins when time is zero. How long should you have it on for it to acquire its maximum possible value? Practically speaking, if you wanted it fully charged, how long would you leave it connected to the source?

8. What would the decay constant be if a 100-ohm resistor had been used instead of a 10,000-ohm resistor? Explain any difficulties you might encounter in collecting data for this case.
9. What would be the maximum current in the circuit if the 10,000-ohm resistor were replaced by a 100-ohm resistor?
10. To be correct in our calculations, we should have included the resistance of the milliammeter. Why were we able to neglect it?
11. Based on your observations in this experiment explain how a capacitor becomes a current conductor when connected to an ac power source.
12. In the equation $\tau = RC$, substitute the appropriate electrical units for R and C, and show that the product RC will have units of seconds.
13. In Step 5 you were asked to find the time constant from only the discharge section of the voltage graph. What difficulty do you encounter if you attempt to find the time constant from the charging section of the voltage graph? *Hint:* Equation [7] for the voltage during charging does not have the same form as Equation [9].

NAME *(Observer)* _____ Date _____

(Partner) _____ Course _____

RECORD OF DATA AND RESULTS

Experiment 32—A Study of Capacitance and Capacitor Transients

STEP 2—CURRENT TRANSIENT DATA

Charging current (mA)						Discharging current (mA)					
Time (sec)	Trial					Time (sec)	Trial				
	1	2	3	4	Avg.		1	2	3	4	Avg.

STEP 3—DETERMINATION OF CAPACITANCE FROM CURRENT TRANSIENTS

Process	I_1	$0.368I_1$	τ_1	I_2	$0.368I_2$	τ_2
Charge						
Discharge						
Avg. $\tau =$		C(Calc) =		C(Mfg) =	Percent discrepancy	
3a.	slope =	$\tau =$		C(Calc) =	Percent discrepancy	

Steps 4 and 6—VOLTAGE AND CHARGE TRANSIENT DATA

| Time (sec) | Potential difference (V) | | | | | Charge (C) |
| | Trial | | | | | |
	1	2	3	4	Avg.	

Step 7—COMMENTS ON GRAPHS

Step 5—DETERMINATION OF CAPACITANCE FROM VOLTAGE TRANSIENTS

Process observed	V_1	$0.368 V_1$	τ_1	V_2	$0.368 V_2$	τ_2
Discharge						

Average $\tau =$	C(calc.) $=$	C(Step 3) $=$	Percent difference $=$

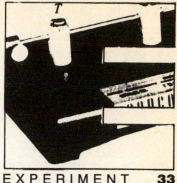

A STUDY OF MAGNETIC FIELDS

NOTE TO INSTRUCTOR:

This experiment may be to long for some students to complete in a laboratory period unless they move along at a steady pace. Using two laboratory periods or omitting selected procedures will ease this pace.

SPECIAL APPARATUS:

Two small compasses, meter stick, bar magnet, horseshoe magnet, several sheets of paper, dip needle, large compass, protractor, iron filings, reversing switch, tap key, small copper wire (about 2 m long), iron rod, mounted conductor (see Figure 33-4).

GENERAL APPARATUS:

Boards and cardboard for placing over magnets, source of direct current (about 6 V), rheostat (about 6 ohms).

THE PURPOSE OF THIS EXPERIMENT

is to investigate the nature of magnetic fields produced by permanent magnets and electric currents.

INTRODUCTION

Magnetic fields can be produced in two ways. One is by the presence of permanent magnets and the other is by the presence of moving charged particles. Permanent magnets are ferromagnetic materials which show strong magnetic effects and are capable of maintaining their magnetism under normal conditions. The most common magnetic fields are those produced by electric currents in conductors which are caused by moving electric charges. Modern theories of magnetism have moving electric charges as the origin of all magnetic fields. Thus, magnetism in matter is produced by atomic charges moving around in the atoms of the material.

Since a compass needle serves as a suitable detector of the presence of a magnetic field and is capable of indicating direction, *we shall define the direction of a magnetic field* at any point as the direction in which the north pole of a compass needle points when placed at the point in question.

If one should move a small compass from point to point, by letting the north pole of the needle dictate the direction to be moved, the path resulting from the motion would reveal the direction of the field at all points along which it moved. A line which one might draw on a piece of paper, or an imaginary line drawn through space, which indicates the direction of a magnetic field at every point, is called *a line of force*. The number of such lines that might be drawn in a particular region is purely arbitrary, but a series of such lines spread over the region in question will reveal the general nature of the magnetic field of the region. The relative spacing of the lines of force at different points indicate the relative intensity of the magnetic field at these points.

The magnetic field of the earth is always present and must be reckoned with in the analysis of any other field. When more than one magnetic field exists in a particular region, the individual magnetic fields are superimposed upon each other and a compass needle placed in the region will indicate the direction resulting from the interaction of the fields. Since a magnetic field has both magnitude and direction, the resulting field intensity at each point in the field is the vector sum of the components at that point.

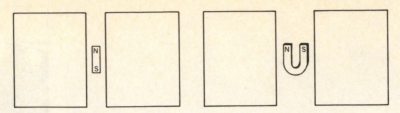

Figure 33–1 Arrangement of Magnets and Boards.

DESCRIPTION OF APPARATUS

In order to study the magnetic fields of various arrangements of magnets it is best to have a selection of boards which can be placed around the magnets and support a piece of cardboard (Figure 33-1). This arrangement holds the magnets in place and allows the experimenter to place a sheet of paper over the magnet and observe the associated magnetic fields. It is also posible to simply tape the magnets to the underside of a piece of cardboard.

The wiring diagram for Procedure B is shown in Figure 33-4. *R* is a rheostat for protecting the power source and for adjusting the current in the circuit. This experiment requires some high currents and the rheostat can get hot if the current is on continuously. Therefore the tap key *K* is placed in the circuit to keep the circuit open when observations are not being made.

The reversing switch *S* is used in the circuit to reverse the direction of the current in the wire *W*. Read Paragraph 2 in Appendix C concerning the use oi a reversing switch, before attempting to build your circuit.

PROCEDURE

Your mapping worksheet will serve as a data record.

A. Magnetic Fields of Permanent Magnets

1. Tape a piece of plain paper to the cardboard and center it over a bar magnet. Once you start your mapping do not move your setup. Note the geographic directions along the edges of the paper to indicate the orientation of your setup in the laboratory room. Label the magnetic poles and sketch the position of the magnet.

2. Determine the polarity of your compass by holding it 1 or 2 m away from the magnet board. Repeat the procedure now and then during the course of the experiment, since the polarity may change due to the proximity of strong magnets in the board. The direction of the magnetic field you are investigating will be determined by the direction in which the north-seeking pole of your compass points. Select a pencil which does not attract the compass needle.

3. Make a dot on the paper at the place you choose to start, and set your compass next to the dot in such a position that one end of the needle points toward the dot. Mark the position of the other end of the needle with a second dot (Figure 33-2). Then move the compass beyond the second dot and continue the process until you have a line of dots that lead to the pole of a magnet or to the edge of the paper. Connect this series of dots by a smooth curve. This line represents a line of force and should have one or more arrows on it to represent the direction of the magnetic field

along the line. Continue this process until you have the entire sheet mapped, spacing the lines so that they are about 2 to 3 cm apart in the regions away from the magnets. Each partner should do half of the mapping with the appropriate name on the corresponding side of the sheet. These mappings should be part of your report.

4. Repeat Steps 1 through 3 with a horseshoe magnet.

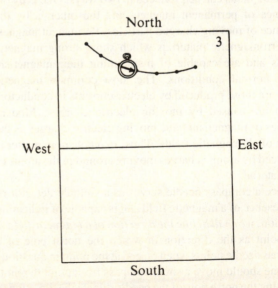

Figure 33–2 Method of Plotting Magnetic Fields.

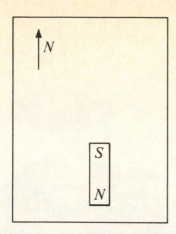

Figure 33–3 Locating the Neutral Point.

5. Ask your instructor if you are to map any other arrangements of magnets.

6. Reconstruct the configuration from Step 1 with a clean sheet of paper. Gently sprinkle iron filings over the entire sheet and tap the cardboard lightly. Compare the pattern with your mapping and describe simularities and differences. Repeat for other configurations you have mapped. Be sure to return your iron filings to the original container.

7. Determine the direction of magnetic north with your compass and construct the arrangement shown in Figure 33-3. The south pole of the magnet should be about 5 cm from the south end of the sheet. Draw a line around the magnet and label the south pole.

8. Plot the general outline of the magnetic field in the region of the south pole of the magnet and the section of the paper to the north of this end of the magnet. Do you find any points where the compass acts in a strange manner? If so, take a careful look at the field pattern and see if you can explain your observations.

9. Now let us take a closer look at the earth's magnetic field. The small mapping compass could be used here but a larger compass, or a mounted magnetic needle, would be more suitable. Place your compass (or magnetic needle) where it will not be influenced by other magnets or magnetic materials and make an estimate of the angle between the direction of the needle. and *true geographic north*. This angle is called the *magnetic declination* for your geographic location. Look at a magnetic map for your section of the country and see if your observation seems reasonable. *Record the observations from Steps 9, 10, 11, on the remaining unused part of your sheet.* Follow the pattern of other data forms in this book.

10. Set a dip needle in the same area as used in Step 9 and orient it so that the plane of the protractor scale is parallel to the direction of a compass needle. Note the direction the needle points and record the angle of dip of the north pole end below the horizontal. This angle is called the *angle of inclination (or angle of dip)*.

11. Now rotate the plane of the dip-needle scale so that it is perpendicular to the north–south magnetic meridian indicated by the compass. What is your observation? Can you explain it? Hold the dip needle with the plane of the scale parallel to the table top and note the result. Record these observations as a part of your report.

QUESTIONS—PROCEDURE A

1. Briefly sketch the general nature of the lines of force in a magnetic field between two unlike poles.

2. Do the same for the field between like poles.

3. Examine your drawing and note whether or not lines of force ever intersect.

4. Can you tell, by examining your drawing (or map), where the regions of strong field intensity are located? Explain.

5. In Step 8, where you mapped the field in the region of a single bar magnet, did you observe any neutral points? If so, explain the reason for it. What can you say about the magnitude and direction of the magnetic field intensity at a neutral point?

6. Suppose a bar magnet is brought near a magnetic needle in a way that does not disturb the latter's position in the magnetic meridian. What can be said concerning the direction of the field due to the magnet? If you are not sure

of the answer, try it out.

7. If the position of the magnet in Step 8 were reversed (end for end), would neutral points still exist at any place in the area around the magnet? If so, where?

8. If you had used a stronger magnet in Step 8, what effect would it have had on the position of the neutral point?

9. Explain why a dip needle points down rather than horizontal as does a compass needle. Why does a compass needle not indicate a dip angle?

10. Does the dip needle give any indication of the relative magnitudes of the horizontal and vertical components of the earth's magnetic field in your locality? Explain, using vectors.

11. Explain the action of the dip needle in Step 11 in terms of the horizontal and vertical components of the earth's field.

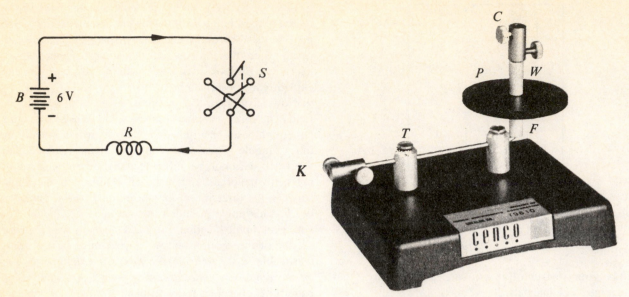

Figure 33–4 Circuit for Studying the Magnetic Field About a Wire Carrying a Current (Courtesy of Central Scientific Co.).

B. Magnetic Effects of a Wire Carrying a Current

1. *Conductor Vertical*. Connect your equipment as shown in Figure 33-4, using some type of dc power source of about 6 V, and join the wires at *C* with a connector. The current for this part should be 8–10 A.

2. *Test with Iron Filings*. Place a section of a sheet of paper on the circular platform *P*, and sprinkle iron filings *evenly and lightly* on the paper. Close the switch, tap the platform with your pencil, and note the pattern arrangement of the iron filings. Do not keep the switch closed longer than necessary, since this is injurious to the battery. In your report describe and make a clear sketch of the pattern of iron filings around the vertical wire. When you have finished, remove the paper and pour the iron filings back into the container.

3. *Test with a Compass*. Since a compass indicates the direction of a magnetic field, three situations present themselves for investigation: (1) no current in the wire, (2) current direction up the wire, (3) current direction down the wire. Trace the current through the reversing switch and then determine which way to close the switch to get the current direction up or down in the vertical portion of your test wire. The symbols $\odot$ and $\otimes$ may be used to represent the cross section of the wire in which the directions of the current are toward the observer and away from the observer, respectively.

With the switch open, place a small compass on the platform near the vertical wire and note the direction of the north pole of the needle as you move the compass around the wire. In recording this on your data sheet indicate the direction of the compass needle by a symbol such as a circle with an arrow, "$\oplus$" to show the direction of the north pole.

Also indicate the orientation of the platform by designating the geographic directions along the edges of the diagrams on the data sheet.

Now repeat the above procedure for the two cases with the current direction up and down the wire and again record the directions of the compass needle around the wire.

From the observations which you have just made, can you relate the directions of the current and the associated magnetic fields to the relative positions of the thumb and fingers of either of your hands? When you are able to make this association, use it to record a summary statement of your findings thus far.

4. *The Extent of the Field*. Replace the coil, *R*, with a rheostat (5- or 6-ohm range) in the circuit of Figure 33-4, and adjust for as high a current as possible without heating the rheostat significantly. You will use the rheostat for the remainder of the experiment.

Now suppose we investigate the extent of the magnetic field in the general region of the vertical wire. Starting with the compass against the wire, move it slowly away from the wire in a horizontal plane in a direction north of the wire and note the action of the compass. Repeat by moving in each of the other three geographic directions. In each case move away from the wire until the compass seems to not be influenced by the current, and record the distance from the wire. These points may best be determined by opening and closing the switch and noting the action of the compass needle.

Reduce the current to about one-fourth (by increasing the resistance about fourfold), repeat the above procedure, and record your observations.

5. *Conductor Horizontal*. Adjust the orientation of the current frame so that, with no current, the horizontal portion of the conductor (or wire) is parallel to the compass needle. Now investigate the direction of the magnetic field, both *above* and *below* the horizontal wire. While holding the compass still, in a specific position, increase and decrease the current, by means of the rheostat, and make note of your observations.

6. *Magnetic Field in the Region of Two Parallel Wires.* Remove the platform frame from your circuit and connect the two wires together with a connector clip. Arrange them, one above the other, along the table top in a north-south direction such that, with no current, they are parallel to a compass needle placed between them. Then with the current in opposite directions in the two wires, adjust the rheostat until the compass shows a large deflection, and record the approximate angle and direction of the compass deflection.

Now, without changing the current or the positions of the wires, investigate the intensity of the field directly above the upper wire and below the lower wire and record your observations. Analyze your setup and see if you can account for the results.

Next arrange the two wires so that the currents are in the same direction and repeat the above procedure.

7. *Effect of Loops.* This test can best be made with 1.5 to 2 m of small-size wire and a large compass with a calibrated circular scale. With the compass directly under one strand of wire, align the wire with the compass needle with no current in the wire. Then close the switch and adjust the rheostat until the deflection is about 20°. Now make one loop around the compass by folding the wire back underneath and note the deflection. Next go to two loops, then three, and so on until you have a quite large deflection.

8. *Magnetic Properties of a Solenoid.* With one section of your wire, make a solenoid (or helix) of 20 or more turns by

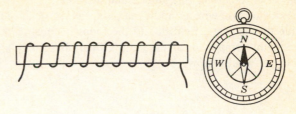

Figure 33–5 Solenoid and Compass.

winding the wire around your pencil. Remove the pencil and, with the helix replacing the platform in the circuit of Figure 33-4, arrange the solenoid (helix) in a direction at right angles to a compass needle (Figure 33-5). For all measurements which follow, *it is necessary* to use a long wire in the battery section of the circuit to remove the rheostat to a considerable distance from the compass.

Close the switch and adjust the rheostat until a small deflection is obtained. The size of the deflection can also be controlled by changing the distance between the compass and the solenoid. Reverse the current and note the effect for both directions of the current. Now insert a large soft iron nail in the solenoid and make note of the effect. Do you note any similarities beteen the solenoid and a bar magnet?

Now trace the current through the reversing switch, then through the solenoid as you note the polarity of the ends of the solenoid for both directions of current. See if you can formulate a rule, using your thumb and fingers, for the relation of the current around a solenoid and the polarity of the ends such that you could predict the polarity for any solenoid carrying a current. Write the rule as part of your report, along with a drawing to indicate your prediction.

QUESTIONS—PROCEDURE B

1. With no current in a vertically oriented wire, did your compass indicate that it was influenced by a magnetic field? Explain the reason for the action of the compass.

2. From observation of the pattern of iron filings around the vertical wire, what relation seems to exist between the wire and the pattern of iron filings? Did all positions of the compass needle show this same relation? If not, explain.

3. In attempting to determine the extent of the magnetic field about a vertical wire, did you find the distance to be the same in all directions? Explain. Theoretically, how far do you think the field should extend? Why does your compass not indicate this?

4. As you examined the magnetic field about a single horizontal conductor, did the direction of the field with respect to the wire correspond to your findings with a vertically arranged wire? Did the compass needle

deflection show the same direction for all values of the current? If not, does the direction of the field due to the current in the wire depend upon the amount of current? Also check with the current reversed. Explain fully.

5. Suppose two insulated parallel wires, carrying equal currents in opposite directions and aligned with the magnetic meridian, are placed against each other and laid directly over a compass. What results would you expect? Explain.

6. When investigating the magnetic field in the area of two parallel wires, what were your observations? Explain the likely reason for the results. Discuss both arrangements, parallel currents and oppositely directed currents.

7. As you increased the number of loops around the compass, what was the result? Does the magnetic field change direction inside a loop as the number of turns is increased? Does it change in magnitude? Explain both aspects of the effect (magnitude and direction).

8. With loops of wire around a compass, can you determine when the magnetic field due to the loops is equal to the horizontal component of the earth's field? If so, explain how you would know.

9. What was the effect of putting a soft iron nail inside the solenoid while near the compass? Give a possible explanation for the effect produced.

10. What advantage has the electromagnet over the permanent magnet? What disadvantage?

NAME *(Observer)* _____   Date _____

        (Partner) _____   Course _____

RECORD OF DATA AND RESULTS

Experiment 33—A Study of Magnetic Fields

B. MAGNETIC EFFECTS OF A WIRE CARRYING A CURRENT

The sketches below show the appearance of the iron filings and positions of compass needles as actually observed, when viewed from above with wire vertical.

| Iron-filing pattern about a vertical wire carrying a current | Compass arrangement, no current in the wire | Compass arrangement, current up the wire | Compass arrangement, current down the wire |

The results of all the above observations may be summarized as follows: _ _ _ _ _ _ _ _

_ _

_ _

The Extent of the Field

Direction from the wire		North	West	South	East
Distance from wire to apparent edge of field effect	Large current				
	Small current				

Comments:

Conductor Horizontal — Results and Comments

Magnetic Field in the Region of Two Parallel Wires

Current arrangement	Position of compass	Angle of deflection	Direction of deflection
Currents in opposite directions	Between wires		
	Above wires		
	Below wires		
Currents in same direction	Between wires		
	Above wires		
	Below wires		

Comments:

Effect of Loops

Number of loops around the compass	Amount of deflection

Comments on loop effects:

Magnetic Effect of a Solenoid

Description of effects: (1) Coil alone

(2) Coil with core

Rule for polarity:

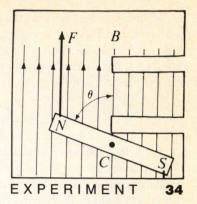

MEASUREMENT OF THE EARTH'S MAGNETIC FIELD

SPECIAL APPARATUS:

Milliammeter (0–100 mA), single switch, reversing switch, bundle of wire, small bar magnet (5 cm long or less), meter stick, watch with second hand (or stop watch), graph paper, regular bar magnet.

GENERAL APPARATUS:

dc current supply, (1–2V), rheostat (22 ohms), solenoid, supply of thread.

THE PURPOSE OF THIS EXPERIMENT

is to determine the horizontal component of the earth's magnetic field by means of an oscillating magnet in the field of a solenoid.

INTRODUCTION

Many electrical instruments employ coils which rotate in a magnetic field as the result of a torque interaction related to the current in the coil. The properties of charged particles are studied by observing their motion while under the influence of magnetic fields. The motion of electrons in electron microscopes and television tubes is controlled by magnetic fields. Often, the earth's magnetic field must be considered in analyzing the motions of coils and charged particles. Coils can be mounted on a vertical axis, and the influence of the earth's magnetic field determined by the action of its horizontal component on the coil

Since the horizontal component of the earth's magnetic field is a function of the geographical latitude and the nature of the earth's surface, its value at any particular location must be determined experimentally. In this experiment we shall determine its value in the region where your laboratory is located. One method of making this determination is to utilize the effect of a magnetic field on the period of a vibrating magnet suspended in the field.

If a bar magnet, which acts as a magnetic dipole, is placed in a magnetic field of flux density B (Wb/m²), the magnetic forces exert a torque τ on the magnet (Figure 34-1) given by

$$\tau = MB \sin \theta \qquad [1]$$

where M is the magnetic moment of the bar magnet. The value of the magnetic moment is dependent upon the

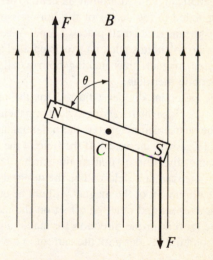

Figure 34–1 Torque Action on a Magnet in a Magnetic Field.

strength of the magnet and the length of the bar. If released in a position such as shown in Figure 34-1, it will vibrate back and forth about the axis C and across a line parallel to the field. If θ is small, the motion will be simple harmonic and the period T will be given by

$$T = 2\pi\sqrt{\frac{I}{MB}} \qquad [2]$$

where I is the moment of inertia of the oscillating magnet.

This arrangement is best accomplished by suspending a small magnet in a solenoid which provides an approximately uniform field at its center. The magnetic field at the center of such a solenoid, of N turns and length L, is given by

$$B_s = \left(10^{-7}\right)\frac{4\pi Ni}{L} \qquad [3]$$

where i is the current in amperes in the solenoid. The direction of the field can be determined by applying the right-hand rule to the current loops, or checking with a compass. If the circuit is arranged as shown in Figure 34-2, the solenoid may be aligned with the earth's magnetic meridian so that the field B_s of the solenoid is either parallel or antiparallel with the horizontal component of the earth's field B_e. Since the fields are vector quantities, we can write

$$B = B_s + B_e \qquad [4]$$

where B is the resultant flux density. If B_s is antiparallel to B_e, it will carry a negative sign. The suspended magnet is then under the influence of the resultant field of the earth and the solenoid. If the magnitudes of B from Equation [2] and B_s from Equation [3] are substituted into Equation [4], we have

$$\frac{4\pi^2 I}{M}\,\frac{1}{T^2} = \left(10^{-7}\right)\frac{4\pi Ni}{L} + B_e. \qquad [5]$$

An inspection of Equation [5] reveals that all quantities involved are constants except the period T of the vibrating magnet and the current i in the solenoid. Considering the variables as $1/T^2$ and i, the equation has the form $ay = bx + c$ and thus predicts a linear relationship between the two variables, both of which can be easily measured. If we set $1/T^2$ (or y) $= 0$, the intercept of this line on the X axis determines B_e from the relation

$$B_e = -\left(10^{-7}\right)\frac{4\pi N}{L}\,i_0 \qquad [6]$$

where i_0 is the current designated by the intercept on the current axis.

PROCEDURE

1. Connect the elements of the circuit as shown in Figure 34-2. You will note that the rheostat is to serve as a potential divider rather than a device to control the current (see Paragraph 7 in Appendix C). With a 22-ohm rheostat, the battery need not be more than 1 to 2 V.

2. With a small thread tied around its center, the magnet can be suspended in the center of the hollow core of the solenoid by means of a small piece of tape. When this is done, align the solenoid so that its axis is parallel to that of the suspended magnet, thus making both parallel to the earth's magnetic meridian.

3. With the potential divider adjusted for the minimum output voltage, close the reversing switch and then note the action of the suspended magnet as the potential applied to the solenoid is slowly increased. If the field of the solenoid is in the same direction as that of the earth's horizontal component, the resultant field will be the direct sum of the two and the alignment of the magnet will not be disturbed. Reverse the current and note the effect on the magnet.

4. With the current in the direction for stable alignment, adjust the potential divider until the current in the solenoid

circuit is 20 mA.* The suspended magnet may be set into oscillation by disturbing its alignment with a larger bar magnet as it is brought up near the outside of the solenoid and then quickly removed.

5. As the magnet oscillates, determine its period by taking the time for about 100 vibrations. Make at least three trials of this measurement and record the information along with the current in the solenoid.

6. Repeat this procedure for other values of the current, say steps of 10–20 mA, until you have five or more sets of values.

7. If your magnet is small enough to reverse direction inside the solenoid, repeat the above steps for some four or five values of a reversed current in the solenoid. These currents may be considered as negative.

8. Now let us determine if our experimental data confirms the predictions of the theory by plotting $1/T^2$ on the Y axis and the current on the X axis. Does it show a straight line as predicted by Equation [5]? If so, extrapolate to the X axis and determine the intercept value on the current scale used. If you performed Step 7 with negative currents, plot these

*If the magnet, solenoid, and ammeter at your disposal suggest different values for the current, the instructor will designate a different range of current values to be used. The field will be more uniform for currents below 200 mA.

data also. Record the intercept i_0.

9. Measure and record the length of the solenoid L and count the number of turns on the top layer; if the solenoid has more than one layer of wire, ask the instructor or check the manufacturer's specifications for the number of layers. With this information and the value of the intercept i_0, compute the magnitude of B_e from Equation [6] and compare with the value from a handbook for your local geographic region. Look under the heading *Horizontal Intensity of the Earth's Field*. From the range of values given for your state, you may have to estimate for your location within the state. If a conversion of units is needed, use the conversion factor: 1 Wb/m^2 (or 1 tesla) = 10^4 gauss. Record both values and find the percent discrepancy.

10. With *no current* in the solenoid, set the magnet into vibration and determine its period, and record along with your other data. Would you expect this period to bear any relation to the graph you have plotted? Examine the Y intercept.

QUESTIONS

1. Look at the individual values determined for the period of the magnet for a given current and compare them with their average. What sort of accuracy does this indicate?

2. Examine the trend of points along the line drawn on the graph. What does it indicate about the accuracy of your data? Do you notice any correlation between the size of the period and the uncertainty of its measurement? Explain in terms of your data.

3. Does Equation [5] predict the same slope and intercept for equal magnitudes of negative and positive currents? Discuss this point and relate it to your results.

4. As the current increased, what change did you observe in the period of the magnet? Is this effect predicted by the theory? If so, explain how the prediction is indicated.

5. If, when the current is reversed in the solenoid, the magnet tends to reverse direction, what does this action indicate about the magnitude of the magnetic field B_s in the solenoid? If, for some given current, the magnet does not tend to reverse, how would its period likely compare with that for the same current in the direct (or positive) direction? Why?

6. The magnet's motion was assumed to be of the simple harmonic type. Does Equation [1] predict angular harmonic motion? If so, explain how the condition for simple harmonic motion is predicted.

7. When you measured the period of the magnet with no current in the solenoid, what relation did this value of the period bear to the graph you had plotted? How is this period related to B_e? Compute T from the Y intercept and compare with measured value from Step 10.

8. Now that you know the value of the horizontal component of the earth's field in your locale, explain how you could use a vibrating magnet to determine the field intensity of some unknown magnetic field not associated with a solenoid.

9. When the current in the solenoid is zero, what relation does equation [5] predict for the magnitude of B_e? How is this related to the intercept on the Y axis? Can you find B_e from this intercept? If not, perhaps we can make use of the slope of the line for the determination. Write Equation [5] in the form of $y = mx + b$, using $1/T^2$ as y and i as x. If the solution is not yet apparent, carry out the steps in the next question.

10. Combine the expression found for m (or slope) in the above question with the expression for B_e in terms of the intercept y_0 and see if you can determine the unknown quantities from the resulting relation. Find the slope m and the intercept y_0 from your graph, and utilize them in evaluating B_e.

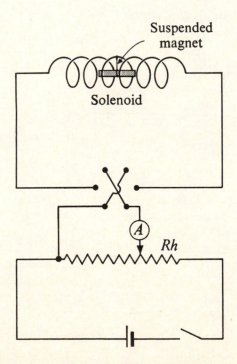

Figure 34–2 Circuit for Vibrating Magnet.

NAME *(Observer)* _____   Date _____

      (Partner) _____   Course _____

RECORD OF DATA AND RESULTS
Experiment 34—Measurement of the Earth's Magnetic Field

Direction of current	Amount of current	Number of vibrations counted	Total elapsed time	Average value of T	$\frac{1}{T^2}$
Positive		1. 2. 3.			
Negative					

RESULTS FROM GRAPH AND COIL DATA — STEPS 8 AND 9

Intercept on X axis i_o (A)	Length of coil L (m)	Number of turns N	B_e Calculated (Wb/m²)	B_e Handbook (Wb/m²)	Percent discrepancy

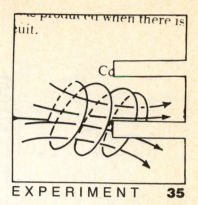

ELECTROMAGNETIC INDUCTION*

SPECIAL APPARATUS:

Table galvanometer, ammeter (0–2 A), coil system consisting of primary and secondary coils, bar magnet, brass rod, iron rod, high resistance (20,000 ohms), rheostat,† switch (tap key type), bundle of connecting wires.

GENERAL APPARATUS:

Source of low-voltage dc current.

THE PURPOSE OF THIS EXPERIMENT

is to study some of the phenomena of electromagnetic induction and to determine the relation between induced current and magnetic flux change.

INTRODUCTION

In Experiment 33 you examined the nature of the magnetic field about a wire carrying a current. In this experiment we shall study some of the effects produced when there is relative motion between a magnetic field and a closed electric circuit.

When magnetic lines of force thread through a coil as shown in Figure 35-1, we say that there is a magnetic flux linkage through the coil. The arrows indicate the directions of the current and the direction of the resulting flux lines. When the flux linking a coil of wire changes, because of either a change in the current in the coil or some outside influence, a potential difference appears between the ends of the coil. We call this potential difference an *emf* for *electromotive force*. The flux linking the coil could come from an external permanent magnet, the earth, or a nearby coil carrying an electric current.

In order to detect the presence of an induced emf, it is generally necessary that the change of flux take place within a coil which is a part of a closed electrical circuit. In this case, the presence of the emf produces an induced current which can be detected by a galvanometer. The galvanometer may not be calibrated to give the exact amount of current but it will indicate the direction and the relative magnitude. The zero current position on the scale is usually in the center, and the scale is calibrated to indicate deflections in either direction from the zero reading.

Coil

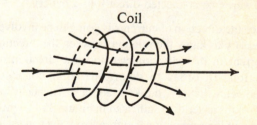

Figure 35–1 Illustration of Flux Linkage.

*This experiment can be done with almost no previous knowledge of induced currents on the part of the students. In the experiment students are led into the various effects and phenomena.
†The rheostat will not be needed if a variable dc voltage supply is to be used.

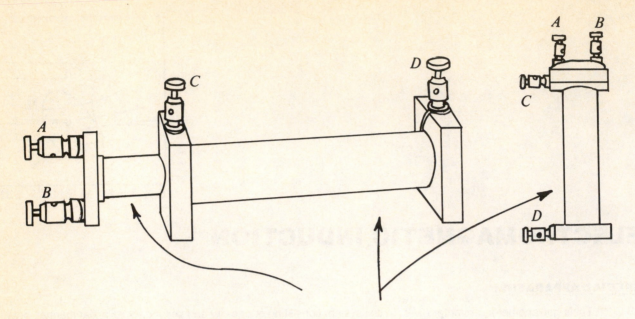

Figure 35–2 Primary–Secondary Coil Set.

DESCRIPTION OF APPARATUS

One form of a primary–secondary coil set for this type of experiment is shown in Figure 35-2. The primary, with a relatively small number of turns of large wire, has terminals at A and B. The secondary is wound on a hollow tube with many turns of small wire and has terminals at C and D. The primary coil can be moved back and forth inside the secondary and can also accommodate a small rod inside its own core. The windings may be covered with a nonconducting material for protection. Most other forms of coils will have these features.

PROCEDURE

On the data form, the numbered sections correspond to the numbered steps of the procedure. The galvanometer which you are to use is very sensitive and is designed for use with very small currents. It should never be used with currents with magnitudes that are wholly unknown and, likewise, should *never* be connected directly to a battery.

1. Since the direction of currents is going to be involved, it is necessary to know the relation between the direction of the current in the galvanometer and the direction of its deflection. In order to determine this, connect the galvanometer, battery, tap key, and a resistance of about 20,000 ohms in series as indicated in Figure 35-3. Trace the direction of the conventional current from the battery to the galvanometer and, when the switch is momentarily closed, note whether the needle deflects to the right, or the left, when the current enters at the right-hand terminal. Repeat with battery reversed.

2. Connect the terminals of the secondary coil to the galvanometer as shown in Figure 35-4. Examine the wires leading from the coil proper to the coil terminals (the apparatus itself, not the drawing), and determine which direction the winding goes around the coil.

Now thrust a bar magnet into the secondary coil (Figure 35-4), first with the north pole entering ahead, and then the south pole ahead. Note the direction of the galvanometer deflection for each case and then determine the direction of the current around the coil, clockwise or counterclockwise,

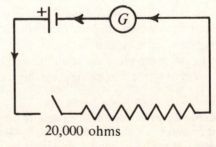

20,000 ohms

Figure 35–3 Circuit for Determining the Direction of Current Flow.

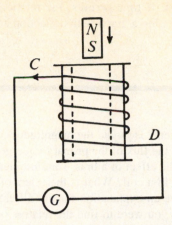

Figure 35–4 Galvanometer-Coil Circuit.

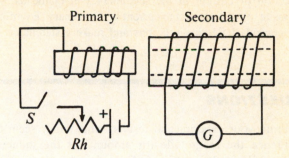

Figure 35–5 Primary–Secondary Circuit Connections.

looking down from above. Also check the directions as the magnet is pulled out of the coil.

As you move the magnet in and out of the coil make observations for the following and record them as a part of your report:

a. The effect of a change of speed of the moving magnet.

b. The effect of a change in polarity of the magnet.

c. The polarity produced in the coil by the induced current for each type of motion of the magnet. Consider the coil as an electromagnet and determine the polarity from the direction of the induced current.

d. The possible appearance of magnetic forces due to the motion. (Do not consider the force of gravity.) If forces do exist, when are they attractive and when are they repulsive? The force effect might best be detected by suspending the secondary coil horizontally as a pendulum by attaching strings to the two terminals, C, and D. This arrangement essentially eliminates the interfering forces of gravity and friction.

3. Line up the two coils so that the axes are parallel and the windings are in the same direction, either both clockwise or both counterclockwise (Figure 35-5). Set the rheostat at a value that will prevent the primary coil from overheating when the switch is closed. With the switch closed, move the primary coil back and forth along the axis as you did the magnet in Step 2. If the galvanometer needle deflects off the scale, reduce the current in the primary. Observe the directions of the current in the secondary, and, as a part of your report, draw two diagrams (omit the rheostat) similar to Figure 35-5. In each diagram show the following:

a. The direction of the current in the primary.

b. The direction of motion of the primary (toward secondary in one diagram, and away from secondary in the other; use arrow as shown in Figure 35-4).

c. The direction of the induced current in the secondary (use arrows) for each case.

Repeat these observations by moving the secondary instead of the primary and make a statement concerning the result.

4. With the primary coil near the secondary (inside, if need be), make and break the current in the primary and observe both the direction and relative magnitude of the deflection. Record your observations on two diagrams in a manner similar to the suggestion in Step 3, except that one is to show the switch being closed and the other is to show it being opened. To illustrate *make* and *break*, respectively, use the symbols:

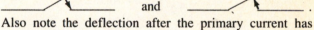

Also note the deflection after the primary current has reached a steady value with switch closed.

5. Insert a brass (or aluminum) rod through the cores of both coils while connected and positioned as shown in Figure 35-5; then repeat Step 4 and make a record of the results. Next replace the brass rod with an iron rod and repeat, recording the results. Compare the galvanometer deflections for both kinds of rods with the deflections in Step 4 without the rods.

6. With the iron rod inside the primary coil only, orient the primary in several different positions with respect to the secondary, make and break the current in each position, and find the orientation for a minimum deflection. Hold the coils as close to each other as possible during this procedure. Make a drawing of the position found for minimum deflection, and illustrate the relative directions of the flux linkages.

7. Place the coils, with the iron rod in the cores, back in the position shown in Figure 35-5 and insert an ammeter in the circuit of the primary coil. Adjust the rheostat for a current of 0.2 A and note the galvanometer deflection for a make and break of the primary circuit. If the deflection is more than one-fifth of the maximum scale reading, partially withdraw the iron rod. Increase the current in steps of 0.2 A up to 1.0 A and record the deflections (switch closed and open) for each 0.2 A increment of the current. Also note the direction of current in the secondary coil (same as, or opposite to) with respect to current in the primary. Examine the current–deflection relations and make a statement of your findings. If time permits, the relations can best be shown by plotting a graph with deflection as a function of the current.

QUESTIONS

1. From your experimental results, what things seem to influence the magnitude (or amount) of the induced current? What observations indicate these?

2. When using the bar magnet, what things affected the direction of the induced current?

3. Did you decide whether a magnetic force resulted from the motion of the magnet? If so, write a statement which relates the direction (attraction or repulsion) of the force to the direction of the induced current.

4. As you moved the coils (with current in the primary) relative to each other, when was the induced current in the same direction as the primary, and when in the opposite direction? Did you observe any relation between these results and those with the moving bar magnet? If so, what?

5. When two coils (one carrying a current) undergo relative motion with respect to each other, which one do you move to get the most induced current in the secondary? Explain.

6. Did you find the "make" and "break" of current in the primary to produce the same or different directions of induced current? How did the magnitudes compare? What change does the flux of the primary undergo in each case?

7. What is the effect of a brass (or aluminum) core on the flux linkage in a coil? What is the effect of an iron core? Can you give an explanation for the difference.

8. In Step 6 you were to find the relative positions of the coils for minimum deflection. Use the flux linkage concept to explain the reason for this minimum.

9. What effect did increasing the current in the primary have on the amount of induced current? Explain.

10. Is there any induced current in the secondary after the current in the primary has reached a steady value? Explain.

11. In this experiment you changed the flux linking the secondary by several methods. In one simple sentence, write a summary statement relating the *amount* of the induced current to the flux linkage in the secondary.

12. If you do not already know Lenz's law, look it up in your text, and then cite some part of your experimental observations which verifies the law. Explain how.

NAME *(Observer)* _____ Date _____

       (Partner) _____ Course _____

RECORD OF DATA AND RESULTS
Experiment 35—Electromagnetic Induction

1. Observation regarding direction of current through the galvanometer and the direction of the deflection.

   _

2. Effects of moving magnet in and out of the coil.

   (1) _

   (2) _

   (3) Polarity produced in the secondary by:

       (a) inserting N pole _ _ _ _ _ _ withdrawing N pole _ _ _ _ _ _ _ _ _ _ _

       (b) inserting S pole _ _ _ _ _ _ withdrawing S pole _ _ _ _ _ _ _ _ _ _ _

   (4) Explanation for a magnetic force _

   _

       (a) condition for attractive force _

       (b) condition for repulsive force _

3. Diagrams and results for relative motion between primary and secondary.

4. Diagrams and results of making and breaking primary circuit.

5. Results for the metal rods in the coil cores.
   (1) Brass (or aluminum)

   (2) Iron rod

6. Diagram for orientation of coils for minimum interlinkage of flux.

7. Effect of change of primary current on induced current.

Primary current (A)	Direction of current in secondary with respect to current in primary		Galvanometer deflection	
	Make	Break	Make	Break
0.2				
0.4				
0.6				
0.8				
1.0				

THE OSCILLOSCOPE

SPECIAL APPARATUS:

Oscilloscope, function generator, step-down transformer* (120 V to between 6 and 12 V), resistance box (0 to 20,000 ohms), diodes (4), capacitor (1 to 15 μF), bundle of connecting wires (with clip leads or some system for connecting components).

GENERAL APPARATUS:

None.

THE PURPOSE OF THIS EXPERIMENT

is to familiarize the student with the oscilloscope and the basic types of measurements that can be made with it.

INTRODUCTION

One of the more versatile laboratory measuring tools is the cathode ray oscilloscope, frequently just called an oscilloscope (Figure 36-1). The display part of an oscilloscope is the cathode ray tube (CRT), the general construction of which is shown in Figure 36-2. The electron gun consists of a number of components including a source of electrons, high-voltage accelerating electrodes, and electron focusing devices. The basic purpose of the electron gun is to produce a very well defined beam of constant velocity electrons traveling down the center of the CRT.

The key to the versatility of the CRT is two sets of electron deflection plates which deflect the electron beam in the vertical and horizontal directions as a function of the voltage applied to each set of plates. By selecting the appropriate applied voltages to these plates it is possible to deflect the beam so that it strikes any point of the screen. In practice the applied voltage on the plates is usually varied as a function of time causing the beam to sweep in some pattern across the screen. The most common technique is to cyclically apply a linearly increasing voltage to the horizontal plates causing the beam to sweep from left to right across the screen. Without any vertical signal this appears as a continuous line across the screen, because of the slow decay of the fluorescent screen and the persistence of vision of the human eye. When a signal is applied to the vertical plates the beam will be deflected vertically as it sweeps left to right causing a wave pattern to form (Figure 36-3).

There are generally three distinct sets of controls on any oscilloscope. One set controls the electron gun, one set controls the internal and external inputs to the horizontal plates, and one set controls the external inputs to the vertical deflection plates. The following discussion will include only the basic controls in each set. All oscilloscopes will contain the controls discussed below; however, the labels on the controls may be different and some additional controls may be present to operate special functions or to perform the usual functions more precisely.

---

*Be sure that the low-voltage side of the transformer is isolated from the high-voltage side. Many variable transformers are voltage dividers and do not give isolation, which could make possible 120 Volts to ground even though the terminals have a voltage of 6 V. Inexpensive transformers are available at electronic supply stores such as Radio Shack.

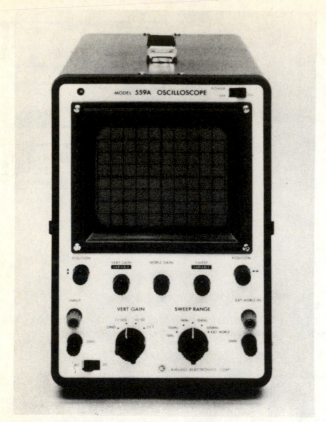

Figure 36–1 Oscilloscope (Courtesy of Central Scientific Co.).

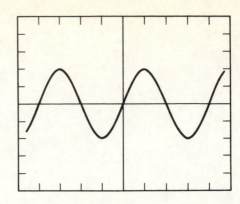

Figure 36–3 Cathode-Ray-Tube Screen Wave Pattern.

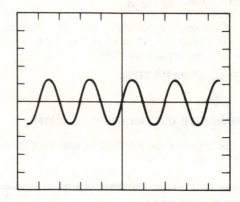

Figure 36–4 Sine Wave.

Electron Gun Controls

Power: Turns the power to the oscilloscope *on* or *off*. May be combined with another control.

Intensity: Controls the brightness of the spot or trace on the screen. This brightness should be kept low to increase the life of the CRT.

Focus: Controls how sharp the spot or trace is on the screen.

Horizontal Plate Controls

Horizontal Position: Controls the center position of the beam in the horizontal direction.

Horizontal (Sweep) Selector: This control allows the operator to control the horizontal sweep rate to one of several ranges from the internal sweep generator. The internal sweep rates are usually given in seconds per division. Thus, if the setting is 2 msec the signal takes 2 msec to sweep across one division on the horizontal axis.

Associated with this control there may be others such as *horizontal gain* and *sweep vernier*. The horizontal gain determines the amplitude of the horizontal sweep and the

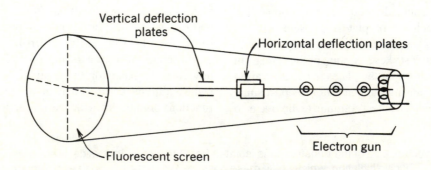

Figure 36–2 Cathode-Ray-Tube Diagram.

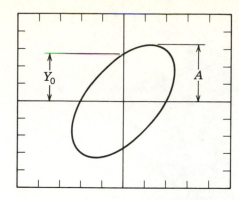

Figure 36–5 Frequency Match Pattern.

sweep vernier permits fine control of the internal sweep generator. Additional controls may allow the operator to select 60-Hz line sweep triggering or external signal sweep frequencies.

Vertical Plate Controls

Vertical Position: Controls the center position of the beam in the vertical direction.

Vertical Input: Other names for this control include *attenuator* and *volts per division*. This control selects the vertical sensitivity of the oscilloscope. It is used to determine the vertical beam deflection as a function of the vertical input signal. Thus for a setting of a 1 V/div, a 5-V input would deflect the beam five vertical divisions.

Perhaps the simplest measurement one can make with an oscilloscope is ac or dc voltage. The oscilloscope becomes a high-resistance voltmeter simply by calibrating the vertical deflections to some known volts per division of deflection and then connecting the unknown voltage to the *vertical (Y)* input. The voltage is then easily determined by reading the deflection on the CRT. Note that the oscilloscope measures instantaneous voltage. Thus, the maximum voltage is from zero to a peak or peak-to-peak divided by 2.

The frequency of an ac source can be determined by calibrating the horizontal sweep rate and then connecting the unknown source to the Y input. The periodic sweep of the beam causes the input pattern to retrace itself and the pattern appears to stand still (see Figure 36-4). By determining the number of X divisions required for one complete cycle of the waveform, the period is easily determined. The sine wave in Figure 36-4 covers two divisions for one complete cycle. If the sweep rate were 2 msec per division, the period would be 4 msec; so the frequency would be 250 Hz. The frequency is the reciprocal of the period.

The relative phase between two ac signals having the same frequency can also be measured with an oscilloscope using a Lissajous pattern. If one signal is input to the horizontal plates and the other to the vertical plates, a pattern similiar to Figure 36-5 will appear on the screen. If the horizontal and vertical amplifiers are adjusted so that the amplitudes of the two signals are the same, the equations of the two waveforms on the screen would be

$$X = A \sin 2\pi ft \qquad [1]$$

and

$$Y = A \sin(2\pi ft + \phi) \qquad [2]$$

where A is the amplitude, f is the frequency, and ϕ is the phase difference. From trigonometric identities Equation [2] can be expanded as

$$Y = A(\sin 2\pi ft \cos \phi + \cos 2\pi ft \sin \phi). \qquad [3]$$

When the X deflection is zero, $\sin 2\pi ft = 0$ and $\cos 2\pi ft = 1$; so

$$Y_0 = A \sin \phi \qquad [4]$$

or

$$\sin \phi = Y_0/A. \qquad [5]$$

Both Y_0 and A can be read from the screen and thus ϕ is determined.

DESCRIPTION OF APPARATUS

In addition to the oscilloscope this experiment requires a function generator (Figure 36-6). This instrument is a variable oscillator designed to output various waveforms with varying frequencies and amplitudes. The controls generally allow one to choose between various waveforms including sine waves and square waves. The output amplitude and frequency are usually selected by several controls which are self-explanatory. If you have difficulty understanding how to set your function generator, ask your instructor for assistance.

Electronic components used in this experiment include a 120-V ac stepdown transformer with a 6- to 12-V ac output, diodes, capacitor, and resistors. The diodes are solid state devices which will allow electric current in only one direction.

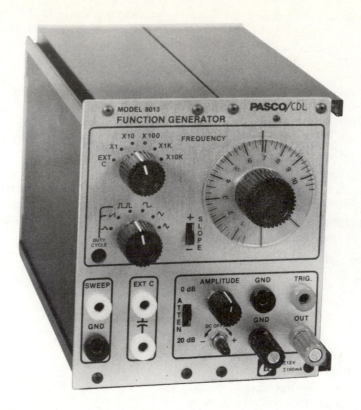

Figure 36–6 Function Generator (Courtesy of Pasco Scientific Co.).

PROCEDURE

1. Identify the controls discussed in the Introduction on your oscilloscope. Your instructor may have some special instructions for the oscilloscope you are using.

2. Turn the intensity and amplifier gains to zero before turning the power on. Also place the focus, horizontal positioning, and vertical positioning controls to the center of the range. Turn the power on and allow 30 sec for the instrument to warm up.

3. Advance the intensity control until the beam is visible on the screen. Adjust the focus to give a sharp point. Adjust the horizontal and vertical positioning controls to center the spot on the screen. Set the internal sweep generator to 60 Hz (*line*) and advance the horizontal gain control until the trace on the screen covers about two-thirds the width of the screen.

4. Set the function generator to output a 60-Hz sine wave and connect it to the vertical input of the oscilloscope (This includes a GND connection). Adjust the vertical gain until a stationary sine wave appears on the screen. Determine and record the voltage and frequency of the input. Determine the percent difference between the frequency

given on the function generator and that measured by the oscilloscope. Draw the waveform on your data sheet.

5. Change the output of the function generator to a higher frequency (i.e., 400 Hz). Again determine the voltage and frequency on your oscilloscope and the percent difference. Draw the waveform on your data sheet.

6. Repeat Step 5 using a square wave instead of a sine wave.

7. Connect the output of your transformer to a 200-ohm resistor, and then to your oscilloscope as shown in Figure 36-7. Determine and record the output frequency of the

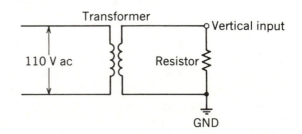

Figure 36–7 Circuit for Transformer Connections.

transformer. Draw the wave observed on the screen. Also measure and record the voltage across the resistor. Note that the oscilloscope measures peak-to-peak (PP) voltages and outputs are usually given as root-mean-square (RMS) values. The conversion factor is given by

$$V_{RMS} = V_{PP}/2\sqrt{2} = 0.3535V_{PP}. \qquad [6]$$

Compute V_{RMS} and compare to the listed transformer output by computing the percent difference.

8. Insert a diode into your circuit used in Step 7 so that the circuit looks like Figure 36-8. Connect the oscilloscope to point 2. Draw a diagram of what you see on the screen and write a brief description. How does this compare to that measured in Step 7? This is called a half-wave rectifier. Connect the oscilloscope to point 1 and note any differences.

9. Using four diodes, construct the circuit shown in Figure 36-9, and connect to your oscilloscope. Draw a diagram of your screen and explain. How does the voltage compare to that measured in Step 7? This is called a full-wave bridge rectifier.

10. Using your transformer, resistance box, and capacitor construct the circuit shown in Figure 36-10. The resistance should be comparable in magnitude to the reactance of the capacitor. (See Experiment 37, Equation [3].) Connect the *horizontal* input of the oscilloscope across AG and the *vertical* input across BG. **Caution:** *Be sure to use point G as the common ground.* Turn on the transformer and adjust the *horizontal* and *vertical gain* controls until you obtain an elliptical pattern like that in Figure 36-5. Read and record A and Y_o and compute ϕ from Equation [5]. For this arrangement ϕ is the phase angle between the voltage and the current in the circuit. See Experiment 37 for a discussion of this phenomenon.

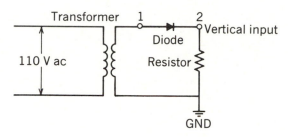

Figure 36–8 Half-Wave Rectifier.

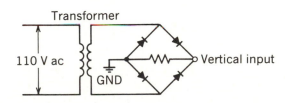

Figure 36–9 Full-Wave Bridge Rectifier.

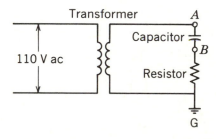

Figure 36–10 Circuit for Measuring Phase Due to Capacitance and Resistance.

QUESTIONS

1. Draw a diagram of two parallel conducting plates like those in a CRT and label the top plate positive and the bottom plate negative. Draw arrows indicating the direction of the electric field and the direction of the force on an electron.

2. Sketch a graph of voltage as a function of time across the CRT deflecting plates that would cause the beam to sweep uniformly across the screen.

3. What is the input frequency of a signal that covers 4 divisions when the horizontal sweep is set to a 4 msec?

4. Describe what you would expect to see on the screen if a

6-V battery were connected to the vertical input. What changes if the connections are reversed? Can an oscilloscope be used as a dc voltmeter?

5. What is the difference between peak-to-peak voltage and RMS voltage? Derive Equation [6].

6. What would you expect to see in Step 8 if the vertical and ground inputs had been reversed? In Step 9?

7. In Step 10 explain why the *AG* connection gives the phase of the voltage and the *BG* connection gives the phase of the current.

RECORD OF DATA AND RESULTS
Experiment 36—The Oscilloscope

OSCILLOSCOPE

Manufacturer _____

Model _____

FUNCTION GENERATOR

Manufacturer _____

Model _____

Step 4

Oscilloscope

Sec/div = _____ f = _____

Volts/div = _____ V = _____

Function generator

f = _____

Percent diff. = _____

Step 5

Oscilloscope

Sec/div = _____ f = _____

Volts/div = _____ V = _____

Function generator

f = _____

Percent diff. = _____

Step 6

Oscilloscope

Sec/div = _____ f = _____

Volts/div = _____ V = _____

Function generator

f = _____

Percent diff. = _____

Step 7

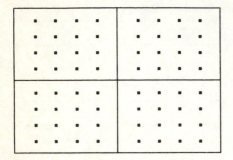

Oscilloscope

Sec/div = _____ f = _____

Volts/div = _____ V_{PP} = _____

V_{RMS} = _____

Transformer

V_{RMS} = _____ Percent diff. = _____

Step 8

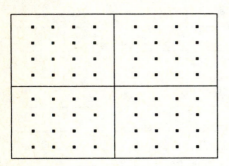

Oscilloscope

Sec/div = _____ f = _____

Volts/div = _____ V = _____

Description

Step 9

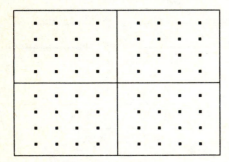

Oscilloscope

Sec/div = _____ f = _____

Volts/div = _____ V = _____

Description

Step 10

Oscilloscope

Sec/div = _____ f = _____

Volts/div = _____ V = _____

A = _____ Y_0 = _____

ϕ = _____

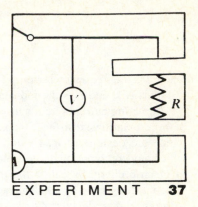

EXPERIMENT **37**

A STUDY OF ALTERNATING CURRENT CIRCUITS

SPECIAL APPARATUS:

Step-down transformer* (120 V to between 6 and 12 V), high-impedance ac voltmeter (such as a Simpson meter), oscilloscope[†] resistor (100–3000 ohms), capacitor (1–15 μF, nonpolarized), inductor (0.2–10 henrys), bundle of connecting wires (with clip leads or some system for connecting components), dc voltmeter, dc ammeter, switch.

GENERAL APPARATUS:

Storage battery or other 6-V dc source.

THE PURPOSE OF THIS EXPERIMENT

is to become familiar with the basic alternating current circuit elements and their properties.

INTRODUCTION

The voltage output (V) of an ac source usually varies with time (t) according to the relationship

$$V = V_0 \sin(2\pi f t) \qquad [1]$$

where V_0 is the maximum voltage and f is the frequency. This means that the polarity of the source changes twice each cycle, and therefore the current will change its direction of flow twice each cycle.

Alternating current voltmeters and ammeters measure the root-mean-square (RMS) voltages and currents. The root mean square is the square root of the average value of the sum of the squares of the instantaneous values. It can be shown that this value, V_e, is related to the maximum value, V_m, by the equation

$$V_e = V_m/\sqrt{2}.$$

Throughout the remainder of this experiment we will always use root-mean-square values, also called effective values. Ohm's law for resistors in ac circuits is still

$$V = IR \qquad [2]$$

which is valid for instantaneous or effective values.

Capacitors and inductors are very important elements in ac circuits. Depending on their physical characteristics and the frequency of the applied voltage, they can effectively show high or low resistance to ac. Both elements show Ohm's law-type relationships where the ratio of applied voltage to the current produced is a constant called the *reactance*. The capacitive reactance and inductive reactance are designated X_C and X_L, respectively, and their units are ohms. The expressions for these reactances have the form

*Be sure that the low-voltage side of the transformer is isolated from the high-voltage side. Many variable transformers are voltage dividers and do not give isolation, which could make possible 120 V to ground even though the terminals have a voltage of 6 V. Inexpensive transformers are available at electronic supply stores, such as Radio Shack.
[†]Only the optional procedures require an oscilloscope. However, the oscilloscope can be used in place of the voltmeter in all procedures if desired. If the oscilloscope is used the student should first perform Experiment 36.

$$V_C/I = X_C = 1/(2\pi fC) \qquad [3]$$

$$V_L/I = X_L = 2\pi fL \qquad [4]$$

where f is the frequency in hertz (Hz), C is the capacitance in farads (F), and L is the inductance in henrys (H).

Alternating current circuits usually contain two or more of three quantities: resistance, capacitance, and inductance; which are designated R, C, and L. The total resistance which any combination of these elements offers to the flow of electric charge is called the impedance of the circuit which is designated by a Z. The ratio of the applied voltage to the current produced is Z so that Ohm's law for ac circuits is given by

$$V = IZ. \qquad [5]$$

When a pure resistance carries an alternating current, the potential difference across the resistor is in phase with the current. This means that when the voltage is zero, the current is zero, and when the voltage is at its maximum value, the current is at its maximum value. This is not true for capacitors and inductors. The voltage across a capacitor is zero when the current is a maximum, and when the voltage is a maximum, the current is zero. We say that the voltage lags the current and is out of phase by a quarter cycle or 90°. A similar phenomenon occurs with inductors except the voltage leads the current by 90°. It is important to note that these statements apply to pure capacitance and pure inductance. Consult your textbook for illustrations of the above-described phase relations.

Because of these phase relationships, the voltages of the *RCL* elements must be added vectorially. Figure 37-1a shows the vector relationship for the voltages across the three ac elements when connected in series to an ac source such as shown in Figure 37-6. Similarly, Figure 37-1b shows the vector relationship for the resistance and reactances of the same three elements when connected in series. Note that V_L does not indicate the voltage across the inductor but rather the voltage due to inductance. (All inductors have resistance.) The resistance component of an inductor must be included as part of R and the voltage due to the inductor's resistance must be included in V_R.

The voltages, resistance, and reactances are added as three vectors which have lengths proportional to the magnitudes of the quantities and which have directions like those shown in Figure 37-1. Figure 37-2 shows the vector addition for the three vectors of Figure 37-1b. From this diagram we see that the circuit impedance is given by

$$Z^2 = R^2 + X^2 = R^2 + (X_L - X_C)^2. \qquad [6]$$

In a similar way the total voltage across these three elements is found to be

$$V^2 = V_R{}^2 + (V_L - V_C)^2. \qquad [7]$$

Note that in Equations [6] and [7] it is possible for some of the factors to be zero when the circuit is without one or more of the three elements.

From Figure 37-2 we see that the phase angle ϕ between the voltage and current is given by

$$\tan \phi = X/R \qquad [8]$$

where X is the total reactance and R is the total resistance. See Experiment 36 for a discussion of how ϕ can be measured with an oscilloscope.

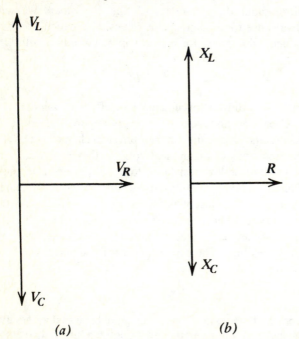

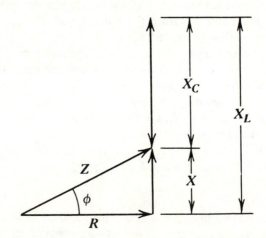

Figure 37–1 The Vector Relations of Voltages and Reactances in *RCL* Circuits.

Figure 37–2 Vector Addition of Reactance in *RCL* Series Circuits.

DESCRIPTION OF APPARATUS

There are commercially available circuit boards suitable for this experiment. Such boards can be constructed easily from readily available components, or the components can be used separately without constructing a circuit board.

The choice of values for capacitance, inductance, and total resistance (resistor plus the resistance of the inductor) is important in order to achieve meaningful results in this experiment. The capacitive reactance, inductive reactance, and total resistance should all be comparable in magnitude. For example: a 5-μF capacitor (X_C = 531 ohms), a 1-H inductor (X_L = 377 ohms), and a total resistance of 400 ohms would be a reasonable choice of components. Magnitude ranges of compatible components are given in the *Special Apparatus* list.

PROCEDURE

A. Pure Resistance Circuit

1. Using the dc source and dc meters connect the circuit as shown in Figure 37-3. Record the voltmeter and ammeter readings and use these to determine the resistance of your resistor. Replace the resistor in the circuit first by the capacitor and then by the inductor. Record the voltmeter and ammeter readings in each case. Use the appropriate values to determine the resistance of the inductor which will be needed later.

B. Circuit with Resistance and Capacitance in Series

2. Connect the circuit as shown in Figure 37-4. *The current source in this case will be the ac step-down transformer*. If you are using a multimeter, be sure that all switches are in the correct position for ac readings before making measurements. Using the ac voltmeter; measure and record the voltage across the capacitor, next the resistor, and then across both elements. Open the switch after completing your measurements. Using Ohm's law, determine the current through the resistor which is the current through the circuit since all elements are in series. Is there any significant difference between the current through the capacitor in this case compared to what was found in Step 1? Now using Equation [3] calculate the capacitive reactance of the capacitor and its capacitance. (The frequency should be 60 Hz unless otherwise stated by your instructor.)

3. Using the data from Step 2, check to see if the voltage V_T across both elements is equal to the sum of the voltages across individual elements. Recall that, in ac circuits, the total voltage across more than one series element is not the arithmetic sum of the voltages but rather the vector sum. Now use Equation [7] to calculate what the total voltage across both elements should be, and find the percent difference between this value and your measured value.

4. Use Equation [6] with the reactance found in Step 2 and the resistance from Step 1 and calculate the impedance of the two elements. With this value and the current measured in Step 2, calculate the voltage across both elements using Equation [5]. Calculate the percent difference between this value of the voltage and your measured value V_T from Step 2.

5. Optional. The optional procedures require an oscilloscope to measure the phase angle between the voltage and the current. Connect the vertical input across *AG* and the horizontal input across *BG* (Figure 37-4). Be sure *G* is a common ground. Adjust the gain controls until an elliptical pattern fills most of the screen. Record the values of *A* and Y_0 (see Experiment 36, Figure 36-5). Using these numbers compute ϕ and compare the value obtained using Equation [8] by computing the percent difference.

C. Circuit with Resistance and Inductance in Series

6. Connect the resistor and inductor to the ac source as shown in Figure 37-5. Measure and record the voltage across the inductor, then the resistor, and finally across

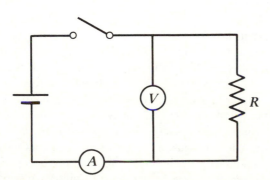

Figure 37-3 Direct Current Resistance Circuit.

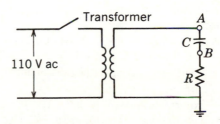

Figure 37-4 Alternating Current Circuit with Resistance and Capacitance in Series.

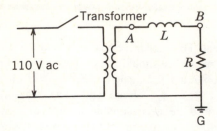

Figure 37–5 Alternating Current Circuit with Resistance and Inductance in Series.

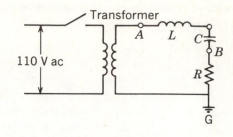

Figure 37–6 Complete *RCL* Series Circuit.

both elements. Open the switch when not making measurements. Calculate the current in the circuit by applying Ohm's law to the resistor. Then find the impedance of the circuit from Equation [5] using the total measured voltage.

7. In Step 1 we found the resistance of the inductor and of the resistor. The total resistance R_T in the circuit is now the sum of the resistances of the inductor and the resistor. Using this value for R_T and the value of Z from Step 6, solve Equation [6] for X_L. *Then find the inductance, L, using* Equation [4].

8. Use Equation [4] along with the values of X_L obtained in Step 7 and the current in Step 6, and calculate the voltage V_L due to the inductance. Why is this not the same as the measured value of the voltage across the inductor in Step 6? If R_L is the resistance of the inductor, the voltage across it is given by

$$V_L = I\sqrt{R_L^2 + X_L^2} \ . \qquad [9]$$

Compute the percent difference between this value and the measured value.

9. Optional. Using the oscilloscope connect the vertical input across *AG* and the horizontal input cross *BG* (Figure 37-5). Be sure *G* is a common ground. Adjust the gain controls until an elliptical pattern fills most of the screen.

Record the value of *A* and Y_0 (see Experiment 36, Figure 36-5). Using these numbers compute ϕ and compare the value obtained using Equation [8] by computing the percent difference.

D. Complete *RCL* Series Circuit

10. Connect all three components in a series circuit as shown in Figure 37-6. Measure and record the voltage across each component and then the total across all three components. Use Ohm's law to find the current through the resistor, which is the same as that through the entire circuit.

11. Using the total measured voltage across all three components and the current from Step 10, calculate the impedance of your circuit from Equation [5]. Use your previously determined values for X_C, X_L, and R to calculate Z from Equation [6]. Calculate the percent difference between these two values for Z.

12. Optional. Using the oscilloscope connect the vertical input across *AG* and the horizontal input across *BG* (Figure 37-5). Be sure *G* is a common ground. Adjust the gain controls until an elliptical pattern fills most of the screen. Record the values of *A* and Y_0 (see Experiment 36, Figure 36-5). Using these numbers compute ϕ and compare the value obtained using Equation [8] by computing the percent difference.

QUESTIONS

1. What is the maximum or peak value of the voltage across the source in Step 2? Review the second paragraph of the introduction.

2. What is the maximum or peak value of the current in the resistor in Step 2?

3. Explain clearly how a capacitor can conduct ac current but not dc current.

4. In this experiment we used a 60-Hz source. What measured quantities would change in Step 18 if we had used a 600-Hz source? What is the nature of the change:

5. Derive Equation [9] using earlier equations.

6. Insert the measured values for the voltages across the components in Step 10 into Equation [7], and solve for the total voltage in the circuit. Is this the same as the measured value? Figure out how to use Equation [7] correctly for this circuit so that one gets the current value for the source voltage. *Hint*: Look again at Step 8 and note how the resistance of the inductor was corrected for.

7. An inductor is just a long wire in a coil configuration. Explain how a wire in this configuration offers opposition to the current in an ac circuit which it would not offer if used as a long straight wire.

8. If the wire used to wind a coil were doubled back on itself and then wound around some form of core, would this affect the inductance of the coil in an ac circuit? Explain.

9. If the capacitor and the inductor were designed so that X_C and X_L could be made variable quantities, what would be the result in Step 10 of the procedure if adjustments were made so that $X_C = X_L$? How would this affect the current in the circuit of Figure 37-6?

10. The tuner knob on a radio operates a variable capacitor which has air between the plates. As the knob is turned the effective area of the plates changes and thus changes the capacitance. Maximum volume is obtained only when the *RLC* circuit of the radio is tuned to the frequency of the broadcasting station. The voltage for the tuning circuit is supplied by resonance coupling with the incoming high frequency signal. Examine the equations in the Introduction of this experiment, along with the circuit of Figure 37-6, and explain how the frequency of the circuit is changed by the tuning knob to resonate with the incoming signal.

11. Was the percent difference between the measured and computed phase angles larger in Step 5 or 9? Explain why one might expect Step 9 to have the larger percent difference.

NAME *(Observer)* _____ Date _____

   (Partner) _____ Course _____

RECORD OF DATA AND RESULTS

Experiment 37—A Study of Alternating Current Circuits

Frequency of ac Source _____ Hertz

Step	Measured Values	Calculated Values	Calculations and Comments
A. PURE RESISTANCE CIRCUIT			
1	$V_R =$ $I_R =$ $V_C =$ $I_C =$ $V_L =$ $I_L =$	$R =$ $R_C =$ $R_L =$	
B. CIRCUIT WITH RESISTANCE AND CAPACITANCE IN SERIES			
2	$V_R =$ $V_C =$ $V_T =$	$I =$ $X_C =$ $C =$	
3		$V =$ Percent Difference $=$	
4		$Z =$ $V_T =$ Percent Difference $=$	
5	$A =$ $Y_0 =$	$\phi = \sin^{-1}(Y_0/A) =$ $\phi = \tan^{-1}(X/R) =$ Percent Difference $=$	
C. CIRCUIT WITH RESISTANCE AND INDUCTANCE IN SERIES			
6	$V_R =$ $V_L =$ $V_T =$	$I =$ $Z =$	
7		$R_T =$ $X_L =$ $L =$	
8		V_L (eq 4) $=$ V_L (eq 8) $=$ Percent Difference $=$	
9	$A =$ $Y_0 =$	$\phi = \sin^{-1}(Y_0/A) =$ $\phi = \tan^{-1}(X/R) =$ Percent Difference $=$	

D. COMPLETE RCL SERIES CIRCUIT

10	$V_R =$ $V_C =$ $V_L =$ $V_T =$	$I =$	
11		Z (eq 5) $=$ Z (eq 6) $=$ Percent Difference $=$	
12	$A =$ $Y_0 =$	$\phi = \sin^{-1}(Y_0/A) =$ $\phi = \tan^{-1}(X/R) =$ Percent Difference $=$	

PART FIVE
LIGHT

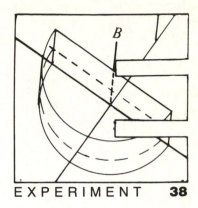

REFLECTION AND REFRACTION OF LIGHT

SPECIAL APPARATUS:

Each individual student should have the following:* plane mirror, mirror support (small wood block), five pins, ruler, protractor, semicircular plastic dish† and/or rectangular glass plate, soft cardboard, rubber band.

GENERAL APPARATUS:

Supply of one or two additional liquids if plastic dish is used.

THE PURPOSE OF THIS EXPERIMENT

is to study some phenomena and laws of reflection and refraction by tracing the paths of light rays falling on reflecting and transmitting surfaces.

INTRODUCTION

When a ray of light OC (Figure 38–1) strikes a reflecting surface MZ at C, it is reflected along the direction CA. Since light may be emitted in all directions from O, many other rays will strike the mirror. One such ray may travel from O to D and be reflected along the path DB. If the normals CN_1 and DN_2 are drawn at the points where the rays strike the mirror, the angles i and i' are called the angles of incidence. You will note that the two incident rays OC and OD diverge from point O, and still diverge after reflection and appear to have originated at point I. This is the position of the virtual image of the object at O.

If light is traveling in air at a velocity v_1, along the direction AC (Figure 38–2), and enters a different medium at C in which the velocity is v_2 (less than v_1), the ray is transmitted along the direction CB, making a smaller angle with the normal CN than did the incident ray. Angles θ_1 and θ_2 are called the angles of incidence and refraction, respectively. If the ray should originate in the second medium at B and emerge in the first medium at C, the refracted ray would take the direction CA, so that the

resulting angle of refraction θ_1 is larger than the incident angle θ_2. If the velocities of light in the two media are known, their relation to the angles take the form

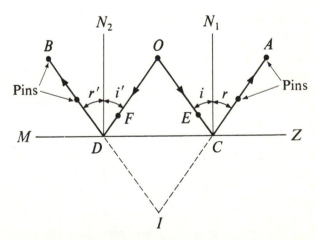

Figure 38–1 Reflection of Light from a Plane Mirror.

---

*If this much equipment is not available, it is recommended that one student work with the reflection equipment while a partner works with the refraction equipment, and then exchange.

†These transparent semicircular plastic dishes are available from scientific supply houses.

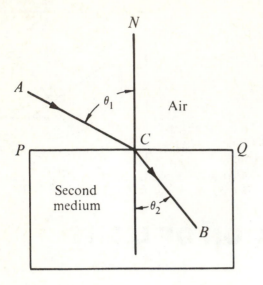

Figure 38–2 Refraction as Light Passes from One Medium to Another.

$$\frac{v_1}{v_2} = \frac{\sin \theta_1}{\sin \theta_2}. \qquad [1]$$

If medium 1 is a vacuum (or air), the ratio $\sin \theta_1/\sin \theta_2$ is called the *absolute index of refraction* n_2 of medium 2. If both media are substances other than air, such as water and glass, then the index is called the *relative index of refraction* and is designated by the relation

$$\frac{n_2}{n_1} = \frac{\sin \theta_1}{\sin \theta_2} \qquad [2]$$

where n_1 and n_2 represent the absolute indices of the two media, respectively. An easily remembered form of Equation [2] is

$$n_1 \sin \theta_1 = n_2 \sin \theta_2. \qquad [3]$$

This equation is easy to apply to the situation, regardless of the direction of travel of the ray of light across the boundary of two media. If medium 1 is air, then $n_1 = 1$, and the index of refraction of medium 2 is written as

$$n_2 = \sin \theta_1/\sin \theta_2. \qquad [4]$$

PROCEDURE

Instructor may choose Section A and B or Section A and C.

Each student should work alone on all parts of this experiment and make his or her own individual report. The drawings are to be made on plain white sheets of paper and constitute a portion of your report along with results recorded on the data form.

A. The Reflection of Light

1. Draw a line across the center of a plain sheet of paper and use it as the base line *MZ* for the position of the mirror (Figure 38–1). Examine your plane mirror and note that the silvered (or reflecting) surface is on the back side. Support the mirror against a wood block by means of a rubber band, and place the reflecting surface along the line *MZ* with the paper resting on a sheet of soft cardboard and held in place with strips of scotch tape.

2. Stick a pin vertically into the paper at some point about 8 cm in front of the mirror to serve as the object. Label it as point *O*. Now with your eye on a level with the surface of the paper, and displaced to one side as you look into the mirror, locate the image of the object pin. You are now looking along the path of one of the reflected rays, in a direction such as *AC*, illustrated in Figure 38–1. While holding your line of sight along the surface of the paper, with your eye at nearly arm's length from the mirror, stick two pins 6 to 8 cm apart along this line so that these two pins

and the image of the object pin stand in a straight line. Now place your eye behind the object pin *O* and adjust your line of sight until the object pin and the *images* of the other two pins are in a straight line. Then set a pin in this line of sight between the mirror and the object at some point, such as *E*. The line along *OE* is the direction of the incident ray. Mark the positions of all pins and, keeping the object pin in place, remove the other pins from the paper.

Now move your eye to other positions (on the opposite side of the object pin would be best), and repeat the above procedure to establish the direction of a second reflected ray. Again align the *images* of these two pins with the object pin and set a fourth pin at some point *F* to establish the direction of the incident ray.

3. Remove the mirror and draw lines along the incident path and reflected path for each of the two arrangements in the preceding setup. While your eye was behind points *A* and *B*, respectively, the reflected rays appeared to come from some point behind the mirror, the image of the object *O*. Since both reflected rays, *CA* and *DB*, appeared to come from the same point, this image point would be at the point of intersection of these lines. You can locate the point by extending these two lines (dotted lines) behind the mirror until they intersect. By means of arrows on the lines, indicate the directions of the rays, draw normals to the mirror at the points where the reflections occur, and

indicate on both the drawing and the summary sheet the sizes of the angles of incidence and reflection, and also the distances of the object and image from the mirror. What conclusions can you draw about the comparison of the angles and of the distances?

4. Hold the mirror up and look at your face as you touch the right cheek with your right hand. Does this appear to be the right, or left, side of the image? What would you say is the relation of the orientation of the image with respect to the object? Consider all directions. What do you observe about the relative sizes? Could you put this image on a screen? Why? Record your conclusions about all of these items on the data form.

5. In order to perform the next step in this study we need to review the meaning of parallax. While holding one pencil (or finger) vertically at arm's length, hold another about 15 cm closer and, as you move your head from side to side, note which way the *nearer* pencil appears to move with respect to your eye. Is it the same direction, or the opposite direction? This apparent relative motion of the two pencils is called parallax. Now move the pencils closer together and determine whether the relative motion between them increases or decreases. Continue to adjust their positions until the apparent relative motion, or parallax, is zero. What are their relatve positions then?

6. Place the mirror on the line *MZ* as before and adjust the height of the mirror on the block so that none of the block extends above the mirror. Stand a pencil (it should be taller than the mirror) at the former object position, *O*, and look at its image in the mirror. Now hold a second pencil behind the mirror so that the part visible above the mirror appears to be an extension of the image of the object pencil. Then move your head from side to side and adjust the position of the pencil behind the mirror until parallax between it and the image is eliminated. Where, in general, with respect to this reflecting face of the mirror, does this observation locate the image of the object pencil? Now look behind the mirror and compare the position of the second pencil with the point of intersection of the dotted lines constructed in Step 3.

7. The next observation may be made by both partners as a team if convenient. Borrow your partner's mirror and set the two mirrors at right angles on a sheet of paper. Hold your pencil as an object somewhere between the mirrors and look at the images. How many do you see? Does moving the object around have any effect on the number of images seen? Now make the angle about 60° and again observe the images. Then change the angle to about 120°. What seems to determine the number of images seen? Record the number of images seen for each of these angles. On your data form draw two lines at right angles to

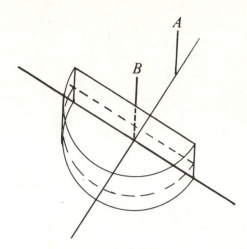

Figure 38–3 Refraction Arrangement.

represent the 90° arrangement of the two mirrors. With some point *O* out in front of the mirror position, use what you learned about ray tracing earlier in this experiment and show that one should expect these images in the positions observed. This drawing need not be exact but it should be approximate.

B. Refraction of Light*

8. Construct a pair of perpendicular lines on a sheet of plain white paper and fasten the paper on the cardboard. Fill a semicircular plastic dish about half full of water and place it on the paper with the center of the semicircle at the intersection of the two lines (Figure 38–3). Stick one pin vertically at *B*, as close as possible against the dish, and a second pin at *A*, 3 or 4 cm away and on the perpendicular line.

Now look at these two pins *through the water* from the curved side of the dish and, when you have lined them up with your eye, place a third pin in the line of sight and against the curved side of the dish. Label this pin position as *C*. Since your eye was in the line of sight, the light from pin *A* must have traveled along the path *ABC*. What is the angle of incidence in this case? For this angle of incidence, what general statement can you make about the direction a ray of light takes after entering a second medium?

9. Move the pin *A* to one side of the normal so that the line *AB* will make an angle of about 20° with the normal. While viewing *through the water* from the curved side of the dish, line up the pins *A* and *B* with a third pin *C* against the curved side of the dish. Label these new positions of the two pins you have moved as A_1 and C_1. Repeat this procedure for incident angles of 40, 60, and 70°, labeling the pin positions for future identification.

---

*If the plastic dish described above is not available, Step 8 and 9 in the procedure can be performed with a rectangular refraction plate as illustrated in Step 11. If the plastic dish is available, the instructor may prefer the combination arrangement outlined in Section C.

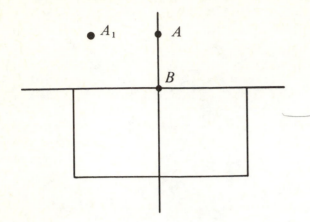

Figure 38–4 Refraction in Glass.

10. Remove the dish and draw lines across the paper for each of the paths when the pins A, B, and C were aligned. Measure and record, on the drawing and on the summary sheet, all of the angles of incidence θ_1 and the corresponding angles of refraction θ_2. Compute and record the sines of each angle and then find the ratio $\sin \theta_1/\sin \theta_2$ for each pair of angles.

Do the values obtained for the ratio $\sin \theta_1/\sin \theta_2$ appear to be related to each other? If so, how? Since the only two media involved are air and water (plastic is about equivalent to water), what relationship would you expect for the ratio of the velocities? This ratio of velocities is called the index of refraction and its value for various materials in comparison to air (or vacuum) can be found in the tables in Appendix B. Compare the table value for water

with your values of the ratio $\sin \theta_1/\sin \theta_2$ by finding the percent discrepancy.

11. Repeat Step 9 for some other liquids in the plastic dish, making only one trial with each for an angle $\theta_1 = 40°$. Set the dish on the same paper as before and label the position of the third pin at D instead of C. Record your observations and calculated results as before.

12. On a new sheet of plain paper place a rectangular glass plate on a set of perpendicular lines as shown in Figure 38–4. With pins at positions A and B, follow the instruction in Step 9, for whatever number of trials your instructor designates, placing the pin C *against* the back side (bottom side of diagram) of the glass plate. Be sure you align the pins while looking *through* (not over) the plate of glass. While your eye is aligned with the pins, set a fourth pin D in the line of sight between the glass plate and your eye. For one of the trials, draw a line along the path $ABCD$ and compare the direction of CD with AB. Compute the index of refraction of glass as described in Step 10 and compare with tabulated values.

13. Set the glass plate back into position as shown in Figure 38–4 and, with the three pins aligned, view the alignment by looking at the pins from the front side of the glass with the eye beyond pin A. The light is now traveling from pin C through the glass and then through the air to your eye. In this case, the angle of refraction is the angle in the air, and the angle of incidence is the angle in the glass. Repeat with a different angle.

Try moving pin C farther and farther away from the normal and see if you can find a position for pin C that would give an angle of refraction of 90°. Record θ_1 and θ_2. In this case the incident angle, θ_1, is labeled θ_c and is called the critical angle.

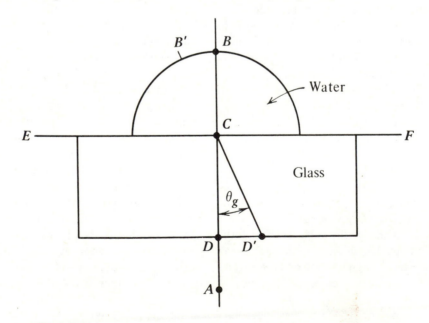

Figure 38–5 Refraction of Light Across Three Media.

C. Tracing the path of Light Through Three Media

14. Construct a pair of perpendicular lines near the center of a sheet of plain white paper, and fasten the paper on a piece of cardboard. For this operation we shall prepare a combination of the equipment in Figures 38–3 and 38–4. Attach a small strip of scotch tape to the flat edge of the plastic dish and, with a ball point pen and a straight edge, mark a vertical line at the midpoint of the diameter to serve in place of a pin for alignment purposes. Now carefully set the dish on the line *EF* so that the foot of the pen mark is at the intersection of the perpendicular lines as shown in Figure 38–5. Then fill the dish about half full of water and place the edge of the glass plate firmly against the flat face of the dish. Now look *through (not over)* the system from the curved side of the plastic dish near *B* and then from the other side of the glass plate near *D* to see if you can locate and identify the ink mark at the boundary of the two media.

15. First establish the position of the normal by aligning pins *B*, *D*, and *A* with the ink mark at *C*. Then, while looking through glass to water, move pin *D* to some point

D' so that θ_g in glass is about 25–27°. Now, with your eye near *D'*, move pin *B* to some point *B'* so that *B'*, *C*, and *D'* appear to be in a straight line. Be sure that your line of sight is *through the water and glass* and not above them. While maintaining this alignment, move pin *A* to some point *A'* until it is also in alignment with the other two pins and the ink mark.

16. Now mark all positions of pins that have been aligned and label as indicated. Then remove the dish and glass plate and draw straight lines between points to represent the complete path of the light ray and show directions of travel by arrows.

17. The line *BCDA* serves as a normal to all surfaces involved. Now draw a normal to the glass surface at *D'*. Measure and record the angles between the normal and the path of the light ray in each of the media—air, glass, and water. Then, by a double application of Equation [3], compute the index of refraction of both glass and water and compare with the accepted values in Appendix B by finding the percent discrepancy.

QUESTIONS

1. What relation do your data show to exist between the angle of incidence and the angle of reflection in a plane mirror?

2. Describe the image formed in a plane mirror as to size, distance, shape, type, and the like, as compared to the object. Is it real? Erect or inverted?

3. If a man should walk toward a plane mirror at the rate of 0.8 m/sec, at what rate would he approach his image?

4. A boy with the letters R.L.S. on his sweater stands in front of a plane mirror. Draw the letters as they appear when seen in the mirror.

5. What do you really mean when you say an image is ''behind the mirror''?

6. In locating the position of the image behind the mirror, is there any advantage gained by tracing three rays rather than just two? If so, what is it?

7. Can you formulate a relation between the angle made by two mirrors and the number of images formed from a single object? *Hint:* Divide the various angles into 360° and study the results.

8. How many images should one expect to see from a single object between two mirrors at an angle of 10°? How many for an angle of 0°? Have you ever observed this situation in a barber or beauty shop?

9. Several angles of incidence and the associated angles of refraction were measured for a given material in this experiment. What rather striking result was found concerning these angles?

10. When the angle of incidence was 0°, what did your experiment indicate about the angle of refraction? Would you expect the situation to be different for other materials? Why? Examine Equation [1] in the light of these angles and comment.

11. When light travels from air into a slab of glass with parallel sides and then emerges into air again, what relation do you find to exist between the directions of the incident and emergent rays?

12. Under what conditions is the angle of refraction larger than the angle of incidence? What special significance is given to the angle of incidence when the angle of refraction is 90°? What happens beyond the point where the angle of refraction is 90°?

13. By using 3.0×10^8 m/sec as the velocity of light in air, calculate the velocity in each of the other media used in this experiment.

14. If you have used both water and glass in the experiment, compute the relative index of refraction for light passing from water into glass.

RECORD OF DATA AND RESULTS
Experiment 38—Reflection and Refraction of Light

A. SUMMARY OF RESULTS ON REFLECTION

Path considered	Angle of incidence	Angle of reflection
First ray path		
Second ray path		
Third ray path		

Distance of object from mirror _ _ _ _ _ _ _ _ _ _ Distance of image _ _ _ _ _ _ _

Conclusions _

_ _

Description of image _

_ _

Distance of image by parallax from geometrically located image _ _ _ _ _ _ _ _ _ _ _ _ _

Multiple images in two mirrors at an angle

    (a)   Effect of moving the object for a given angle _ _ _ _ _ _ _ _ _ _ _ _ _ _ _ _ _

    (b)   Number of images when

        (1)   angle is 120° _ _ _ _ _

        (2)   angle is 90° _ _ _ _ _

        (3)   angle is 60° _ _ _ _ _

        (4)   Drawing for 90° angle

B.   SUMMARY OF RESULTS ON REFRACTION ACROSS ONE BOUNDARY

Kind of medium	Angle incidence θ_1	Angle of refraction θ_2	Sin θ_1	Sin θ_2	$\dfrac{\text{Sin } \theta_1}{\text{Sin } \theta_2}$	Index of refraction (Tables)	Percent discrepancy
Water							
				Average→			
Crown glass							
				Average→			
Glass to air							
$\theta_1 = \theta_c =$	near 90°						

C.   SUMMARY OF RESULTS ON REFRACTION ACROSS TWO BOUNDARIES

Kind of medium	Angle at boundary θ	Sin θ	Ratio $\dfrac{\text{Sin } \theta_1}{\text{Sin } \theta_2}$	Index of refraction n (calc.)	Accepted value of n (Tables)	Percent discrepancy
Air						
Glass			$\text{Sin } \theta_a / \text{Sin } \theta_g$			
Water			$\text{Sin } \theta_a / \text{Sin } \theta_w$			
Water to glass			$\text{Sin } \theta_w / \text{Sin } \theta_g$	Relative index $\dfrac{n_g}{n_w} =$	Computed n	

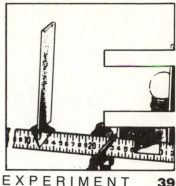

THE FOCAL LENGTH OF A CONCAVE MIRROR

SPECIAL APPARATUS:

Meter stick, two optical-bench supports, concave mirror, mirror holder, light source, object screen, image screen, screen holder, two object and image markers, ruler.

GENERAL APPARATUS:

120-V ac outlets, if the light source is an electric lamp.

THE PURPOSE OF THIS EXPERIMENT

is to study the image-forming properties of a concave mirror, and to determine its focal length by several methods.

INTRODUCTION

The laws of reflection of light for plane mirrors, as given in the preceding experiment, are applicable to curved surfaces if we apply these laws to each individual point of the curved surface where the light strikes. However, when a beam of light strikes a curved surface, the angle of incidence will not, in general, be the same for all parts of the beam, since the curved surface does not make the same angle with all parts of the beam. Hence, when a beam of light is incident upon a spherical mirror, it is either converged to a real focus somewhere in front of the mirror or diverged so as to appear to come from a point (the virtual focus) behind the mirror.

A line drawn perpendicular to the mirror at the central point of its surface is called the *principal axis*. The *principal focus* of a mirror is the point through which a bundle of rays parallel to the principal axis passes, or appears to pass, after reflection. If the incident rays are not parallel to each other or to the principal axis, the image is formed at some position other than the principal focus.

A *real image* is an image formed as a result of the light rays from a point on the object actually converging to a point after reflection from the mirror. A *virtual image* is an image formed at a point through which the reflected light rays only appear to have passed. In the latter case the reflected rays are said to diverge from a focus.

The *focal length* of a concave mirror is the distance from the mirror to the principal focus. The principal focus is approximately halfway between the mirror and the center of curvature, thus making the focal length equal to one-half the radius of curvature.

If we represent the distance of the object from the mirror (the object distance) by D_0, (Figure 39–1), the image distance by D_i, the radius of curvature by R, and the focal length by f, it can be shown that

$$\frac{1}{D_0} + \frac{1}{D_i} = \frac{2}{R} \tag{1}$$

or

$$\frac{1}{D_0} + \frac{1}{D_i} = \frac{1}{f}. \tag{2}$$

A rule for signs to be assigned to numbers being used in Equations [1] and [2] may be summarized as follows:

1. D_0 is always positive.
2. D_i is positive for real images and negative for virtual images.

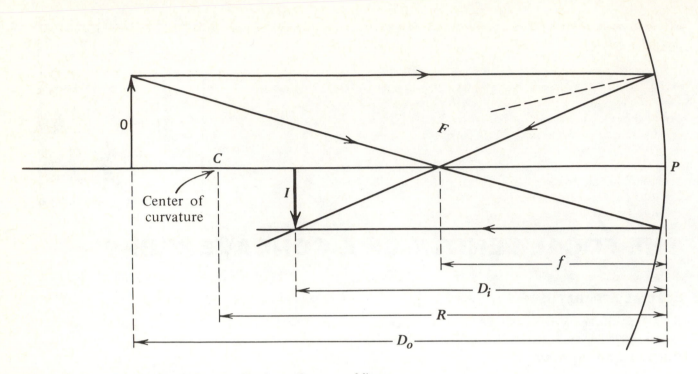

Figure 39–1 Real Image Construction for a Concave Mirror.

3. *R* and *f* are positive when the principal focus is real (concave mirrors), and negative when the principal focus is virtual (convex mirrors).

By examination of the construction in Figure 39–1, it is easily seen that the relation between the image size S_i and the object size S_0 is given by

$$\frac{S_i}{S_0} = \frac{D_i}{D_0}.$$ [3]

This ratio is called the *magnification*.

DESCRIPTION OF APPARATUS

An optical bench is made by clamping each end of a meter stick into optical-bench supports (Figure 39–2). The light source consists of a small incandescent lamp (frosted bulb) which is mounted on a support containing double electrical outlets, and which slides along on the meter stick. The double outlet and several feet of cord make it possible to connect several sets of equipment in parallel electrically from a single 120-V outlet. A cross-wire screen mounted in front of the lamp serves as an illuminated object.

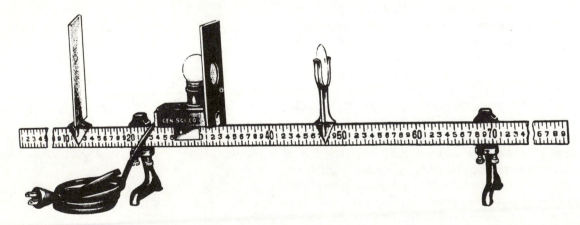

Figure 39–2 Opitical Bench, Showing Arrangement of Light Source and Mirror (Courtesy of Central Scientific Co.).

THE FOCAL LENGTH OF A CONCAVE MIRROR

PROCEDURE

1. Focal Length by Direct Measurement. Mount the mirror and a white cardboard screen on the meter stick and, while holding the mirror toward some distant object such as a cloud, tree, or building, adjust the distance until a distinct image is focused on the screen. Record the image distance. Repeat for two other trials, using a different object each time, and making an independent adjustment for each trial. Determine and record the mean of the three trials.

2. Focal Length from the Radius of Curvature. Set up your illuminated object on the meter stick and adjust the position of the mirror until a real and distinct image of the cross-wire appears on the white portion of the object screen just beside the cross wire. This gives the object and image at the same distance from the mirror. A little thought will show that this distance is equal to R, the radius of curvature of the mirror. Make three independent trials and record the distance. Find the mean and compute the focal length f by Equation [2].

As a means of studying the nature of images formed by a concave mirror, mount the illuminated object near one end of the meter stick and the mirror near the other. Now with the image always in focus on a white cardboard screen, slowly move the object toward the mirror and carefully note the changes which take place in the image until the object almost reaches the mirror. Also note when it is inverted and when erect.

3. Focal Length by Real Image Formation When D_0 Is Greater Than 2f. Mount the mirror at a convenient distance from the object equal to about two and one-half to three times the focal length as found in Observation 1. Make three independent determinations of the image distance and record as D_i. With the third trial adjustment, make careful measurements of the size of the image and the size of the object and record as S_i and S_0, respectively (see the second table on the Record of Data and Results).

Find the mean of the recorded values of the image distances and, using this as the value of D_i, compute and record the focal length, using Equation [2]. Show this computation following the data table as a part of your report.

Determine the magnification from each of the ratios in Equation [3] and compute the percent difference.

4. Focal Length by Real Image Formation When the Object Is between Principal Focus and Center of Curvature ($f < D_0 < 2f$). Mount the mirror at a convenient distance from the object somewhere near $1.5f$, and perform all the operations outlined in Observation 3.

5. Focal Length When Image Is Virtual. Replace the light source with a tall pin (the object marker), which is to serve as the object for this part of the experiment. Make $D_0 < f$, say $0.5f$ to $0.6f$, and mount a second pin (the image marker) behind the mirror so that, while viewing the image of the object pin in the mirror, you can also see the top of the image marker above the mirror. The image is virtual and appears to be at some point behind the mirror. The problem of the experiment is to set the image marker at the position where the image appears to be. This is done by the method of parallax described in the following paragraph.

When two objects nearly in line with the eye are viewed by moving the eye sidewise, the apparent change in their relative positions is called parallax. While viewing two fingers in this manner, note the change in the magnitude of the relative sidewise displacement as the distance between the fingers is changed. Also note which finger appears to move in the same direction as the eye while the eye is moving sidewise. What is the result when the fingers are against each other (at the same position)?

Now with the position of the object pin remaining fixed, move the image pin back and forth until, while viewing one pin in the mirror and the other above the mirror, all *parallax* is eliminated. Record the distance of the image pin as D_i, and compute the focal length by using Equation [2], showing your computation following the data table.

QUESTIONS

Answer each question so that the question asked may be ascertained from the answer.

1. Explain why the method of Observation 1 gives the focal length.

2. What does Observation 2 show about the relation of focal length and radius of curvature?

3. Which of your methods of determining the focal length of your mirror do you think is least accurate? Why?

4. (a) As the object is brought from a great distance toward the center of curvature, what changes take place in the image? (b) As the object is moved from the center of curvature to very near the mirror, describe the changes that take place in the image.

5. In Observation 1, when the object distance was very large, the image distance was used as the focal length. By using Equation [2], substitute ∞ for D_0 and solve for D_i. Does this computation verify the results of Observation 1? Explain.

6. By using Equation [1], determine, in a manner similar to the above, the image distance when the object is at the center of curvature. Does this verify the observations in Step 2? How?

7. Use Equation [2] to locate the position of the image when the object is at the principal focus, that is, when $D_0 = f$. What observation did you make in Step 2 which verifies this result?

8. If this mirror were large enough to use as a mirror for shaving, at what distance from your face would you place it? (A definite distance in centimeters is not required.)

9. Does your nose appear to be enlarged when viewing your face in the mirror? If your nose is 5 cm long and at the same distance from the mirror as was the object in Step 5, compute the size of the image of your nose and its distance from the mirror.

10. Which of the methods used in this experiment would you suggest for finding the focal length of a convex (diverging) mirror? Give reasons for your choice.

NAME *(Observer)* _____ Date _____

(Partner) _____ Course _____

RECORD OF DATA AND RESULTS
Experiment 39—The Focal length of a Concave Mirror

Trial	Image distance, D_i				
	Step 1: Focal length by direct measurement	Step 2: Focal length from the radius	Step 3: $D_0 > 2f$ _____	Step 4: $f < D_0 < 2f$ _____	Step 5: $D_0 < f$ _____
1					
2					
3					
Mean					
f, Calculated					
Mean value of f from Steps 1 and 2					Percent Difference
Mean value of f from Steps 3, 4, and 5					

MAGNIFICATION OF MIRROR

Setup	D_i	D_0	S_i	S_0	D_i/D_0	S_i/S_0	Percent Difference
$D_0 > 2f$							
$f < D_0 < 2f$							

PROPERTIES OF CONVERGING AND DIVERGING LENSES

SPECIAL APPARATUS:

Meter stick, two optical-bench supports, two converging lenses (different focal lengths), one diverging lens with f between those of the converging lenses, lens holder, light source, object screen, image screen, screen holder, two object and image markers, plane mirror, mirror support.

GENERAL APPARATUS:

Electric lamp.

THE PURPOSE OF THIS EXPERIMENT

is to study the image-forming properties of a converging lens and to determine the focal length of a single lens and also of a lens combination.

INTRODUCTION

A converging lens is one which is thicker at the center than at the periphery and converges incident parallel rays to a real focus on the opposite side of the lens from the object. A diverging lens is thinner at the center than at the periphery and diverges the light from a virtual focus on the same side of the lens as the object.

The *principal axis* of a lens is a line drawn through the center of the lens perpendicular to the face of the lens. The *principal focus* is a point on the principal axis through which incident rays parallel to the principal axis pass, or appear to pass, after refraction by the lens.

The *focal length* of a lens is the distance from the optical center of the lens to the principal focus. The reciprocal of the focal length in meters is called the *power of a lens* and is expressed in diopters. If two thin lenses with focal lengths of f_1 and f_2, respectively, are placed in contact to be used as a single lens, the focal length F of the combination is given by the equation

$$\frac{1}{F} = \frac{1}{f_1} + \frac{1}{f_2}.$$ [1]

If D_0 is the distance of the object from the optical center of the lens (Figure 40–1a), D_i the distance of the image from the lens, and f the focal length, it can be shown that

$$\frac{1}{D0} = \frac{1}{D_i} + \frac{1}{f}.$$ [2]

In the special case of focusing on a distant object, D_0 may be considered equal to infinity. When such is the case, Equation [2] gives $D_i = f$.

Examination of Figure 40–1a shows that a diminished image, I, is formed just a little beyond the principal focus. If the object were placed at the position of the small image, a reversal of the direction of the light rays would give an enlarged image O. These two interchangeable positions of object and image are called *conjugate foci*. However, the same result can be accomplished by moving the lens from position a (Figure 40–1b), a distance d, to position b and leaving the position of the object, O, unchanged. With the lens at position b, an enlarged image I' will be formed on the screen, and the values of D_0 and D_i will change accordingly. Hence, there are two positions between the screen and the object, namely, position a and position b (Figure 40–1b), at which a converging lens may be placed and give a real image without any change in the positions of the screen or object.

If L represents the fixed distance between the screen and the object, and d is the distance between the two possible positions of the lens, it is easily seen from Figure 40–1a, when compared with Figure 40–1b, that

$$D_0 = \frac{L}{2} + \frac{d}{2} = \frac{L + d}{2} \qquad [3]$$

or

$$D_i = \frac{L}{2} - \frac{d}{2} = \frac{L - d}{2}. \qquad [4]$$

By substituting these values of D_0 and D_i in Equation [2], we get

$$f = \frac{(L^2 - d^2)}{4L}. \qquad [5]$$

This method of finding the focal length is especially valuable when using thick lenses, since the position of the optical center need not be known.

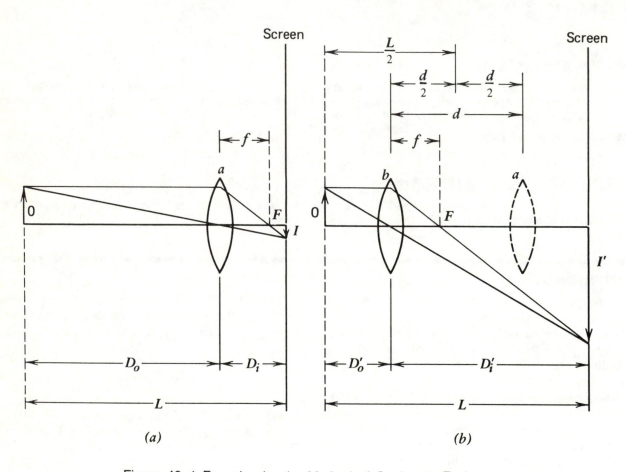

Figure 40–1 Focusing by the Method of Conjugate Foci.

DESCRIPTION OF APPARATUS

An optical bench is made by clamping each end of a meter stick into optical-bench supports (Figure 40–2). The light source consists of a small incandescent lamp (frosted bulb) mounted on a support containing double electrical outlets and slides along on the meter stick. The double outlet and several feet of cord make it possible to connect several sets of equipment in parallel electrically from a single 120-V ac outlet. A cross-wire screen mounted in front of the lamp serves as an illuminated object.

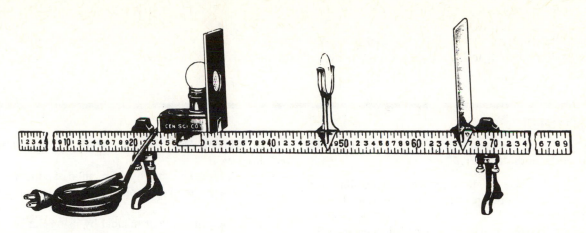

Figure 40–2 Optical Bench, Showing Arrangement of Light Source, Lens, and Screen (Courtesy of Central Scientific Co.).

PROCEDURE

1. *Focal Length by Direct Measurement.* (a) Mount one of the lenses, to be known as lens 1, on the meter stick, together with a cardboard screen. By using three different distant brightly illuminated objects, as viewed through a window, make three independent trials to locate the positions of the real images thus formed. Record the image distances and, from their average (or mean) value and the definition of the principal focus, determine the focal length of lens 1. (b) Repeat the above procedure for lens 2.

2. *Focal Length of a Lens Combination.* (a) Mount the two lenses in a single lens holder. A strip of paper about 2 cm wide placed around the periphery of the pair of lenses will help hold them in the lens holder. Handle the lens combinations very carefully to avoid dropping one of the lenses and breaking it. Follow the procedure in Step 1a and determine and record the focal length of the lens combination. (b) Repeat the above procedure using a combination of a diverging lens with the converging lens of shorter focal length. Note that you have already measured the focal length of the converging lens in Step 1. With this focal length known, and that of the combination which you have just measured, use Equation [1] to compute the focal length of the diverging lens. Record in the data form as f_3.

3. *Focal Length by Coincidence.* Select either of the two converging lenses just used for this part and all the remaining parts of the experiment. Mount a plane mirror about 5 cm from the lens, using a small piece of paper in the mirror support to prevent scratching the silvered surface. Mount the illuminated object on the opposite side of the lens and adjust its position until a distinct image of the cross-wire shows on the object screen just beside the object. The light is now returning nearly along its original path. This requires that the rays be parallel on the opposite side of the lens from the object, and that they strike the mirror perpendicular to its surface. The point at which the parallel rays form an image is the principal focus. The

object and image are nearly coincident and their distance from the lens is the focal length. Make three independent trials, record the image distances, and determine the mean value of the focal length.

4. *Focal Length by Method of Conjugate Foci.* Use the same lens here as was used in Step 3. Mount the illuminated object near one end of the meter stick, and place the screen at some convenient distance toward the other end so that the distance between them is about 10 cm more than $4f$. Adjust the lens at the two positions for image formation, and determine whether both the enlarged and the diminished images are clear and distinct. If not, change the distance between the screen and the object until they are distinct, and then record their positions (meter-stick readings). This setup is that of Figure 40–1. Now make three independent trials for each position of the lens and record the positions of the lens, the object, and the screen, together with the image distance for each. Some time may be saved by alternating positions between trials. From the means of the values for D_0 and D_i compute the focal length of the lens by Equation [2], making the computations a part of your report. Determine the values of L and d, compute the focal length from Equation [5] and include it in your report.

5. *Focal Length When Image Is Virtual.* (a) Remove the illuminated object and the screen and mount a tall pin (the object marker) at a distance less than f, say $(2/3)f$, from the lens used in Step 4. Place your eye on the opposite side of the lens from the object marker and look at it through the lens in the same manner as you would use a reading glass. Place a second pin (the image marker) at the apparent position of the vertical image and, by careful adjustment, locate the best position of the image marker by the parallax method. Read Step 5 of the procedure for Experiment 39 for a discussion of the parallax method of locating virtual images. Keep the object position fixed and make two or three independent trials. Record the object and image distances, and find the mean of D_i. Compute the focal length from Equation [2] and add to your report. (b) Repeat above procedure with the diverging lens, using $D_0 > f_3$ (Step 2b).

QUESTIONS

Answer each question so that the question asked may be ascertained from the answer.

1. Calculate the focal length of the lens combination from Equation [1] and compare with the experimental value found in Step 2.

2. Compute the power of each lens and the power of the combination in diopters from the experimental focal lengths. What relation do you notice?

3. In which parts of the experiment did you find the image inverted? Erect? Real? Virtual? Enlarged? Reduced? Make a general statement which will embody all of these answers and also include the conditions for producing these results.

4. What do your data from the method of conjugate foci in Step 3 show concerning the relation of the object and image distances for the two positions of the lens?

5. Follow the suggestion given in the Introduction, and derive Equation [5].

6. In Step 4 you were told to make L, the distance between object and screen, about 10 cm more than $4f$. If you had used $L = 4f$, what changes, if any, would this have made in your experiment? *Hint: Substitute 4f for L in Equation [5] and solve for d.*

7. When locating the image by the method of parallax, how do you tell when the image marker is too near or too far away?

8. Read some print in your manual by using the lens as a reading glass. (a) What observations do you make concerning the image? (b) Make an image construction for this observation, showing the position of the eye, the lens, the printed page, and the image.

9. Assume that the method of conjugate foci is used with the combination lens of Step 2, and the object and screen are the same distance apart as you used. Compute the distance d between the two positions of the lens at which images are formed.

10. Based on observations made in this experiment, describe two methods by which the focal length of a concave (diverging) lens could be measured.

11. Does either method described in the answer to Question 10 put any limitations on the focal length that could be measured? If so, explain.

NAME *(Observer)* _____ Date _____

(Partner) _____ Course _____

RECORD OF DATA AND RESULTS

Experiment 40—Properties of Converging and Diverging Lenses

FOCAL LENGTH BY DIRECT MEASUREMENT AND COINCIDENCE METHOD

Trial	Step 1a: Image Dist. Lens 1	Step 1b: Image Dist. Lens 2	Step 2a: Image Dist. Lens Comb. 1	Step 2b: Image Dist. Lens Comb. 2	Step 3: Image Dist. by Coincidence Lens No. _____
1					
2					
3					
Mean	$f_1 =$	$f_2 =$	$F =$	$f_3 =$	$f =$

FOCAL LENGTH BY CONJUGATE-FOCI AND PARALLAX METHODS

	Step 4: Image Enlarged. Position of object _____ Position of screen _____			Step 4: Image Reduced. Position of object _____ Position of screen _____			Step 5a: $D_0 < f$ D_0 _____	Step 5b: $D_0 > f$ D_0 _____
Trial	Position of lens	D_0	D_i	Position of lens	D_0	D_i	D_i	D_i
1								
2								
3								
Mean								
$d =$				$L =$			$f =$	$f =$
$f(\text{Eq }[2]) =$				$f(\text{Eq }[5]) =$				

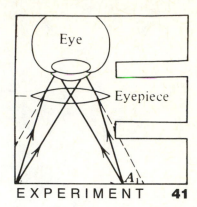

OPTICAL INSTRUMENTS EMPLOYING TWO LENSES

SPECIAL APPARATUS:

Two meter sticks, two rulers, two optical-bench supports, light source, object screen, cardboard screen, screen holder, five lenses (convex with foci of 5, 10, 15, and 30 cm, respectively, and concave with focus of 10 cm), three lens holders

GENERAL APPARATUS:

Electric lamp.

THE PURPOSE OF THIS EXPERIMENT

is to observe the operation of microscopes and telescopes when set up with simple lenses, and to measure their magnifying power.

INTRODUCTION

As the term is generally used, an optical instrument is an instrument designed to form on the retina of the eye an image of the object larger than would be formed if the object were viewed with the naked eye. The magnifying power is defined as the ratio of the angle subtended at the eye by the enlarged image of the object when viewed through the instrument to the angle subtended at the eye by the object when viewed directly. In this calculation, the angles should be expressed in radians.

The *compound microscope* consists essentially of a short-focal-length objective lens and a longer-focal-length eyepiece lens (Figure 41-1), both lenses being convex. The object AB, just outside the principal focus F_0, of the objective, gives a real enlarged image A_1B_1 just inside the principal focus F_e of the eyepiece. This real image serves as an object to give a magnified virtual image A_2B_2, which is viewed by the eye at the distance of most distinct vision.

The magnifying power M_0 of the objective lens is the ratio A_1B_1/AB, which is also given by

$$M_0 = \frac{D_i}{D_0} \qquad [1]$$

where D_i and D_0 are the distances of the lens from A_1B_1 and AB, respectively. If we treat the image A_1B_1 (Figure 41–1) as an object being viewed through a simple magnifier (the eyepiece), the normal eye sees the virtual image A_2B_2 most distinctly at some distance d, usually about 25 cm from the eye (or lens). Since the eye sees a virtual image, d is negative, and, if we use $d = -25$ cm in the general lens equation, it takes the form

$$\frac{1}{D_0} - \frac{1}{25} = \frac{1}{f_e} \qquad \text{or} \qquad \frac{25}{D_0} - 1 = \frac{25}{f_e}$$

But the magnifying power of the eyepiece is

$$M_e = \frac{D_i}{D_0} = \frac{25}{D_0}.$$

When this value of M_e is substituted in the above equation, it takes the form

$$M_e = \frac{25}{f_e} + 1 \qquad \text{or} \qquad M_e = \frac{d}{f_e} + 1 \qquad [2]$$

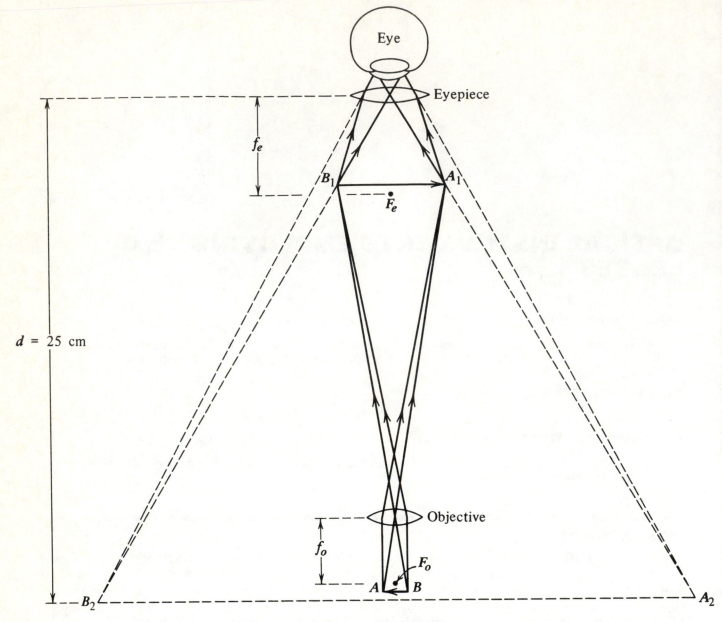

Figure 41-1 Compound Microscope.

where d is the distance of most distinct vision, and f_e is the focal length of the eyepiece. Hence, the total magnifying power of the microscope is

$$M = (M_0)(M_e) = \frac{D_i}{D_0}\left(\frac{d}{f_e} + 1\right).$$ [3]

The objective of a *simple telescope* gathers light from a distant object and gives a real, inverted image AB (Figure 41–2) very near the principal focus of the objective. This real image, which is just slightly inside the principal focus of the eyepiece, acts as the object for the eyepiece, which gives an enlarged virtual image A_1B_1. The angle α is half of the angle subtended at the eye by the image, and β half of

that subtended by the original object, assuming that the eye is at the objective when viewing the object directly. In practice the angles are quite small so that the magnifying power, which is by definition the ratio of the angles, may be considered equal to $\tan \alpha / \tan \beta$. By examination of Figure 41–2, it may be seen that this ratio gives the magnifying power as

$$M = f_0/f_e$$ [4]

where f_0 and f_e are the focal lengths of the objective and eyepiece, respectively. This does not mean that the image is actually larger than the object. The image is merely brought close to the eye and is larger than the object appears to be when viewing it from a great distance.

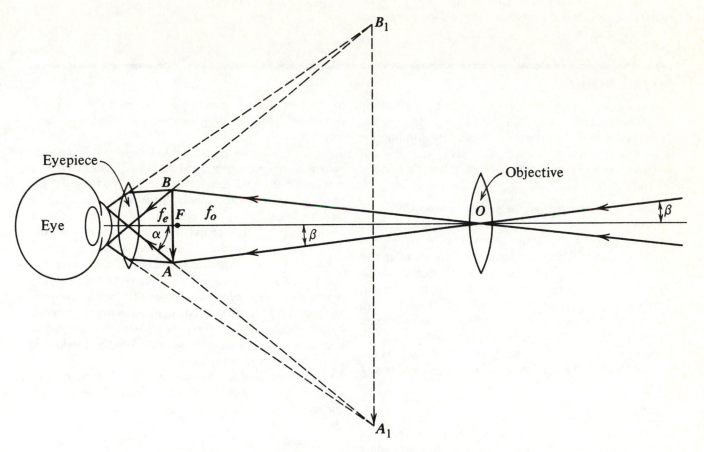

Figure 41–2 Astronomical Telescope.

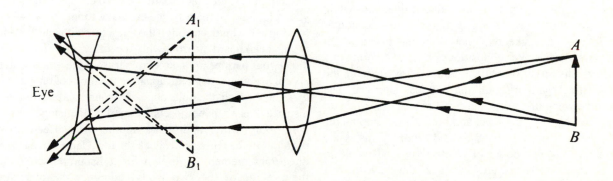

Figure 41–3 Opera Glass (Galilean Telescope).

The *opera glass* (or *Galilean telescope*) (Figure 41–3) employs a concave (diverging) lens as the eyepiece and gives an erect virtual image A_1B_1 of the object AB. This instrument is designed to give an enlarged image of a not too distant terrestrial object. Its magnifying power is also given by the relation $M = f_0/f_e$, the same form as for the astronomical telescope.

PROCEDURE

You are to use simple lenses in constructing each of the optical instruments described on the preceding pages. Answer all assigned questions relative to each instrument at the time you are performing the operations. Be sure to put all lenses back in the proper envelope before checking in the equipment.

1. *Compound Microscope.* Select a short-focus (about 5 cm) double-convex (converging) lens for the objective and a slightly longer-focus (about 10 cm) double-convex lens for the eyepiece. Check the focal length of each lens by directly viewing a distant object and measuring the image distance on a small screen and use it as the focal length. Then mount the eyepiece a few centimeters from the zero end of a meter stick and, while holding this end vertically down against a printed page of your book, adjust the position of the lens for the most distinct image of the printed matter. Observe the magnification as compared to that of the naked eye.

While viewing printed matter through the center of the lens, also examine the appearance of the letters while looking through the lens near the periphery. Do you notice any difference? Move the lens as necessary to get a better image through the periphery. Which way did you have to move it?

2. Now replace the printed page with a ruler lying flat on a blank sheet of paper. Adjust the lens for a sharp focus on a small portion of the scale at the ruler, then measure and record the distance from the lens to the ruler. Now view the centimeter scale through the lens with one eye, and, while looking with the other eye (unaided) at a nearby section of the blank paper, make two marks on the paper which correspond to the apparent positions of the end points of a centimeter division (or half centimeter) as seen in the eyepiece. Each partner should make at least one trial. The distance between these two marks is the size of the image. Record the size of both object and image and compute the magnification from $M_e = S_i/S_0$ and record. This is the observed magnifying power of your microscope eyepiece.

3. Move the eyepiece upward to about 25 cm above the end of the meter stick, then mount the objective lens at a position slightly more than the focal length above the end of the meter stick. You now have the arrangement for a compound microscope as illustrated in Figure 41–1. With the end of the meter stick against a ruler, adjust the position of both lenses for the most distinct image of the ruler scale. Now select some number of scale divisions within the field of view of your microscope and, by following the procedure in Step 2, measure the magnifying power and record the observed value.

4. Without changing the positions of the lenses, set up your system as an optical bench and place an illuminated object at the position on the meter stick occupied by the ruler. By use of a cardboard screen placed between the lenses, adjust its position until you get a distinct real image A_1B_1 (Figure 41–1). Record the object and image distances and compute the magnifying power of your objective lens by use of Equation [1]. Then, from the focal length of the eyepiece and the distance d (Figure 41–1), compute the magnifying power of the eyepiece from Equation [2]. Finally, compute and record the magnifying power of the entire microscope by use of Equation [3]. Show the computation as a part of your report and compare it with the observed value by computing the percent difference.

5. *The Simple (or Astronomical) Telescope.* In constructing a simple telescope select a short-focus converging lens (about 5 cm) for the eyepiece and one of about 30-cm focal length for the objective, and mount on a meter stick. While viewing some distant object (a tree or a building), adjust the relative positions of the lenses until a distinct image is seen. Make a general statement concerning the relative size of the image as seen in the instrument and the size of the object as seen with the unaided eye.

To measure the magnifying power, make two horizontal chalk marks on the blackboard about 10 cm apart and focus your telescope on them from across the room. Now, while looking at the marks through the telescope with one eye and at the blackboard nearby with the other eye (unaided), direct your partner to make two marks on the board which appear to coincide with the apparent positions of the images as seen through the telescope. Measure and record the distances between both sets of marks. Exchange places with your partner and make another trial. Record the positions of the lenses and check the focal length of each by direct measurement while viewing a distant object. From the average of the two trials, compute the observed magnifying power and compare with the value obtained by use of Equation [4].

Now focus on some object across the street, such as a brick wall, then substitute a 15-cm-focus lens for the objective and make a comparison of the magnifying power in the two cases. See Question 9.

6. *The Opera Glass (Galilean Telescope)*. By using the same 30-cm-focus objective as was used with the simple telescope, construct an opera glass by using a 30-cm-focal-length diverging lens as the eyepiece. Repeat Step 5 with the setup in Figure 41–3.

QUESTIONS

The Compound Microscope

1. In using each lens as a reading glass, what relationship do you observe between focal length and magnifying power?

2. What movement of the lens was necessary to get a better image when viewing through the periphery? Explain the reason for this.

3. Describe the imperfections present in the microscope image.

4. What names are given to the defects observed in the images?

5. What is the purpose of an eyepiece in a microscope?

6. In setting up the microscope, where must the object be placed with respect to the objective lens?

7. Where is the image formed by the objective with respect to the eyepiece?

8. What changes could be made to construct an instrument with greater magnifying power?

The Simple Telescope

9. When viewing a distant brick wall or tree, where does the image appear to be with respect to the space between the eye and the object?

10. Did you see any chromatic aberration in the telescope? How can you tell? Which lens is most likely to contribute to it?

11. How does the distance between the two lenses compare with the sum of their focal lengths?

12. If cross hairs are to be put in the telescope, at what point should they be placed? Why?

13. Did you notice any change in magnification when the focal length of the objective was changed? What other change could be made to increase or decrease the magnification?

14. What changes would be necessary in constructing a terrestrial telescope using convex lenses throughout? Construct an image diagram similar to Figure 41–2.

15. What is the advantage of making the objective lens of an astronomical telescope of large diameter?

The Opera Glass (or Galilean Telescope)

16. Is any chromatic aberration observed?

17. How does the magnifying power compare with that of the astronomical telescope?

18. What relation seems to exist between the two focal lengths and the distance between the two lenses? What advantage does this relation give to the instrument which the astronomical telescope does not have?

19. Is there a real image formed in the system? If so, locate its position on a screen.

20. Would an instrument which produces an erect image be of any advantage as an astronomical telescope? Why?

NAME *(Observer)* _____ Date _____

(Partner) _____ Course _____

RECORD OF DATA AND RESULTS
Experiment 41—Optical Instruments Employing Two Lenses

Kind of optical instrument / Observations and measurements	Compound microscope		Simple telescope		Opera glass (Galilean telescope)	
Focal length of eyepiece by direct measurement					Use labeled value	
Focal length of objective by direct measurement						
Position of eyepiece on meter stick						
Size of object being measured, S_0						
Size of image measured, S_i	Trial 1	Trial 2				
Magnifying Power of eyepiece −observed	Average					
With two lenses −position of eyepiece on meter stick						
Position of objective on meter stick						
Size of object being measured, S_0						
Size of image measured, S_i	Trial 1	Trial 2	Trial 1	Trial 2	Trial 1	Trial 2
Magnifying power of complete instrument (observed = S_i / S_0)	Average		Average		Average	
Distance of object from objective lens, D_0			Approx. _____		Approx. _____	
Distance of image from objective, D_i						
Magnifying Power of objective (calculated)						
Magnifying Power of eyepiece (calculated)						
Magnifying power of complete instr. (calc.)						
Percent Difference observed *vs* calculated						

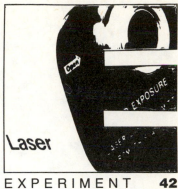

EXPERIMENT 42

THE LASER

NOTE TO INSTRUCTOR:

This experiment is divided into four parts each of which can be done independently. It may be necessary to omit one or two parts if the student is to complete the experiment in one laboratory period. Also note that Procedure B of Experiment 44 uses a laser and could be combined with this experiment.

SPECIAL APPARATUS:

He–Ne laser (1 mW or less), cylindrical lens, beam splitter, plane mirror, mirror holder (wood block), glass plate.

GENERAL APPARATUS:

Meter stick string, tape, plain paper, plumb line.

THE PURPOSE OF THIS EXPERIMENT

is to familiarize the student with lasers and laser techniques.

INTRODUCTION

The term *laser* is an acronym for *light amplification by simulated emission of radiation*. This statement is actually a concise description of the internal mechanism which produces laser light. Light is produced when excited atoms undergo transition to lower energy states and emit the energy as radiation. In this respect a laser is no different from many other light sources. Figure 42-1 shows a He–Ne laser, which is the type most commonly used in beginning laboratory studies.

The basic feature of a He–Ne (helium–neon) laser is shown in Figure 42-2. When the power supply is turned on a high-voltage electric discharge occurs in the He–Ne gas, producing a plasma of ionized atoms. This discharge causes collisions between the He atoms and the Ne atoms which causes many of the Ne atoms to go to an excited relatively stable state. When these Ne atoms undergo transition to a lower energy state they emit light of wavelength 632.8 nm (nanometers). When light of 632.8 nm passes by one of the excited Ne atoms it increases the probability that the Ne atom will undergo this same energy transition. This is called simulated emission of radiation.

Many different wavelengths of light are emitted in the He–Ne plasma. By carefully aligning the mirrors to be an integral number of wavelengths apart, it is possible to select the 632.8-nm wavelength light. This technique sets up standing waves for the selected wavelength. As this light moves back and forth between the mirrors it amplifies the 632.8-nm light by increasing the number of Ne atoms that undergo this transition. The transmission mirror actually transmits a small percent of the light which is the red beam observed from a He–Ne laser.

A laser light source has four very special properties which are not found in ordinary light sources. These are decribed below:

1. *Directionality*—Because only the light moving along the axis defined by the two mirrors escapes, the beam is highly directional. This gives a high concentration of light in a given direction.

2. *Coherence*—All of the light waves emerging from the laser are in step or all the waves are in phase. Coherence is produced by the integral wavelength separation of the mirrors which sets up the standing waves.

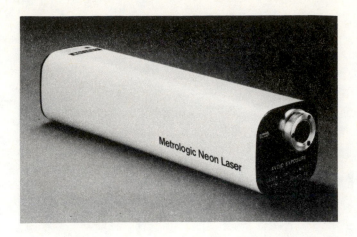

Figure 42–1 He–Ne Laser (Courtesy of Metrologic Instruments Inc.).

3. *Monochromaticity*—The laser beam consists of a single wavelength of light.

4. *Intensity*—Properties 1–3 combine to make the energy concentrated in a laser beam very high. A 1-mW laser beam is actually 100 times brighter than the sun. There is no better source than the sun for lighting large areas but as a compact, spectrally pure light source, the laser stands alone.

Because of these properties the laser has become a very useful tool for studying properties of light. It makes available many observations and measurements which cannot be done with other light sources. It has also become a very important scientific and industrial tool. Before working with the laser you should read the laser safety section in Appendix E.

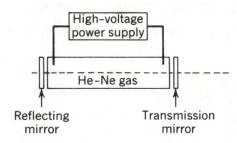

Figure 42–2 Basic Components of He–Ne Laser.

PROCEDURE

Be sure you have read Appendix E on laser safety. Turn on the laser only long enough to make measurements and note carefully where it is pointed before turning it on. In particular, be sure that the beam does not shine in anyone's eyes.

A. Triangulation

1. Point the laser toward a wall several meters or more away. Turn the laser on and identify the approximate location of the red spot on the wall. With the laser off tape a piece of paper with an X in the center to the wall at the approximate location identified for the red spot. Turn the laser on long enough to adjust it to hit the X on the paper.

2. Tape a meter stick firmly to the laboratory table with one end lined up with one edge of the laser as shown in Figure 42-3. Measure and record the angle, α, between the laser beam direction and the baseline defined by the meter stick.

3. Move the laser to point *B* so that the edge as before lines up with the other end of the meter stick. Turn the laser on and adjust it so that the beam strikes the *X* on the wall. Turn the laser off. Measure and record the angle β. Also record the length, d, of the baseline.

4. By using trigonometry it can be shown that the distances a and b are given by

$$a = d\,\frac{\sin(180 - \alpha)}{\sin(\alpha - \beta)} \qquad [1]$$

$$b = d\,\frac{\sin\beta}{\sin(\alpha - \beta)}\ . \qquad [2]$$

Calculate and record these distances.

5. Measure and record the distances a and b by measuring the length of a string stretched between AC and BC, respectively. Compare these values with those obtained in Step 4 by computing the percent difference.

B. Triangulation Using Beam Splitter

6. Set up the system as shown in Figure 42-4. Turn the laser on for brief periods to adjust the beam splitter so that the reflected beam strikes the mirror. Tape the meter stick to the laboratory table so that one end is even with the point where the beam strikes the mirror. Adjust the laser so that the beam transmitted through the splitter strikes the point X on the wall. Then rotate the mirror so that the reflected beam strikes the point X.

7. Measure and record angles α and β. The direction of the beam can be identified on the table by hanging a small plumb line with the string in the beam. This arrangement is essentially the same as before. Thus the distances a and b can be determined by using Equations [1] and [2]. Also, these distances can be measured as before using the string.

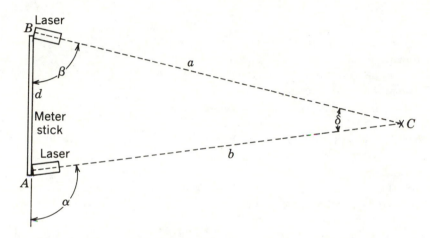

Figure 42–3 Laser Method for Measuring Distance.

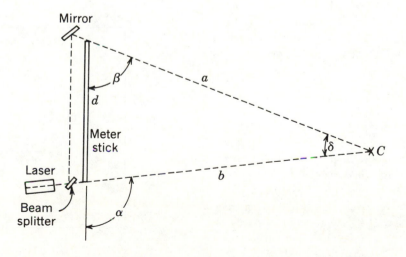

Figure 42–4 Beam Splitter Method for Measuring Distance.

Record all values in your data table and compare the laser values with the measured values by computing the percent difference.

C. Measurement of Rotation

8. Spread out the laser beam by using a cylindrical lens as shown in Figure 42-5. A glass, Lucite, or plexiglass rod with diameter between 2 and 4 mm will work well. One way of mounting this lens is to cement it to a doughnut-shaped magnet which makes it easily removable. This reduces the beam intensity on any point; however, it is still important to practice laser safety. The remaining parts of this experiment can be done without this lens but they will work better with the lens.

9. Draw perpendicular lines AB and CD across the center of a sheet of paper as shown in Figure 42-6, and label the point of intersection O. Draw a third and fourth line on the sheet which pass through O and make angles of approximately 10 and 20° with respect to the line CD. Label the two new lines EF and GH. Also draw lines perpendicular to EF and GH and label points on the three perpendiculars X, Y, and Z.

10. Support the mirror with a rubber band and a wood block along the line CD. It may be necessary to place the paper and mirror on a board or piece of cardboard and to elevate it on a book or other object to get it to the level of the beam. Shine the spread out laser beam on the mirror at point O with an incident angle between 30 and 40°. Label the incident and reflected beams from the mirror I and R, respectively. Now leave the laser and paper fixed but rotate the mirror so that it lies along the line EF. The incident beam should still be defined by point I but the reflected beam has changed. Label the reflected beam R_1. Repeat for line GH and label the reflected beam R_2.

11. Turn off your laser, remove the mirror, and draw dashed lines IO, OR, OR_1, and OR_2. Measure, record, and compare angles COE and ROR_1. Note that these two angles represent the mirror rotation and the reflected beam rotation. Repeat for angles COG and ROR_2. What conclusions can you draw? Does this suggest a way to determine small rotations? Explain. Attach your diagram to your report.

D. Measurement of the Index of Refraction

12. Lay a glass plate on a sheet of plain paper and carefully draw a line around it (Figure 42-7). Turn on the laser and adjust the beam to shine along the paper so you can see it before and after entering the glass plate. With a sharp lead pencil mark and label the points I, O_1, O_2, and R.

13. Turn off the laser and remove the glass plate. Construct perpendiculars P_1P_1' and P_2P_2' to the two surfaces at points O_1 and O_2, respectively, and draw line O_1O_2. Measure and record the angles P_1O_1I, $P_1'O_1O_2$, $P_2'O_2O_1$, and P_2O_2R. Recall the law of refraction given by

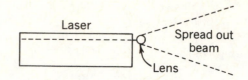

Figure 42–5 Cylindrical Lens Arrangement.

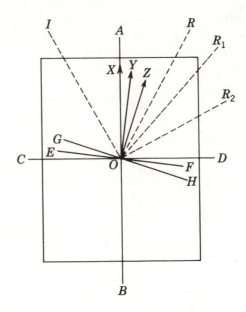

Figure 42–6 Mirror Arrangement.

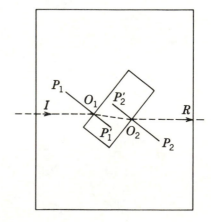

Figure 42–7 Index of Refraction Using a Laser.

$$n_1 \sin \theta_1 = n_2 \sin \theta_2 \qquad [3]$$

where n_1 and n_2 are the indexes of refraction of the media in which the angles θ_1 and θ_2 are measured. θ_1 and θ_2 are the angles between the normal to the refracting surface and the beam of light. Use this relation to compute the index of refraction of your glass plate. How do angles P_1O_1I and P_2O_2R compare?

QUESTIONS

1. Was the length of the laser a factor in the distances determined in Step 4? Explain.

2. What were your most likely sources of error in Step 4? Step 5? Which measurement do you think was the most accurate? Why?

3. Compare the sources of error in the laser measurements in Steps 1–4 with those in Steps 6 and 7. Which of these two measurements do you think is the most accurate? Give reasons for your answer.

4. Using the law of sines derive Equations [1] and [2].

5. Would the laser technique used in this experiment be a useful way of measuring large distances? Small distances? What are the limiting factors in this technique?

6. In Steps 9–11, is there any advantage to locating points R, R_1, and R_2 as far as possible from the mirror? Why or why not?

7. What are the limiting factors on the size of the rotation angle that can be determined by the technique in Steps 9–11?

8. Could the technique used in Steps 12 and 13 be used to measure the index of refraction of a plate of glass that was shaped like a triangle?

NAME (Observer) _____ Date _____

(Partner) _____ Course _____

RECORD OF DATA AND RESULTS

Experiment 42—The Laser

Step	Measured Values	Computed Values
2–4	α = β = d =	a = b =
5	a = b =	Percent diff. = Percent diff. =

Step	Measured Values	Computed Values
7	α = β = d =	a = b =
	a = b =	Percent diff. = Percent diff. =

Step 11: Angles

COE = _____ ROR_1 = _____ Comparison:

COG = _____ ROR_2 = _____ Comparison:

Relation discovered:

Step	Angles	Index of refraction of glass plate
12 & 13	P_1O_1I = $P_1{}'O_1O_2$ =	n =
	$P_2{}'O_2O_1$ = P_2O_2R =	n =
	Percent diff. between P_1O_1I and P_2O_2R =	

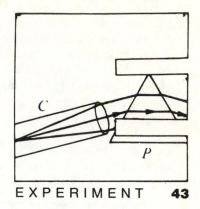

INDEX OF REFRACTION WITH THE PRISM SPECTROMETER

SPECIAL APPARATUS:

Small electric light and cord, ruler, graph paper, prism.*

GENERAL APPARATUS:

Spectrometer, source for spectral lines (sodium flame or neon-glow lamp), clamp for electric light.

THE PURPOSE OF THIS EXPERIMENT

is (1) to measure the angle of a prism and its minimum deviation for sodium (or neon) light, (2) to determine the index of refraction of the prism, and (3) to calibrate the spectrometer for spectral measurements.

INTRODUCTION

A spectroscope is an instrument for producing and viewing spectra. When the instrument is graduated for measuring angles for refracted and diffracted beams, it is called a spectrometer. It consists of three essential parts (see Figure 43-1): a collimator C to produce a parallel beam of light, a prism P to disperse the light into a spectrum, and a telescope T with which to examine the spectrum. (Figure 43-1 illustrates an arrangement for monochromatic light.) The collimator is a tube with a convex lens in one end and a slit S in the other. The distance between them is the focal length of the lens. The source of light, L, is placed just beyond the slit. An image of the slit is formed in the telescope, and cross hairs are provided for adjusting the position of the telescope for measurement on any part of the spectrum.

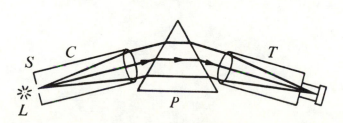

Figure 43–1 The Essential Parts of a Prism Spectrometer.

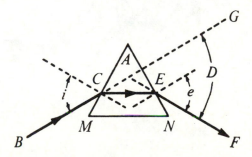

Figure 43–2 Illustration of the Angle of Minimum Deviation.

*If, in Step 10, the instructor desires to use a second prism with a different density instead of different light sources, then other prisms will be needed in the equipment list.

Light incident on a prism in the direction *BC* (Figure 43-2) will be refracted along some line *CE* and then emerge from the prism in the direction *EF*. Angle *A* is called the angle of the prism, and angle *D* is called the angle of deviation, with line *CG* being the original direction of the beam, and *EF* being the direction of the deviated beam. If the angle of incidence *i* is varied, angle *D* also varies, but it has a minimum value for one particular angle of incidence. This minimum value of angle *D* is called the angle of minimum deviation of the prism for the wavelength of light being studied. When the prism is set for angle *D* to be a minimum, the index of refraction *n* is given by

$$n = \sin \tfrac{1}{2}(A + D)/\sin \tfrac{1}{2}A. \qquad \textbf{[1]}$$

A major portion of this experiment is concerned with the measurement of the angle of the prism *A* and the angle of minimum deviation *D*, from which the index of refraction is calculated.

DESCRIPTION OF APPARATUS

Two forms of student spectrometers are shown below in Figure 43-3. In the force-table model, the collimator is shown at the left and consists of a slit and lens. The other movable arm on the right carries the telescope, and is supplied with a vernier scale which will permit angles to be read to 0.1°. When set up for operation, it is suggested that the collimator arm be set at 0° on the circular scale.

The standard model in Figure 43-3*b* has essentially the same features as a force-table model except that its components are enclosed, and therefore it does not require a darkened room. If the spectrometer assigned to you is equipped with a special prism table and the conventional type of telescope and collimator, please do not disturb any of the adjustments until you are sure that it is necessary. Much time is required to restore such adjustments once disturbed. Study the circular scale and learn to read angular settings to the limit of precision of the instrument. Also examine the slit on the collimator and learn which way to turn the screw to adjust the width, but do not change the length of the collimator tube.

PROCEDURE

1. Adjust the eyepiece of the telescope until the cross hairs are in good focus and then focus the telescope for parallel rays by sighting on a distant object (prism not on table). **2.** Mount a small electric lamp near the slit, the slit being at the principal focus of the collimating lens. While viewing the slit image with the unaided eye, move the telescope in the line of sight and adjust for a sharp, clear-cut image. **3.** Now mount the prism in the center of the prism table and orient it so that the groundglass face *MN* is roughly perpendicular to the collimator tube. This arrangement

(a)

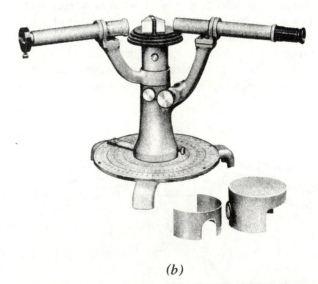

(b)

Figure 43–3 Prism Spectrometer. (*a*) Force-Table Model (Courtesy of Central Scientific Co.). (*b*) Standard Model (Courtesy of Sargent-Welch Scientific Co.).

makes angle A face the collimator tube (see Figure 43–4). In this position the prism splits the parallel beam of light from the collimator and reflects a portion of it from each of the smooth faces.

4. With the unaided eye, locate the image of the slit as reflected from either face of the prism. If you are unable to see either image, try moving the prism a small distance toward or away from the collimator. While viewing the image from face AM (Figure 43–4), move the telescope in front of the eye and adjust the cross hairs on the center of the image. This may best be done with the slow-motion screw. Now narrow the slit, if adjustable, so that the very best setting can be made. Read the angle for position T_1 and record. Repeat the procedure for the image from the other face of the prism.

5. It may be shown that the angle between positions T_1 and T_2 is equal to $2A$. Hence, you have the information for determining the refracting angle A of the prism.

6. Rotate the prism about 120° so that it is in the position shown in Figure 43–5a. With the unaided eye look along direction T_1B until you see the spectrum of colors. Now move the telescope in position and examine the spectrum more carefully, noting how many colors are seen. Do you note any change in the positions of the colors when the telescope is used instead of the unaided eye? Change the width of the slit and note the effect. Set the cross hairs on the yellow portion of the spectrum, and record the angle.

Now replace the electric light with a neon or sodium source and note the position of its spectrum with respect to the cross hairs of the telescope. You may have to readjust the slit width for this observation.

7. While observing the sodium spectrum with the unaided eye, rotate the prism back and forth slowly and follow the movement of the image. Note the reversal of the direction of motion of the image while the prism is still rotating in the same direction. When the prism is set for the position of the reversal of the motion of the image, it is in the position for minimum deviation. Now view the image in the telescope, and when the cross hairs have been set on the image in the minimum-angle position, read and record the position of the telescope.

8. Rotate the prism to the position shown in Figure 43–5b and record the position of the telescope for minimum deviation.

9. Now remove the prism and record the position of the telescope when viewing the image of the slit directly from the collimator along direction EB (Figure 43–5). Determine the angle of minimum deviation D from each of these trials, and record the average as your determination of angle D.

10. Repeat Observations 7, 8, and 9, for any other sources which the instructor may designate or repeat with other prisms of different composition.

11. With the information now at hand, compute the index of refraction of the glass for the sources used by means of Equation [1]. Determine the percent discrepancy compared with tabulated values.

12. Calibration of the Spectrometer (Optional). Place a mercury source near the slit and with the prism position shown in Figure 43–5a, note the spectral lines emitted by mercury. Select a bright line near the center of the spread of colors and adjust the prism for minimum deviation for this line. Without disturbing the position of the prism, measure

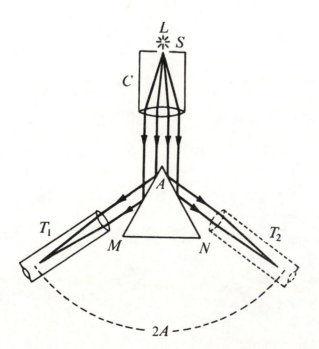

Figure 43–4 Measurement of Angle of Prism.

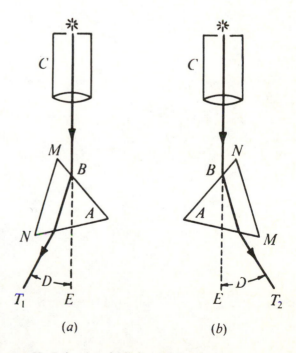

Figure 43–5 Angle of Minimum Deviation.

and record the angular position of each line in the mercury spectrum. Also record the color of each line measured and look up the wavelength of each in Table 12 of Appendix B.

Now plot a graph of wavelength versus angular position of the mercury spectral lines. Start the scale on each axis with numbers slightly less than the smallest values measured. Now use this calibration curve to determine the wavelength of the spectral lines from other sources with the same prism setting from your measurement of the angular positions.

QUESTIONS

1. What distinct difference did you notice between the spectrum of the electric-light filament and the sodium flame?

2. When the angular position of the sodium spectrum is compared with the positions of the various colors in the continuous spectrum, what color in the latter matches the position of the sodium spectrum?

3. Does narrowing the slit cut out some of the colors when viewing the electric-light spectrum? Describe the effect.

4. Do you think the index of refraction of glass for red light or blue light would be different from that for yellow light? What observation did you make which verifies your answer?

5. Which color of light has the greatest speed in glass? What correlation do you observe between the angle of deviation and the speed of light?

6. The handbooks usually give the index of refraction of materials with respect to the light from sodium. Discuss the possible reason for this designation.

7. Compute the velocity of sodium light in the glass used in this experiment.

8. If your prism had had a smaller refracting angle, how would the dispersion have been affected? How would the index of refraction have been affected? Explain.

9. When you were using the electric lamp as a source, why did the image of the slit look different by reflected light (Figure 43–4) than by refracted light (Figure 43–5)?

10. What difference did you observe in the positions of the colors of the continuous spectrum when viewed with the unaided eye? Why is this?

11. In Step 12, a mercury source was suggested for the calibration of the prism. Examine Table 12 in Appendix B and give some possible reasons why mercury might be better than some other elements.

12. If, in Step 10, you used different prisms instead of a variety of light sources, explain the reason for the differences in the index of refraction.

RECORD OF DATA AND RESULTS

Experiment 43—Index of Refraction with the Prism Spectrometer

Angular position of image for angle A		Source used	Material of prism used	Angular position of image for angle of minimum deviation, D			Angle D	Index of refraction n	Percent discrepancy
Left	Right			Left	Direct	Right			
$2A =$									
$A =$									
✕									

CALIBRATION OF PRISM SPECTROMETER

Source used and color of line	Angular position of the line			Wavelength of the line (Angstroms)
	Direct	Left	Angle D	
Mercury violet				

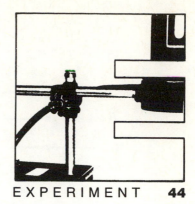

THE WAVELENGTH OF LIGHT

NOTE TO INSTRUCTOR:

Procedure B requires a laser and should be done after performing Experiment 42 or in conjunction with Experiment 42. Also note that Procedure A can be done as an independent experiment and Procedure B lends itself well to laboratory demonstration with all students receiving data from a single demonstration and each student doing the computations.

SPECIAL APPARATUS:

Two-meter stick, one-meter stick, grating, grating support, lamp and cord, enclosure (with slit) for lamp, support clamps, two image markers, optical-bench support.

GENERAL APPARATUS:

Two ring stands, sodium source light, (see Figure 44–2 or 44–3).

THE PURPOSE OF THIS EXPERIMENT

is to measure the wavelength of light with a diffraction grating and to determine the approximate wavelength limits of the visible spectrum.

INTRODUCTION

A transmission diffraction grating is a piece of glass on which have been ruled a large number of equally spaced parallel lines (or grooves) a few wavelengths apart. The transparent portions between the lines act as apertures, whereas the ruled portions diffract light and thereby change the direction of some of the incident rays. The parallel rays A, B, and C (Figure 44–1) which pass through the grating G may be focused at I_0 by the convex lens L if no deviation occurs. However, that portion of the beam which strikes the slits will be deviated. Consider the rays diffracted at an angle θ_1 from the original direction. If these three deviated rays are in phase, an image of the source may be focused at I_1. If the phase difference of adjacent rays is one wavelength, the image I_1 is called the first-order image. If the distance between lines in the grating is d, examination of the diagram in Figure 44–1 will show that the relation

$$\lambda = d \sin \theta_1 \qquad \text{[1]}$$

must exist to get a first-order image at I_1. If the diffracting angle θ is large enough for the phase difference to be two wavelengths, we have

$$2\lambda = d \sin \theta_2 . \qquad \text{[2]}$$

For the general case when the phase difference is n wavelengths, the above relation becomes

$$n\lambda = d \sin \theta \qquad \text{[3]}$$

where n is called the order number. Hence, if the grating space d is known and the angle θ is measured, the wavelength λ may be calculated. **Note:** *n = 1 for the first-order image, and note also that it is the first image either to the right or to the left of the center. Similarly, n = 2 for the second-order images, which are the second images to the right or left of the center.*

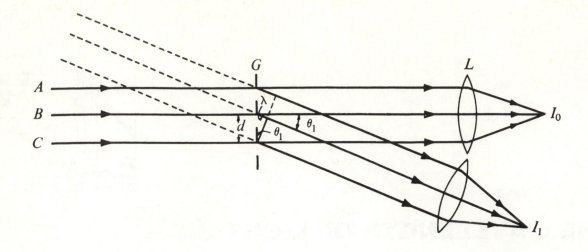

Figure 44–1 Passage of Light Through a Plane Grating.

The angle θ may be measured by either of two methods. If, in Figure 44–1, the grating is placed at the center of curvature of a circular scale, the number of degrees through which lens L must be rotated to produce image I_1 may be read directly from the scale. A second method, and the one to be used in Procedure A, is illustrated in Figure 44–2. Here lens L is replaced by the eye, which sees a virtual image I_1 of the source I_0 displaced from the source by an angular distance θ. From the known distances I_0I_1, I_0G, and I_1G we have

$$\sin \theta = I_0I_1/I_1G \qquad \textbf{[4]}$$

or

$$\tan \theta = I_0I_1/I_0G \qquad \textbf{[5]}$$

Several different sources may be used for this arrangement. A sodium flame produced by mounting a small piece of glass tubing* just above the inner cone of a bunsen flame or a sodium arc lamp is preferred (Figure 44–2b or 44–3). Also required are meter sticks and supports to build the arrangement illustrated in Figure 44–2a.

The diffraction grating provides a very convenient method of measuring the wavelength of a laser. If a laser beam is incident on a transmission grating the diffracted images can be observed on a screen large distances away (Figure 44-4). Because of the extreme intensity of the laser beam it is possible to observe several orders of diffracted images. The geometrical arrangement of Figure 44–4 is the same as that of Figure 44–2a so the Equations [3], [4], and [5] apply to both experimental arrangements. Procedure B uses a grating to measure the wavelength of a laser.

PROCEDURE A

1. Regulate the burner for a clear blue flame and then adjust the position of the sodium attachment to give a distinctly yellow flame. If available, the standard sodium source shown in Figure 44–3 may be preferred. In either case you may want to use a slit near the source to limit the width of the image (Figure 44–2b).†

2. Mount a meter stick AC (Figure 44–2) a few centimeters behind (or in front of) the burner at the same height as the yellow part of the flame. If mounted in front of the burner,

it should be lowered to allow the top part of the flame to show above the scale. Set up a meter stick along the direction BD perpendicular to the first scale at its midpoint (the 50-cm mark).

The plane of the grating should be parallel to that of slit and scale. If the slit images are higher on one side than on the other, the grating rulings are not parallel to the slit. This can be corrected by rotating the grating slightly about an

---

*If a brighter sodium flame is desired, a holder for a sodium salt, such as fused sodium chloride, may be mounted in the flame of a bunsen burner.

†The slit arrangement shown in Figure 44-2b can be made easily from one-half of a coffee can, supported on a block of wood to elevate it. The slit should be about 4 mm wide. If a slit and shield are not available, the burner alone, as already described, works quite well, with images the shape of the flame.

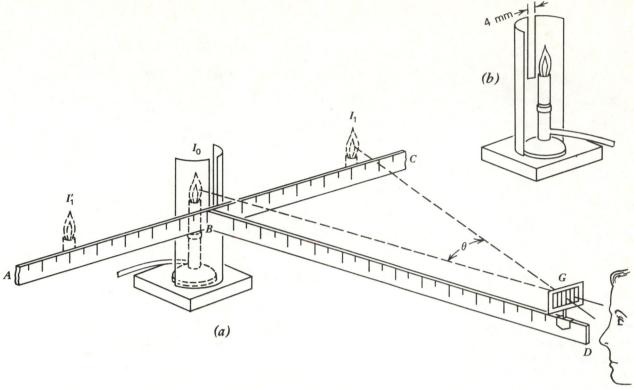

Figure 44–2 Arrangement for Measuring the Wavelength of Light.

Figure 44–3 Sodium Arc Source (Courtesy of Sargent–Welch Scientific Co.).

axis perpendicular to its face. Also check to see that the plane of the grating is vertical.

3. Mount the grating G on the second scale at a distance 50 cm from the transverse scale AC. Now with the eye near the grating, locate the positions of the first- and second-order virtual images of the yellow flame along the scale AC on both sides of the source. Mark the positions with a rider such as an image marker and record the readings. Also record the distance I_0G and the grating constant (the number of lines per meter).

4. Repeat the above procedure for two other positions of the grating, say 100 and 150 cm, from the scale AC. The second-order image may not be visible for this part.

5. Replace the sodium burner by a small electric lamp placed behind a slit. The lamp should be in an enclosure to keep the room dark. The slit may be made in the enclosure. When this source is viewed through the grating, a continuous spectrum may be seen as the image on either side of the source. If an electric lamp is not available, the same result may be obtained by using the bunsen burner with the air supply cut off so as to produce an incandescent flame.

6. With the grating at 50 cm from the scale AC, mark the extremities of the color band for each image in the first order and record the positions. These readings will give information from which the wavelength limits of the visible spectrum can be computed. Examine the position and

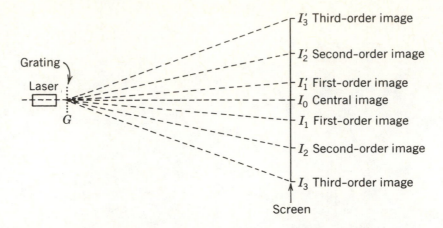

Figure 44–4 Diffraction Grating Images of a Laser.

spread of the second-order image, and record color limit positions.

7. From the left and right positions of the images of each order, determine the average displacement from the central image $1/2(I_1I_1')$ and compute $\tan \theta$ for each arrangment from Equation [5]. From the average of $\tan \theta$ determine $\sin \theta$ for each order.

8. By use of Equation [3], compute the wavelength of sodium light in centimeters for both the first- and second-order images.

9. Compute the wavelengths of the limits of the visible spectrum for both red and violet light.

PROCEDURE B

Before performing this procedure the student should review the Introduction to Experiment 42 and Appendix E.

1. Record the number of lines per meter given for your grating. Then compute and record d by taking the reciprocal of this number.

2. With the laser turned off set up the experimental arrangement shown in Figure 44–4. The grating-to-screen distance should be 1 meter or more if possible. Carefully measure and record the grating-to-screen distance.

3. Turn on the laser and note the diffracted images. Measure and record the distances on the screen between the left and right first-order images (I_1I_1'). Use good technique

and be sure to practice laser safety. Repeat for higher order image pairs. Dividing these values by 2 gives I_0I_1, I_0I_2, etc.

4. Using Equation [5] compute and record the angles associated with each diffracted image. Then compute and record the associated wavelengths using Equation [3]. Do not forget to take the order of the image into account.

5. Compute and record the average wavelength. Compute the deviation of each value from the mean and then compute the average deviation from the mean. Compute the percent uncertainty in your measurement.

6. Record the known wavelength of your laser and compute the percent discrepancy of your measurement.

QUESTIONS—PROCEDURE A

1. Compare your measured value of the wavelength of sodium light with the accepted value given in Table 12 of Appendix B by computing the percent discrepancy.

2. How did the intensity of the first-order image compare with that of the second-order image?

3. Why is there no illumination on the scale between the images?

4. Why did the images from the electric lamp not appear as simple white images of the slit?

5. Find the angles for the first- and second-order images for sodium light. Is the angle for the second order twice that for the first order? Explain.

6. If the grating had had fewer lines to the centimeter, how would the angles have been affected?

7. How did the spread of the second-order image of the continuous spectrum compare with that of the first order image?

8. How did the distance between the images of the first and second order of the continuous spectrum compare with the corresponding distance for sodium light? If the wavelength of violet light were just half the wavelength of red light, what would you be likely to notice concerning the distance between the first- and second-order images? Why?

9. Draw a rectangle showing the general shape of the broadened-slit image for the continuous spectrum, and indicate one end as the violet limit and the other as the red limit. By examination of the wavelength in your data table, show the approximate position of the sodium light in this range of colors by drawing a line across the rectangle.

10. If the meter stick *AC* (Figure 44–2) were replaced with a curved scale, shaped to fit the arc of a circle, what advantages would it have over a linear scale? What disadvantages would a curved scale have in this experiment?

QUESTIONS—PROCEDURE B

1. How did the intensity of the various order images compare?

2. Is there a linear relation between the angles associated with each order image? Explain.

3. If the grating contained fewer lines per meter, what differences would you expect?

4. How does your percent uncertainty compare to your percent discrepancy? Is it possible that some quantity used in determining the wavelength may not have been accurate? What would be the most likely quantity?

5. Perhaps it would be better to assume that the wavelength of the laser is known and use this technique to determine the grating space, *d*. Explain.

NAME *(Observer)* _____ Date _____

(Partner) _____ Course _____

RECORD OF DATA AND RESULTS

Experiment 44—The Wavelength of Light—Procedure A

Grating Data: Number of lines per m _____ Grating space _____ m

Source used	Grating distance	Image positions		Mean displacement from zero order	Tan θ	Sin θ	Wave-length λ
		Left	Right				
First order, sodium light							
Second order, sodium light							
First order, violet limit Electric bulb							
First order, red limit Electric bulb							
Second order, violet limit Electric bulb							
Second order, red limit Electric bulb							

Experiment 44—The Wavelength of Light—Procedure B

Laser Data

Manufacturer _____ Wavelength _____

Grating data

Number of lines per m _____ Grating space _____ m

Screen–grating distance I_0G _____ m

Image order	Image separation	I_0I_n	θ	Wavelength λ	Deviation
$n = 1$					
$n = 2$					
$n = 3$					
$n = 4$					
			Average		

Percent uncertainty _____

Percent discrepancy _____

EXPERIMENT **45**

A STUDY OF SPECTRA WITH THE GRATING SPECTROMETER

SPECIAL APPARATUS:

Small electric light, ruler, diffraction grating, mercury spectral tube, other spectral tubes for unknowns.

GENERAL APPARATUS:

Spectrometer, high-voltage source for exciting the spectral tubes, support for lamp, sodium arc lamp.

THE PURPOSE OF THIS EXPERIMENT

is to observe the nature of continuous and bright-line spectra and to measure the wavelength of some of the prominent spectral lines of both known and unknown elements.

INTRODUCTION

When light is dispersed by a prism or diffracted by a grating so that the different wavelengths of which it is composed are separated, the array of colors is called a spectrum. When solids, such as carbon particles in a Bunsen flame or the filament in an electric lamp, are heated to incandescence, all wavelengths are produced, each color merging gradually into the next. Such a spectrum is called a *continuous spectrum*. If a gas, such as neon or sodium vapor, is heated to incandescence, distinct wavelengths separated from each other are produced and, emerging from the slit arrangement of a spectroscope, show up as bright lines. These lines are called *bright-line spectra*.

The Diffraction Grating. A transmission diffraction grating is a piece of glass on which has been ruled a large number of equally spaced parallel lines a few wavelengths apart. The transparent portions between the lines act as apertures, while the rule portions diffract light and thereby change the direction of some of the incident rays. The parallel rays A, B, and C (Figure 45–1) which pass through the grating G may be focused at I_0 by means of the converging lens L, if no deviation occurs. However, that portion of the beam which strikes the slits will be deviated. Consider the rays diffracted at an angle θ_1 from the original

direction. If these three deviated rays are in phase, an image of the source may be focused at I_1. If the phase difference of adjacent rays is one wavelength, the image I_1 is called the first-order image. If the distance between lines in the grating is d, examination of the diagram in Figure 45–1 will show that the relation

$$\lambda = d \sin \theta_1 \qquad [1]$$

must exist to get a first-order image at I_1. If the diffracting angle θ is large enough for the phase difference to be two wavelengths, we have

$$2\lambda = d \sin \theta_2. \qquad [2]$$

For the general case, when the phase difference is n wavelengths, the above relation becomes

$$n\lambda = d \sin \theta \qquad [3]$$

where n is called the order number. Hence, if the grating space d is known and the angle θ is measured, the wavelength may be calculated. *Note: n = 1 for the*

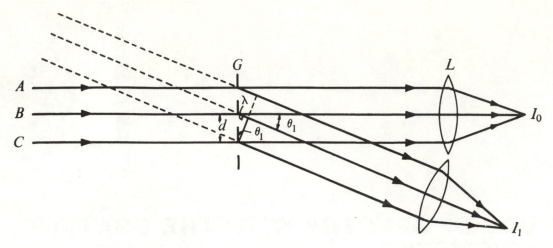

Figure 45–1 Passage of Light Through a Plane Grating.

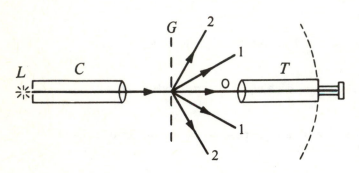

Figure 45–2 Essential Parts of a Grating Spectrometer.

first-order image, and note also that it is the first image either to the right or to the left of the center. Similarly, n = 2 for second-order images, which are the second images to the right or left of the center.

The grating spectrometer is one of the types of instruments used to observe spectra and to determine wavelength. It consists of a collimator, C (Figure 45–2), for producing a parallel beam of light from the source L, a grating G for diffracting the light, and a telescope T for observing the spectrum. The zero, first, and second orders of the diffracted beam are indicated in the figure. The telescope is mounted on a movable arm so that it may be rotated along a circular arc for observing the various orders of the spectrum.

DESCRIPTION OF APPARATUS

Two forms of student spectrometers are shown in Figure 45–3. In the force-table model, the collimator is shown at the left and consists of a slit and lens. The other movable arm on the right carries the telescope, and is supplied with a vernier scale which will permit angles to be read to 0.1°. When set up for operation, it is suggested that the collimator arm be set at 0° on the circular scale.

The standard model in Figure 45–3b has essentially the same features as the force-table model except that its components are enclosed, and therefore it does not require a darkened room. If the spectrometer assigned to you is equipped with a special prism table and the conventional

type of telescope and collimator, please do not disturb any of the adjustments until you are sure that it is necessary. Much time is required to restore such adjustments once disturbed. Study the circular scale and learn to read angular settings to the limit of precision of the instrument. Also examine the slit on the collimator and learn which way to turn the screw to adjust the width, but do not change the length of the collimator tube.

Ask the instructor if the collimator and telescope have been adjusted for parallel light. If not, ask for instructions concerning the adjustment.

PROCEDURE

In this experiment, we shall use a known wavelength of light to calibrate the grating and then use the calibrated

grating to determine the wavelength of other spectral lines. On the data form, under grating data, record the number of

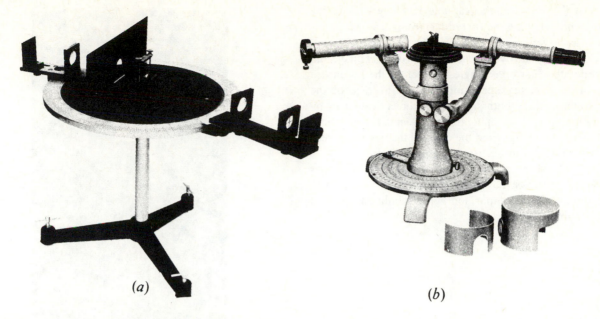

Figure 45–3 Grating Spectrometer. (*a*) Force-Table Model (Courtesy of Central Scientific Co.). (*b*) Standard Model (Courtesy of Sargent-Welch Scientific Co.).

lines per meter indicated by the manufacturer for reference purposes only (also compute the grating space *d*).

1. Adjust the eyepiece of the telescope until the cross hairs are in good focus and then, being *careful not to touch* the grating surface with your fingers, mount the grating in a vertical position on the table in the center of the spectrometer. Then, adjust the grating so its plane is perpendicular to the direct line of sight as shown in Figure 45–2.

2. Set a sodium light source near the slit and, with the telescope in the direct line of sight, adjust until you have a good focus of the slit with a clean image of the cross hairs centered on the slit. Be sure to eliminate parallax between image and cross hairs (see Figure 45–3).

3. Now move the telescope to the positions of the first- and second-order images on each side of the central line and record the angular position of each.

4. In the table of spectral lines in Appendix B, Table 12, you will note that sodium has two lines with wavelengths 5890 and 5896 Å, respectively. These are two yellow lines very close together and may not be separated by your grating. If not, use 5893 Å, or 5.893×10^{-7} m, as the wavelength for calibrating purposes.

Now apply Equation [3] to the data you have and compute the grating space *d*. Record the average value in the space indicated above the data table, and use this value for future wavelength determinations.

NOTE:

The room must be made as dark as possible for all the remaining observations in this experiment. Use a small light for reading the angular scale.

5. Now replace the sodium light with a small electric lamp and, with the unaided eye, observe the first-order image of this spectrum, and note the nature and sequence of the colors. Then move the telescope into your line of sight and check to see if there is any change in the sequence (or order) of the colors. Also observe the second-order image and note its characteristics.

6. In order to determine the wavelength limits of the continuous spectrum, set the cross hairs of the telescope on the red and violet extremities of the color band of the first-order image, and read and record the angle for each. Repeat for the first-order image on the other side of the direct, or zero-order, image.

For comparison purposes, also read and record the angles on both sides for the brightest part of the yellow region of the first-order image.

7. Now replace the electric lamp with a mercury spectral tube, such as that shown in Figure 45–4.* Adjust the height so that the small portion of the tube is in front of the spectrometer slit. Record the number of lines seen and also their respective colors. Observe the second order and compare the appearance and intensity with that of the first

---

*If the transformer shown is not available, a midget high-frequency tesla coil or some form of spark coil can be used.

order. If the slit is of the variable width type, vary the width, while viewing the first order, from fairly wide to as narrow as possible, and note the effect on the image.

8. Set the slit as narrow as practical for a distinct image and determine the first-order angle for each of the prominent lines in the mercury spectrum. Use the average of the angles measured on both sides of the direct beam and, by using Equation [3], compute the wavelength for each line. Record the wavelength and the color of the line measured, and compare with those listed in Appendix B.

9. Repeat the above procedure for unknown spectral tubes which the instructor may assign to you and try to identify the elements by comparison with tabulated spectral wavelengths in Appendix B.

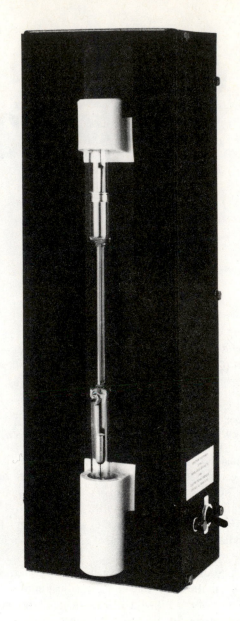

Figure 45–4 Spectral Tube and Excitation Transformer (Courtesy of Central Scientific Co.).

QUESTIONS

1. The grating you used is a replica of some standard glass grating used as a pattern. The replica is made by placing a film of Collodion on the standard and then, where the impression of the lines is made, the film is fastened to a sheet of plastic or glass. With this information, give a reason why it might be desirable to calibrate the grating with a known wavelength.

2. Was the sequence (or order) of the colors the same when viewed with the unaided eye as when viewed through the telescope? Explain. Which color, violet or red, is diffracted through the smallest angle as seen with the unaided eye?

3. Describe the characteristics of the first-order spectrum when using the electric light as a source. To which type of spectra does it belong?

4. What differences did you notice between the second-order image and the first-order image in the spectrum of the electric light?

5. Why are there no images in the region between the zero order and the first order?

6. Describe the spectrum produced with the mercury discharge tube. To what type of spectra does it belong?

7. What effect does a change in the width of the slit have on the images in the mercury spectrum?

8. What comparison did you find between the wavelength of the yellow line in the mercury spectrum and the bright part of the yellow region of the spectrum of the electric light? What, in general, does this reveal about the relation of the wavelengths of the continuous spectrum and those of bright-line spectra?

9. What did you find to be the wavelength limits of the visible spectrum? If a film had been used and a picture of the spectrum obtained, what difference, if any, do you think you might have obtained in the wavelength range on the exposed part of the film? Explain.

10. At what position in the system should a film be placed if a good photograph of a bright-line spectrum is desired?

11. Compute the angular spread of the second-order image of the visible spectrum. *Hint:* Determine the angle for the violet and red limits.

12. The manufacturers are now listing grating descriptions in terms of grooves per millimeter. In terms of this notation, what advantage would a grating of 1000 grooves/mm have over using one with 600 grooves/mm? Explain clearly.

13. What problem might one encounter in attempting to use a grating with 2000 grooves/mm? Would there likely be any problem in the manufacture of it? Compute d for this case; then use λ for sodium light and compute θ from Equation [1].

14. More grooves per millimeter for a grating means a smaller value for d and, hence, a greater resolving power for viewing spectral lines which are close together. Also, the angle θ is greater, resulting in a lower intensity.

Examine the spectrometer you used and make a guess at the largest angle you think might be practical for viewing and measuring. Now, with this angle and the wavelength of sodium light, use Equation [1] to compute d; then determine the grooves per millimeter to specify in selecting the proper grating.

15. In measuring the wavelength of light with a grating, what would be the advantages of using the second-order image rather than the first-order image? What would be the disadvantages? Cite experimental evidence to support each answer.

16. In this experiment, you have viewed two types of spectra, continuous and bright line. What atomic mechanism accounts for these two types being so different? You may need to consult your textbook under the heading "spectra."

NAME *(Observer)* _____ Date _____

(Partner) _____ Course _____

RECORD OF DATA AND RESULTS

Experiment 45—A Study of Spectra with the Grating Spectrometer

Grating Data: Number of lines per m _____ Grating space, *d* _____ m

Source used and color observed	Order number, *n*	Angular position of image		Average angle, θ	Sin θ	Wavelength, λ	Percent Discrepancy
		Left	Right				
Sodium, yellow	1 2						
Electric lamp, red limit							
Electric lamp, violet limit							
Electric lamp, yellow region							

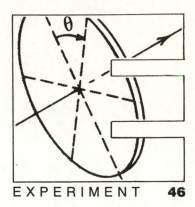

POLARIZED LIGHT

SPECIAL APPARATUS:

Photodetector with microammeter, two Polaroid disks each mounted so it can rotate, glass plate, microscope slide, thin piece of mica, piece of cellophane, protractor.

GENERAL APPARATUS:

Optical bench, white light source (preferably with collimator).

THE PURPOSE OF THIS EXPERIMENT

is to study the polarization of light by observing polarization phenomena.

INTRODUCTION

There are two ways to transmit energy between two points: by waves and by particles. Waves may have longitudinal vibrations (parallel to the propagation direction) or transverse vibrations (perpendicular to the propagation direction). Some waves, such as water waves, may have combined longitudinal and transverse vibrations. The phenomena of interference and diffraction of light show that light behaves like a wave. The fact that light may be polarized demonstrates that light waves are transverse, because longitudinal waves cannot be polarized.

Visible light is an electromagnetic wave in which the electric and magnetic fields oscillate in size and direction. The electric and magnetic field vectors at any point and instant are perpendicular to each other and perpendicular to the direction of propagation. These two field vectors oscillate in phase and are proportional to each other, so we usually represent the wave by referring only to the electric vector.

While much of this model of light is "classical" in form, one quantum aspect of light is essential for an understanding of polarization. That is the fact that light energy is not emitted in a continuous stream, but is emitted in "bursts" of energy called *photons*.

Ordinary light sources emit *unpolarized* light, which means that the photons they emit have their electric vectors oriented in random directions, uncorrelated with other photons. If we could keep track of the directions of the electric vectors of photons over a very large number of photons, we would find that they have no preferred orientation direction. This may be schematically represented as shown in Figure 46–1, where the diagram shows that in a light beam consisting of very many photons all possible electric vector directions are represented.

Several means exist for producing light in which all photons have their electric vectors oriented in the same plane. Such light is called *plane polarized*. This may be schematically represented as shown in Figure 46–2. The direction of the electric field vector is called the direction of polarization. Some lasers emit plane polarized light. Plane

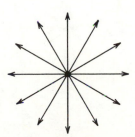

Figure 46–1 Unpolarized Vibrations.

Figure 46–2 Plane Polarized Vibrations.

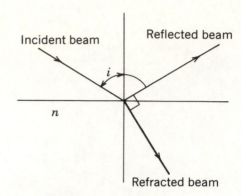

Figure 46–3 Brewster Angle.

polarized light may also be produced by reflection or by passing light through certain nonisotropic materials.

When light is incident on a transparent medium part of the beam is reflected and part is refracted. The reflected beam will show partial polarization at all reflected angles. At one particular angle, called Brewster's angle, the reflected beam will be completely polarized with the plane of polarization being parallel to the surface. When light is incident at the Brewster angle the angle between the reflected beam and the refracted beam is exactly 90° (Figure 46–3). Using this fact, it can be shown that the relationship between the Brewster angle i and the index of refraction of the material n is given by

$$n = \tan i. \qquad [1]$$

The second way to polarize light is to pass it through certain nonisotropic materials, called "doubly refracting" or "birefringent" materials. Ordinary light passing through such materials may be thought of as consisting of two component light beams. Each component beam is plane polarized and the planes of polarization are perpendicular to each other. Many crystals found in nature, such as calcite, are doubly refracting. Some manufactured materials made up of oriented long-chain molecules also have this property. Cellophane is an example.

Some of these materials also have the property of absorbing one of the two beams, so only one plane polarized beam emerges from the material. This is useful for producing plane polarized light and for this purpose is manufactured in the form of sheets called "polarizing sheets." Polaroid brand polarizers are a well-known example. The *polarization axis* of such a sheet is an imaginary line drawn on the sheet parallel to the polarization axis of the light it produces.

A common way to study polarization phenomena is to pass a beam of light through two polarizing sheets as shown in Figure 46–4. The first sheet through which the light passes is called the "polarizer," for its function is to produce polarized light. The second sheet through which the light passes is called the "analyzer," for its function is to detect the polarization of the light and to determine its axis (direction) of polarization. If the transmission axes of the sheets make an angle of 90° with each other, very little light intensity gets through the analyzer. When the transmission axes make an angle θ with respect to each other the transmitted intensity I is given by

$$I = I_0 \cos^2\theta \qquad [2]$$

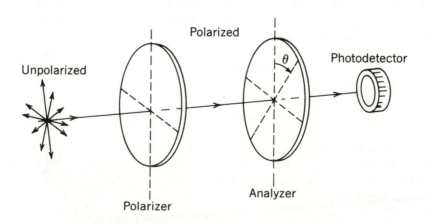

Figure 46–4 Polarizer–Analyzer Arrangement.

where I_0 is the intensity when $\theta = 0$. This equation is known as the law of Malus.

Plane polarized light represents a special and relatively simple type of polarization. Consider two plane polarized waves of the same frequency propagating along the same axis with their planes of polarization perpendicular to each other. The resultant plane of polarization is determined by the vector sum of the two field vectors. At any point along the propagation axis this resultant plane of polarization will vary in time and amplitude depending on the relative phase between the two waves. In the general case the amplitude vector will sweep out an ellipse. This is called elliptically polarized light. Several of the phenomena you will observe in this experiment are a result of elliptical polarization. If you have studied this phenomenon in class try to explain your observations. If you have not studied this phenomenon, make careful observations and ask your instructor for explanations or refer to your textbook for assistance.

PROCEDURE

1. Set up your polarizer–analyzer system similar to that shown in Figure 46–4. Your equipment may be different from that shown but should contain the same general features. Check with your instructor concerning any special considerations for your equipment. Turn on the light source and adjust the two Polaroid sheets to allow maximum-intensity light to strike the photodetector. If possible adjust your microammeter for maximum reading at this point.

2. When the light intensity is a maximum the angle θ between the transmission axes of the polarizer and analyzer is 0. Note what happens when this angle is changed from 0 to 180°. Now start at $\theta = 0$ and increase the angle in 10° increments up to 180° and record the microammeter reading for each angle.

3. Note that $\cos^2\theta$ is the same for $\theta = 80$ or $100°$, 70 or $110°$, etc. Thus the microammeter readings should be the same for these angle pairs if your measurements were perfect. Compute and record the average values for each pair of angles.

4. Plot microammeter reading versus $\cos^2\theta$. Use the average values of the angle pairs. Do the plotted points verify Equation [2]?

5. Remove the photodetector and set the analyzer for minimum intensity with the Polaroids several centimeters apart. Place a piece of mica between the Polaroids and write down what you see through the analyzer. What effect does the mica seem to have on the polarized light?

6. Repeat Step 5 using a microscope slide.

7. Repeat Step 5 using a piece of cellophane. Fold the cellophane into several thicknesses and again record your findings. Crumple the cellophane and record your findings.

8. Remove the analyzer and mount a glass plate so that light from the polarizer is incident on it at an angle of approximately 45°. Observe the image of the source reflected from the glass plate through the analyzer. Is the reflected beam polarized? Adjust the orientation of the glass plate or the analyzer or both until the reflected beam shows maximum polarization. Determine the angle of the incident or reflected light for this case which is the Brewster angle for the glass. Use Equation [1] to determine the index of refraction of the glass plate. If the accepted value of the index of refraction is available compute your percent discrepancy.

9. Make a dot on a piece of paper. Observe the dot through the calcite crystal and write down your observations. Rotate the crystal and record your observations. Examine the light coming through the crystal from the dot with a Polaroid and record your observations. Include a discussion of the direction of vibration of the light beams coming through the crystal.

QUESTIONS

1. Explain how polarization phenomena indicate that light is a transverse wave.

2. Derive Equation [1] using the condition that the angle between the reflected and refracted beam is 90°. You will need the laws of reflection and refraction to do this.

3. If an unpolarized beam of light passes through a Polaroid, what percent of the light is transmitted? Look at a beam with and without a Polaroid and estimate this percent. You might also want to try measuring it with your photodetector. Give an explanation of your results? *Hint:* A component of each light wave is transmitted.

4. What advantage do polarizing sunglasses have over tinted sunglasses?

5. Light from an ordinary source is unpolarized. What does this say about the source?

NAME (Observer) _____ Date _____

(Partner) _____ Course _____

RECORD OF DATA AND RESULTS
Experiment 46—Polarized Light

LAW OF MALUS

θ	Microammeter 10^{-6} A	θ	Microammeter 10^{-6} A	Average 10^{-6} A	$\cos^2\theta$
0		180			1.000
10		170			0.970
20		160			0.883
30		150			0.750
40		140			0.587
50		130			0.413
60		120			0.250
70		110			0.117
80		100			0.030
90		90			0.000

OBSERVATIONS

Step 5

Step 6

Step 7

Step 8   **Step 9**

Brewster angle _____

Index of refraction

Calculated _____

Accepted _____

Percent discrepancy _____

PART SIX
NUCLEAR PHYSICS

6

THE CHARACTERISTICS OF A GEIGER TUBE*

SPECIAL APPARATUS:

Pair of split-disk radioactive sources, two rulers, items furnished by the student: graph paper and watch with sweep second hand.

GENERAL APPARATUS:

Geiger tube with cable and mount, scaler (or ratemeter), electric timer (watch can be substituted).

THE PURPOSE OF THIS EXPERIMENT

is to learn how to operate a radioactivity indicator and to determine the current–voltage characteristic and the resolution time of a Geiger tube.

INTRODUCTION

All radioactive radiations, whether they be charged particles or gamma rays, have the property of ionizing air and many other gases. This ionizing property is utilized in the operation of devices designed for the detection of radiation. One of the most common of these detecting devices is the Geiger–Müller (G–M) counter (Figure 47–1). The G–M tube consists of a positively charged wire W mounted along the axis of a negatively charged cylinder, the whole being enclosed inside a glass tube. This glass tube contains the gas to be ionized by the incident radiation. Very thin windows have to be used in the tube wall if alpha particles are to be detected, their penetrating power being very low. Gamma photons may be distinguished from beta particles by shielding the tube with aluminum to prevent the beta particles from entering. Other types of tubes have a thin mica window in the end, and radiation enters through the end rather than the side.

When a beta particle or gamma photon enters the G–M tube, some of the energy of this incident particle (or photon) may be transferred to a gas molecule within the tube. The absorption of this energy by the gas molecule results in ionization, whereby an outer electron is ejected from the molecule. The ejected electron, being negatively charged, is attracted to the positively charged wire of the G–M tube, and the positive ion is attracted to the negatively charged cylinder (Figure 47–1). If the potential difference between the wire and the cylinder is sufficiently high (800–1000 V), these primary ions acquire a high speed within a very short distance and, on their way toward the electrodes, collide with other gas molecules and thus produce additional (or secondary) ions. These secondary ions, as well as the primary ions, may produce other ions and these, in turn, still other ions before reaching the electrodes. This cascading effect produces what is called an "avalanche" of ions. The arrival of this avalanche of ions at a charged electrode results in a change in the charge of the electrode. This change in charge causes a sudden change in the potential difference between the wire and the cylinder, which, in turn, causes a current pulse to be sent through the circuit. This current pulse causes a difference

*This experiment, along with the next two on radioactivity, may be done as a demonstration experiment if insufficient equipment is available for smaller groups of students.

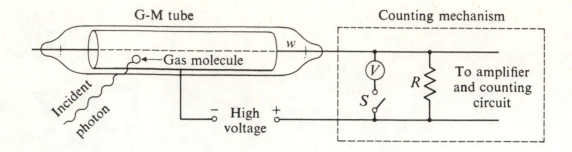

Figure 47–1 Schematic of a Geiger–Müller Counter Circuit.

of potential across the ends of the resistance R. When amplified, this current pulse may actuate a loudspeaker or a neon light, either of which may serve as a counting device. Signals of this nature would indicate the entrance of a radioactive particle (or photon) into the Geiger tube. If the pulse frequency is too fast to count, the ion current may be registered on a meter which has been calibrated to read in counts per minute.

Cosmic rays and radioactivity in the air usually cause a Geiger tube to register counts, even though no radioactive source seems to be present. Counts resulting from such stray radiations are called *background counts*.

The G–M Tube Plateau. Because the speed of the ions inside the tube is related to the potential difference applied to the tube, one might suspect that the count rate would bear some relation to the applied voltage. The general nature of the relation is shown in Figure 47–2, but the detailed shape of the curve is one of the characteristics of each individual tube. There is a threshold voltage below which no counts are registered but, within the first 50 V above the threshold, the count rate rises very rapidly to the knee K. The slope of the plateau region KL depends on how the tube is made and the kind of gas in it. In the plateau region, small fluctuations in the line voltage to the tube have a small effect on the count rate. Hence, the best operating voltage for the tube is on the flattest portion of the plateau.

Resolution Time. If the ions resulting from an incident particle have not all reached the collecting electrodes before a second particle (or photon) enters the G–M tube, the current pulse from the second group of ions may be superimposed on the first pulse and only one count will be registered. The shortest time required for the tube to register separate pulses for two successive particles (or photons) is called the *resolution time*. Each tube has its own characteric resolution time and a knowledge of the resolution time makes it possible to correct for the coincidence loss during the operation time. The value of the resolution time for a particular counting system depends on several factors: the construction of the counter, the gas in the tube, the voltage applied to the tube, the counting rate, and others. For a particular counter, and for a low count rate, the resolution time may be assumed to depend only on the applied voltage. If T sec is the resolution time, then the counter is insensitive for a time of $T/60$ min during each count interval that is registered. If there are N counts per minute (cpm) being registered, then the counter is insensitive for $N(T/60)$ min during each minute. Hence, during 1 min, the fraction of the counts lost is equal to the fraction of 1 min the counter is insensitive, or $N(T/60)/1$. If n is the number of counts that should have been registered, then the number lost in 1 min is $nN(T/60)$. Hence the corrected number of counts per minute is given by

$$n = N + nN(T/60) \qquad [1]$$

or $n(1 - NT/60) = N$, which when solved for n gives

$$n = \frac{N}{1 - NT/60}. \qquad [2]$$

$NT/60 << 1$ and, in such cases, the approximation

$$\frac{1}{1 - NT/60} = 1 + NT/60$$

can be used, and we then have

$$n = N + N^2 T/60 \qquad [3]$$

as the value of the corrected count rate.

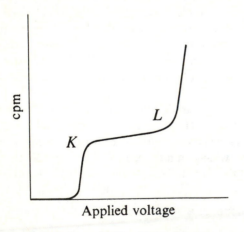

Figure 47–2 Geiger–Müller Tube Plateau.

THE OPERATION OF COUNTING INSTRUMENTS

The directions below are somewhat general and, because they cannot fit every make of equipment, the equipment manual should also be consulted.

Count-Rate Meter. One instrument which registers the pulse rate from a G–M tube in terms of the ion current produced is called a *ratemeter*, and the instantaneous count rate is read directly from a meter on the face of the instrument (Figure 47–3).

The G–M tube shown is the side-window type, but an end-window model will serve equally well. These tubes are both costly and easily damaged. Therefore, they must be handled with care. After you have read Appendix D relative to handling radioactive material, secure a sealed radioactive source from your instructor to use in getting acquainted with the operation of the ratemeter. When you have turned all switches off, plug the ratemeter into a 120-V ac line, and place the source about 15 cm from the side-window tube or on the bottom shelf of an end-window tube mount. Ask the instructor which side of the source should face the G–M tube. It usually will be the side giving the greatest count rate.

Turn the meter selector switch to the position to read volts; then turn on the power switch or the high-voltage knob until it just clicks on, and allow the circuit elements to warm up before increasing the voltage. Transistorized units do not require a warmup period. After the meter needle indicates its initial reading (about 500 V), increase the voltage slowly and watch for the flash of the neon light, which is an indication of the first pulse to be registered. Increase in the volume control will permit pulses to be indicated by clicks in a small speaker.

Increase the voltage about 150 V above the threshold value and then turn the selector knob to the highest range position for counts per minute and note the meter reading.

Figure 47–3 Nuclear Ratemeter and Geiger–Müller Tube (Courtesy of Sargent–Welch Scientific Co.).

The number of counts per minute is the product of the meter reading and the scale factor on the selector dial. Now move the source back and forth from the tube and note the change in count rate. When the count rate is low enough, a lower range may be used on the selector knob. When there is a need to change the voltage, the selector knob must be turned back to volts so that the meter will register volts instead of count rate.

Scalers. This electronic switching mechanism registers the pulses from a Geiger tube circuit and totals the accumulated count for whatever counting time is used. It can also be used for microtiming in mechanics experiments and for pulse counting. Two types of scalers are shown in the diagrams below but all laboratories may not be equipped with these particular models.

The scalers shown in Figures 47–4 and 47–5 are very fast, high-count-capacity instruments designed for a variety of applications, including radioactive emission counting. The accumulated count is displayed in digital fashion on a screen. When counting low-activity samples, these scalers may be left running overnight or for a weekend to obtain high statistical accuracy. The models shown here have built-in timers, but similar models are available to be used with an externally connected timer. The G–M tube shown in Figure 47–4 will operate equally well with any model scaler.

General Instructions Relative to Scalers. Since it is impossible to provide specific operating instructions in this book for all models of scalers on the market, it will be assumed that you have more specific information from the apparatus company which supplied your instrument. Hence, the instructions which follow will serve as a general guide for your procedure. Examine the knobs and switches on the face of the scaler and the plug-in receptacles on the back side of the instrument. You will connect your G–M tube to the receptacle labeled *detector*.

Place a sealed source on the bottom shelf of the tube mount. Turn all switches off and plug the scaler into the 120-V ac outlet. Now turn on the power switch and, after time is allowed for the circuit elements to warm up, turn on the high-voltage switch and allow another warmup period. Models with solid state circuitry require no warmup.

Move the count switch to the reset position and then to the count (or on) position. Now slowly increase the voltage and watch for the first indication of a pulse by the flashing of some of the first set of lights, or by the appearance of numbers on the digital screen. You will need either a watch or an electric timer to indicate the counting time. In order to check the reliability of the scaler it is always desirable to check it against the local 60-cycle ac line pulses. This is done by setting the function switch to the *Test* position and counting for exactly 1 min. If the count is approximately 3600, you are assured that the scaler is operating properly.

Figure 47–4 Scaler and Geiger–Müller Tube (Courtesy of Sargent–Welch Scientific Co.).

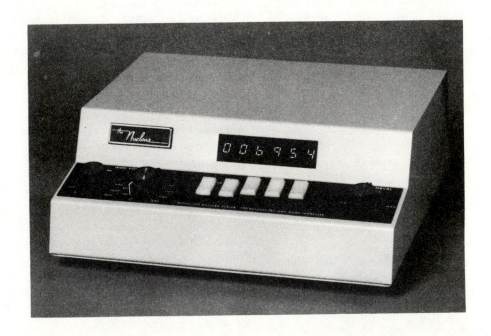

Figure 47–5 Scaler/Timer (Courtesy of the Nucleus Inc.).

With the switch back to the count position and a source on the bottom shelf, increase the voltage to about 150 V above the threshold value and check the number of counts in some given time, say 1 min. This may be accomplished by setting the function switch to preset count and the timer for a 1 min count. Change the position of the source to another shelf and again note the counts in 1 min.

PROCEDURE

Examine the instrument assigned to you to see if it is a ratemeter, or a scaler, and then follow the instructions outlined in the preceding section for your type of instrument. When you have become acquainted with the operation of your counting system, you will use it to study the characteristics of the G–M tube attached to the system.

A. The Count-Rate Plateau

1. When the equipment is warmed up and ready for operation, slowly increase the voltage until you reach the threshold value. Then, decrease the voltage until the counting just barely stops and record this reading as the threshold voltage, the highest possible voltage for zero count rate.

2. By examination of Figure 47–2, you will note that the count rate would be expected to rise rapidly until the knee is reached. Hence, obtain the count rate (counts per minute) for the smallest voltage steps you can estimate on the voltmeter until the rate begins to level off. Then obtain the count rate for about 50-V steps until the meter reads about 300 V above the threshold value. Voltages beyond point L (Figure 47–2) on the curve will *cause a continuous discharge in the tube and ruin it for further use*. When finished with the readings, reduce the voltage to its minimum value while performing the next steps.

3. Plot the count rate as ordinate (zero at the origin) versus the voltage as abscissa with the origin beginning at 50–100 V below the threshold value. Be sure to include the threshold value as one of your plotted points. Then sketch a smooth curve through your series of points.

4. Select a point on the *flattest portion* of the plateau curve *as the operating potential* for the determination of the resolution time in Part B.

B. Resolution Time of a Geiger Tube

5. For this part of the experiment you will need to check out two nearly identical sources. The set may consist of two semicircular disc sources plus a blank piece of the same dimensions.

6. Adjust the voltage for the operating potential for your G–M tube and *do not* change it through this portion of the experiment. Then place source A (or No. 1) alongside the blank at a distance from the tube window that produces about 12,000 cpm. Most of the conventional G–M tubes do not begin to lose a significant amount of counts until the rate exceeds about 10,000 cpm. Make three trials for this arrangement and record the average count rate as N_1.

7. Now remove source A and place the blank in the position from which source A was removed; then place source B in the position previously occupied by the blank disc. The purpose of this exchange is to maintain the same geometrical pattern for both sources. Make three trials for the count rate of source B and record as N_2. These count rates should likewise be 12,000 cpm or more.

8. For a combined count rate of the two sources, replace the blank disc with source A so that you now have both A and B in position to produce pulses. This arrangement of both sources should now push the count rate into the coincidence-loss region and thus provide information from which the resolution time can be computed. Make three trials of the count rate for the combination and record as N_c. If N_c is not somewhat less than $N_1 + N_2$, then all of the above procedures will have to be repeated with the sources closer to the tube to produce higher count rates.

9. It can be shown that an approximate value of the resolution time in seconds is given by

$$T = \frac{(N_1 + N_2) - N_c}{2N_1 N_2} \times 60. \qquad [4]$$

Use this formula to compute the resolution time for your Geiger counter circuit and express the final result in microseconds.

QUESTIONS

1. When the high voltage is first turned on, the voltmeter reads about 500 V but no counts are observed, even with a source near the Geiger tube. Explain why no counts are registered at this voltage, but counts are observed if the voltage is increased by some 200–300 V.

2. Examine your plotted curve and explain why smaller

voltage increments were recommended for the first few sets of data.

3. From an examination of the curve explain the significance of the term "count-rate plateau" curve. Can you think of an explanation for the existence of the plateau? For which voltage increment is the percent change in the counting rate the smallest?

4. What does the shape of the curve indicate about the sensitivity of the count rate to small fluctuations in the applied voltage, if the voltage is set near the center of the plateau region?

5. What procedure in obtaining data for the plateau curve would you suggest might give a smoother alignment of the points?

6. What percent of the counts would be lost when counting at 10,000 cpm with the counter you used? What percent at 20,000 cpm? *Hint:* Refer to Equation [3].

7. If the observed count rate is 10,000 cpm, what is the true count rate corrected for the coincidence loss? See Equation [3].

8. Compute the corrected count rate for each of the observed count rates in Steps 6, 7, and 8, and then compare the corrected count rate for the combination with the sum of the corrected count rates for the two sources individually.

9. If one made no correction for coincidence count loss, would the percent discrepancy be greater for a low count rate or for a higher count rate? Use your data to verify the answer.

10. Use the following hints to derive the expression for the resolution time given in Equation [4] of Step 9. *Hints:* In Equation [3], N represents the observed count rate and n the corrected count rate. Use Equation [3] and write expressions for n_1, n_2, and n_c. Then set $n_c = n_1 + n_2$ and solve for T. Next assume $N_c \cong N_1 + N_2$ and simplify the denominator of your expression for T by setting $N_c^2 = N_1^2 + N_2^2 + 2N_1N_2$.

NAME *(Observer)* _____ Date _____

(Partner) _____ Course _____

RECORD OF DATA AND RESULTS

Experiment 47—The Characteristics of a Geiger Tube

Part A		Part B		
Threshold voltage		Operating voltage of plateau		
Voltage (V)	Count Rate (cpm)	Source designation	Trial	Observed count rate (cpm)
		A + Blank	1	
			2	
			3	
			Average N_1	
		B + Blank	1	
			2	
			3	
			Average N_2	
		A + B	1	
			2	
			3	
			Average N_c	
Resolution Time of Geiger tube = _____ sec _____ μ sec.				

Show steps of calculation below

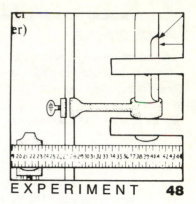

THE NATURE OF RADIOACTIVE EMISSION

SPECIAL APPARATUS:

Radioactive source (encased in capsule), lens holder, burette clamp, graph paper, and watch with sweep second hand.

GENERAL APPARATUS:

Radioactivity demonstrator* with Geiger tube, ring stand, meter-stick-type optical bench (see Figure 48–1).

THE PURPOSE OF THIS EXPERIMENT

is (1) to study the statistical nature of the emission of radiation, (2) to measure the background count on a Geiger counter, and (3) to study the relation of the intensity of radiation to the distance from the source.

INTRODUCTION

Radioactivity is the spontaneous emission of certain charged particles and, in some cases, electromagnetic radiation from the nuclei of certain heavy elements near the upper end of the periodic table. Some radioactive substances emit helium nuclei, called *alpha particles*, whereas others emit high-speed electrons which are called *beta particles*. Certain substances of both of the above types may also emit short-wavelength electromagnetic radiations called *gamma rays*, which are photons similar in nature to X rays. These nuclei do not emit these particles at regular intervals. The rate of emission follows the laws of probability (or chance), and it is only when a very large number of particles are considered that laws can be established for the expected rate of emission.

The atmosphere always contains a few radioactive particles, and the cosmic ray particles and radiations are ever with us. The latter are very penetrating, and it is almost impossible to shield a Geiger counter from them completely. Hence, these ever-present ionizing rays always produce a few counts per minute in a counting

circuit. This low count rate is called *background count* and, when a radioactive sample is being counted, the total count rate includes the background counts which must be subtracted to get the true count of the sample.

A source is said to be an isotropic emitter if the probability of emission is the same in all directions. If N is the number of particles or photons emitted in all directions from an isotropic source in 1 sec, then the intensity I (number of particles per unit area per unit time) at a distance d from the source (assuming none is lost in the air) is given by

$$I = \frac{N}{4\pi d^2} \qquad [1]$$

where $4\pi d^2$ is the surface area of a spherical shell at a distance d from the source. For a relatively long-life source, Equation [1] states that the intensity of radiation is inversely proportional to the square of the distance from the source. This relation is often called the inverse-square law.

*If a scaler is available, the same procedure can be followed except that a watch will be used for timing the registration of counts. An electric timer may also be used.

PROCEDURE

Before beginning any part of the experiment, read the instructions in Appendix D concerning the procedures for handling radioactive material.

A. Emission Fluctuations and Background Count

Caution:

As stated in Experiment 47, a Geiger tube circuit must be handled with respect if it is to last and give reliable results. Handle it with care and Do Not apply more voltage than is suggested in the instructions below or in the manual supplied by the manufacturer. Excessive voltages shorten the life of the tube.

1. Refer to the preceding experiment or to the equipment manual relative to the general operation of the counting system supplied you for this experiment. To ensure that the ionization current (or count rate) will be independent of small fluctuations in the voltage, set the voltage about 100–150 V above the threshold voltage as your operating potential for the entire experiment. Your instructor may have reason to suggest a different operating potential.

2. In preparation for getting the background count, remove all radioactive sources and radium dial watches from within several feet of the G–M tube. Make ten 1-min count trials with no source near the system and record the average as the background count. If you are using a ratemeter, you cannot use the meter indication for such low count rates, but the neon lights or loudspeaker will correctly register the counts.

3. Now bring the radioactive source toward the Geiger tube until the counts are about as fast as you can count them. Keeping the source in a fixed position and using 15-sec time intervals, make ten trials. What do you notice about the count rates for the different trials? Multiply each of your count values by 4 and record the count rate in counts per minute. From the average (or mean) of these trials, determine the deviation of each trial from the mean and record these deviations without regard to algebraic sign. The average of these deviations is the amount that any one counting trial taken at random might differ from the mean value of the several trials you have made. Record this count rate in the form $C \pm D$ cpm, where C is the average of all of your trials and D is the mean deviation.

4. If your instrument is a ratemeter, place the source in the same position as in Step 3; then set the count-rate selector switch on the lowest range and observe the action of the meter for a few minutes. Does it check with the average count rate you recorded in Step 3? Does the reading remain constant? If not, why not?

B. Relations of Radiation Intensity to Distance

5. Set up the equipment as shown in Figure 48–1, using a meter-stick-type optical-bench arrangement, unless your equipment is designed to be used differently. This setup will reduce scattering of radiation from the table top and instrument case and give a more realistic reading. If an end-window tube is used, it could be mounted horizontally or mounted about 1 m above the table with a window down.

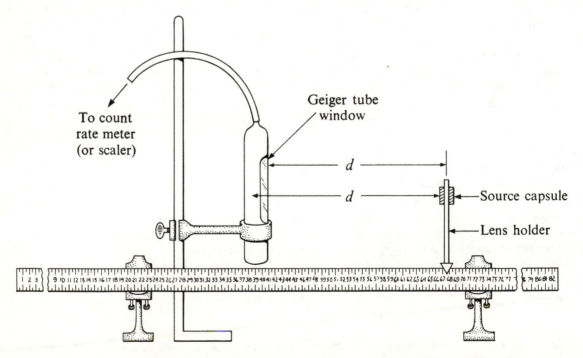

Figure 48–1 Arrangement for Measuring Radiation Intensity Versus Distance.

In the latter case, the source could be moved up and down to vary the distance. *Extreme* care must be taken if an end-window tube is used, since the window is very fragile.

6. If a side-window tube is used, fasten the source capsule in a lens holder with the active face toward the tube, and measure the distance in either of the ways shown in the diagram. If the source is a thin disk, the distance should be measured from the source face to the center of the tube. If an end-window tube is used, measure to the window by setting the meter stick along the side of the tube, *not against the window.*

7. Adjust the position of the source until you get 7000 to 8000 cpm from either a ratemeter or a scaler. Record the distance and the count rate, and then move the source away until the count rate is reduced some 700–800 cpm, and record. Continue moving the source away until you have about 8 or 10 sets of data, and the count rate is near the background level.

8. Subtract the background count determined in Part A from each of the registered count-rate readings, and record it as the net count rate C for each of the corresponding distances. Equation [1] indicates that the intensity should be inversely proportional to d^2 or directly proportional to $1/d^2$. In preparation for checking this relation, compute and record the values for $1/d^2$. The count rate C is proportional to the intensity and thus will bear the same general relation to $1/d^2$.

9. Plot C versus $1/d^2$ and examine the trend of the plotted points. Do they indicate a linear relation? Draw a smooth curve that best fits the points, and, if it is not a straight line as expected, try to give an explanation. Examine your setup and determine what assumptions have been made that may not be exactly correct.

10. Optional. Since the intensity I is defined as the number of particles emitted per unit time per unit area, we must divide the count rate C by the area A of the counter window to convert to intensity units. Hence, $I = C/A$ is the conversion relation, and, in terms of our recorded data, Equation [1] becomes $C/A = N/4\pi d^2$. If written in the form $C = (NA/4\pi) \times (1/d^2)$, it may be noted that $NA/4\pi$ is a constant and becomes the slope of a straight line with C and $1/d^2$ as the variables.

The area of the window can be determined from the dimensions of the opening. If the window has a curved surface as shown in Figure 48–1, measure the dimensions (width and length) of the opening and treat the window as if it were flat. Record at the bottom of the data form.

Select that portion of your graph that most nearly approaches a straight line and extrapolate as a straight line to permit the computation of the slope. Compute the slope S of the extrapolated straight line and, from the relation $S = NA/4\pi$, compute the value of N. This value of N is the total number of photons or particles emitted from the source in all directions. Show this computation as a part of your report.

QUESTIONS

1. What do your data show about the nature of the rate of emission from a given source?

2. Can you suggest a faster way to get the background count without losing accuracy? Explain.

3. Why does the needle on the count-rate meter fluctuate back and forth when it is registering pulses of radiation?

4. Suppose that in Step 3, where you timed the counts with a watch, you had counted for 600 sec each time instead of 15 sec and divided by 10 to obtain the counts per minute. Do you think your results might have been different? Explain.

5. From your observations of the action of radiation on a Geiger counter, what do you think might be the effect of radiation on the molecules of the human body?

6. Which part of your curve deviated most from a straight line, the part for small distances or that for large distances? Consider the size of the Geiger tube window and your distance measurements, and try to correlate your method of getting distances with the nonlinearity of the curve.

7. At small distances you were registering rather high count rates. If the counts were too fast for the circuit to register all of them, relate this to the shape of your curve and explain.

8. It is dangerous to handle some radioactive sources directly with the hands because of the risk of radiation exposure, even though the source may be in a small container. The same source might be carried around without danger of overexposure if kept in a pasteboard box of dimensions $50 \times 50 \times 50$ cm, say. Give an explanation of this, based on information obtained from this experiment.

9. Source strengths are usually measured in curies, where 1 curie $= 3.7 \times 10^{10}$ disintegrations/sec. If time permitted you to carry out Step 10, determine the strength of your source in curies.

10. Examine the source capsule used and see if you think it might be isotropic and emit radiation uniformly in all directions. Explain your conclusions.

NAME *(Observer)* _____ Date _____

(Partner) _____ Course _____

RECORD OF DATA AND RESULTS

Experiment 48—The Nature of Radioactive Emission

Threshold voltage _____ V   Operating voltage _____ V

Background	PART A. EMISSION FLUCTUATIONS			
(cpm)	Time (sec)	Number of counts	Count rate (cpm)	Deviation from mean
	←Sum of all trials→			
I_B	←Average (or mean)→			
			$C \pm D =$ _____ cpm	

PART B. RELATION OF RADIATION INTENSITY TO DISTANCE				
Distance d (cm)	d^2 (cm^2)	$1/d^2$ (cm^{-2})	Registered count rate $C + BC$	Net count rate C (cpm)
Procedure 10 Window Data	Width =	Length =	Diameter if Circular =	Area =

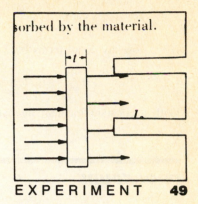

PROPERTIES OF RADIOACTIVE RADIATION

SPECIAL APPARATUS:

Graph paper, radioactive sources (beta and/or gamma).

NOTE:

The source should emit pure beta or gamma, not both (see isotope table, Appendix D).

GENERAL APPARATUS:

Ratemeter (or scaler), Geiger tube and mount, 30 or 40 small sheets of mimeograph paper (or kitchen aluminum foil) for beta source, 8 to 10 thicker (0.4–0.5 mm thick) aluminum sheets for gamma source, electric timer (or watch). Use calibrated absorbers, if available.

THE PURPOSE OF THIS EXPERIMENT

is to study some of the ionizing and absorption properties of radioactive radiation.

INTRODUCTION

As we have already discussed in Experiment 48, *radioactivity* is the spontaneous emission of certain charged particles, and, in some cases, electromagnetic radiation also, from the nuclei of certain heavy elements near the upper end of the periodic table. Some of these radioactive elements emit helium nuclei, called *alpha particles*, while others emit high-speed electrons which are called *beta particles*. Certain substances of both of the above types may also emit short-wavelength electromagnetic radiations called *gamma rays*, which are photons similar in nature to X rays.

There are several methods used for detecting radioactive radiations, each method depending upon some property of the particles or photons. All of the above-named radiations have the property of ionizing the air as they pass through it and of causing the fluorescence of certain substances such as zinc sulfide. The charged particles are also deflected by magnetic and electric fields.

In both the ionization and fluorescence processes the radiation particles (or photons) may be removed from their original path by either deflection or absorption. If I_1 is the

intensity of a beam of radiation as it enters a slab of material of thickness t (Figure 49–1), and I_2 is the intensity of the emerging beam, then an amount $I_1 - I_2$ has been absorbed by the material. The relative value of these two intensities is dependent upon both the thickness t and the kind of material of the absorbing slab. It can be shown that

$$I_2 = I_1 e^{-\mu t} \qquad [1]$$

where $e = 2.718+$, the base of natural logarithms, and μ is the linear coefficient of absorption of the material. The *linear absorption coefficient* may be defined as the fraction of the beam absorbed per unit thickness of material. A decreasing relation, such as Equation [1], is often called an exponential decrease. If one takes the natural logarithm (symbol ln) of both sides of Equation [1], the result is

$$\ln I_2 = \ln I_1 - \mu t. \qquad [2]$$

Since $\ln N = 2.3 \log N$, Equation [2] may be rearranged and expressed in terms of common logarithms by writing

$$2.3 \log I_2 = -\mu t + 2.3 \log I_1. \qquad [3]$$

If we now divide all terms by 2.3, we have

$$\log I_2 = (\mu/2.3)t + \log I_1. \qquad [4]$$

With a constant incident intensity I_1 and a constant μ,

Equation [4] has the familiar linear form $y = mx + b$, where $-(\mu/2.3)$ is the slope and $\log I_1$ the intercept.

Although the true intensities I_1 and I_2 are not actually known, the count rates from a Geiger counter will be directly proportional to the intensities and can be substituted for them. The numerical value of μ will depend on the units used for the thickness t.

PROCEDURE

Caution:

Before beginning any part of this experiment be sure you have read the instructions for handling radioactive sources in Appendix D. If you are not already familiar with the operation of the counting system assigned you, read either the instructions on ratemeters and scalers in Experiment 47, or the manufacturer's instruction manual.

The absorbers required for this experiment may be in any one of three forms: (1) small sheets cut from mimeograph paper and kitchen aluminum foil; (2) sheets of cardboard, aluminum, and lead, furnished with your counting system; or (3) a set of calibrated aluminum and lead absorbers with thickness indicated in milligrams per square centimeter.

1. If the counting system is to give reliable readings, the circuit should be turned on and allowed to warm up before any data are taken. Leave the high-voltage setting at the minimum (about 500 V) until all other preparations are ready; then increase to the operating value. Examine your absorber set and decide on the best method of arranging them with respect to the source and Geiger tube. If you have an end-window tube, the tube mount will contain holders for both the source and the absorber (Figure 47–4).

2. Arrange the source holder and absorber holder in the positions you intend to use and adjust the high voltage in the operating potential for your G–M tube. With all radioactive sources several feet away from the G–M tube, make a 5-min count on the background radiation level, but record as counts per minute to correspond with other data you will be taking.

3. With the source in place, but with no absorbers, make a trial run to determine the initial counting rate. Adjust the

source distance to give a sizable count rate but do not exceed about 10,000 cpm. Keep in mind that this count rate is the total of the source and the background. Now make your initial run for I_1 with $t = 0$, that is, no absorber in the path.

4. Then insert sufficient absorber material between the source and G–M tube to reduce the count rate by about 10% of the value obtained in the Step 3. Record the absorber thickness as numbers of sheets if using uniform sheets, or as milligrams per square centimeter if you are using sheets calibrated in these units. Continue this procedure, reducing the intensity by about 10% of I_1 each time, until you have about ten sets of readings or have reached the background level.

5. Subtract the background count rate (counts per minute) from each intensity reading and record as I_2, the intensity which has penetrated the absorbers. Also record $\log I_2$ for each of these corrected intensity readings.

6. Plot two graphs as follows: (a) plot intensity I_2 as the ordinate versus absorber thickness in sheets (or milligrams per square centimeter), and draw a smooth curve through the series of points to represent an average of the plotted data; (b) plot $\log I_2$ versus absorber thickness and draw a straight line which best represents the average of the plotted points.

7. *Computations from the First Graph.* If Equation [3] is written as $2.3(\log I_1 - \log I_2) = \mu t$, and we assume half of the radiation is absorbed ($I_2 = 1/2 I_1$), then we have $2.3(\log 2) = \mu T$, and the equation simplifies to $0.693 = \mu T$, where T is the half thickness (the thickness to reduce the intensity to half). Determine the half thickness from your first curve and use the simplified form of Equation [3] to calculate μ, the linear absorption coefficient.

8. *Computations from the Second Graph.* Select two points on the straight line, one near each end, and call the ordinate of the upper point $\log I_1$ and the ordinate of the lower point $\log I_2$. The abscissa distance between these two points represents the thickness of absorber necessary to reduce the intensity from I_1 to I_2. With these coordinate points, you can determine the slope of the graph. Then, as indicated in the paragraph below Equation [4], the slope can be equated to $-(\mu/2.3)$, thus μ can easily be

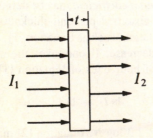

Figure 49–1 Absorption of Radiation.

determined. Note that the slope of your graph is negative. Compute and record the linear absorption coefficient μ, and then find the percent difference resulting from the two methods.

QUESTIONS

Answer each question so that the question asked can be ascertained from the answer.

1. Without timing any counts, what did you observe about the regularity of the background counts? What is the reason for this?

2. What is the best method of preparing absorbing plates (or sheets) so that they will be of approximately uniform thickness?

3. Do you find that the same number of absorbing sheets are required to reduce the intensity by the same amount each time? Justify your answer from your data.

4. In the examination of the first graph, you read from the graph the thickness of absorbing material to reduce the intensity to one-half of the initial intensity I_1. Sketch these values on the first graph so that they can be easily seen. Now find the thickness to reduce this new intensity value to one-half and sketch on the graph. Repeat for another half thickness value. How do the three half thickness values compare with each other?

5. Prove that $2.3(\log I_1 - \log I_2) = \mu t$ simplifies to $0.693 = \mu t$, when $I_2 = 0.5I_1$.

6. What thickness of absorber do you think would be required to reduce the intensity of radiation to zero? Examine your graph No. (1) and also Equation [3].

7. If you have had differential calculus, take the derivative of both sides of Equation [1] and then solve for μ. Formulate a definition of μ from this relation.

8. What would you suggest as a means of experimentally improving the alignment of the points on the graph?

9. If you had used a source which emitted higher energy rays, how would the shape of curve No. 1 have been affected? Illustrate with sketches.

10. By considering observations made in this experiment, along with those from previous experiments, what methods can a person use for protection against radiation, assuming he must work in areas where radioactive sources are used?

11. If your source had been one that emitted both betas and gammas, how would the shape of the second graph be affected? Assume gammas have much higher energy than betas.

NAME *(Observer)* _____ Date _____

      (Partner) _____ Course _____

RECORD OF DATA AND RESULTS
Experiment 49—Properties of Radioactive Radiation

Kind of source used _____ Background count, I_B _____ cpm

Registered count rate $I_2 + I_B$ (cpm)	Source Intensity I_2 (cpm)	log I_2	Absorber thickness t (sheets)	Absorber thickness t (mg/cm²)

ABSORPTION COEFFICIENT FROM THE GRAPHS

Graph No. 1	Graph No. 2
Half Thickness, $T =$	Slope of graph =
Absorption coef., $\mu =$	Absorption Coef., $\mu =$
Percent difference of the two methods for $\mu =$	

MEASUREMENT OF RADIOACTIVE HALF LIFE*

SPECIAL APPARATUS:

Radioactive isotope (indium foil),[†] forceps, graph paper.

GENERAL APPARATUS:

End-window Geiger tube in sample mount, scaler (or ratemeter), timer (watch may be used), neutron source for activating the foil (A) or minigenerators (B).

INTRODUCTION

The decay of a radioactive material is a process which follows statistical laws. There is no possibility of stating the exact time at which a particular atom will decay, but if there is a sufficiently large number of atoms collected together, then it becomes possible to predict how many of them will decay in a given time interval. Experiments with a wide variety of radioactive materials have shown that the rate at which atoms decay is proportional to the number of atoms present, and the rate dN/dt can be expressed by the relation

$$\frac{dN}{dt} = -\lambda N \qquad [1]$$

where N is the number of nuclei present, and the proportionality constant λ is called the decay constant. λ is a positive constant characteristic of the kind of radioactive nuclei under consideration. The negative sign is used because N decreases as t increases. If N_0 is the number of atoms present at time $t = 0$, integration of Equation [1] gives

$$N = N_0 e^{-\lambda t} \qquad [2]$$

where N is the number of atoms left at time t which have not undergone nuclear decay.

At a time $t = T$, when half of the original atoms have decayed, we have $N = N_0/2$, and Equation [2] takes the form

$$T = \ln 2/\lambda = 0.693/\lambda \qquad [3]$$

where $\ln 2$ represents $\log_e 2$. Since this is the time for one-half of the atoms to undergo decay, the quantity T is called the half life.

The half life of a radioisotope may be determined by either of the following methods: (1) measure the activity at periodic intervals and determine the time for the activity (counts per minute) to be reduced to half; (2) measure the decay constant λ and compute the half life from Equation [3].

The activity A, measured in counts per minute by a Geiger tube and scaler assembly at any time t, is directly proportional to the rate of decay dN/dt indicated in Equation [1]. Hence, we may write $A = kN$, and when

---

*Two methods of producing radioisotopes are discussed on the following pages: (A) The Neutron Generator (Figure 50–1), and (B) Minigenerators. If neither of these is available in your laboratory, the isotopes iodine-131 (half life 8.05 days) and phosphorus-32 (14.3 days) are made available by the suppliers several times during the year, and they can be purchased and shipped to you when they are needed.

[†]If indium foils are not available, samples of $NaIO_3$ can be made up and used as described by F. L. Moore in *Am. J. Phys.* **31**:362 (May 1963). Silver coins (dimes) can be used but the half life is only 2.3 min.

$t = 0$, $A_0 = kN_0$. If both sides of Equation [2] are now multiplied by k, the result is equivalent to

$$A = A_0 e^{-\lambda t} \qquad [4]$$

and, if we take the logarithm of both sides, we have

$$\ln A = \ln A_0 - \lambda t. \qquad [5]$$

Equation [5] indicates a linear relationship between $\ln A$ and t, both of which are easily measured. The slope of the line will yield the value of the decay constant λ, thus permitting the half life to be computed from Equation [3].

PRODUCTION OF RADIOISOTOPES

A. Neutron Generator

Many of the radioisotopes which occur in nature have extremely long half lives or belong to a radioactive series and cannot be easily isolated from the decay products. Radioisotopes with half lives suitable for laboratory use can be produced by bombarding certain stable isotopes with neutrons from a neutron generator.

One such source of neutrons is prouced by a mixture of plutonium-239 and beryllium.* The reactions are as follows:

$$_{94}Pu^{239} \longrightarrow {}_2He^4 + {}_{92}U^{235} \qquad [6]$$

$$_2He^4 + {}_4Be^9 \longrightarrow {}_6C^{12} + {}_0n^1 \qquad [7]$$

The isotopes I^{127}, In^{115}, and Ag^{107}, when exposed to a source of neutrons, are converted to the radioactive isotopes I^{128}, In^{116}, and Ag^{108}, respectively, with half lives of 24, 54, and 2.3 min. If we consider indium, the reactions are:

$$_0n^1 + {}_{49}In^{115} \longrightarrow {}_{49}In^{116}$$
$$_{49}In^{116} \longrightarrow \beta^- + {}_{50}Sn^{116}$$

Silver in the form of foils, or coins, contains approximately 51% Ag^{107} and 49% Ag^{109}. Both isotopes absorb neutrons, but Ag^{108} formed from Ag^{107} contributes 98% of the initial beta activity from the activated silver sample.

One form of neutron generator is that shown in Figure 50–1. It contains a Pu–Be mixture encased in a small cylinder fastened to the end of a Lucite rod which is inserted in the center of a paraffin-filled container. The paraffin serves as a shield and also a moderator to slow the neutrons down and increase their probability of being absorbed. Two portholes permit samples to be pushed to the center of the container where they will nearly touch the neutron source when it is lifted to the porthole level.

B. Minigenerators[†]

Radioactive equilibrium provides another means of obtaining short-half-life radioisotopes. For example, $_{55}Cs^{137}$ with a half life of 30 years decays to $_{56}Ba^{137}$ which has a half life of 2.6 min. Thus, any source of $_{55}Cs^{137}$ will have $_{56}Ba^{137}$ present, because it cannot decay before it is formed. A radioactive equilibrium situation results in

Figure 50–1 Neutron Generator.

*It takes about 16 gm of Pu^{239} in the mixture to produce a 1-Ci neutron source. A mixture of Ra–Be will also produce a good neutron source. An NRC license is required to purchase either of these.

[†] Several forms of minigenerators are available at a modest price from physics and chemistry apparatus supply houses. Only one set is needed for an average-size class. Each set is supplied with complete instructions for producing the radioisotopes and performing experiments with them. A ratemeter (or scaler) will also be needed.

which the $_{56}Ba^{137}$ will decay at the same rate that it is being formed by the decay of $_{55}Cs^{137}$. Because the two elements are chemically different, they can be separated by a chemical wash, and a sample of short-half-life $_{56}Ba^{137}$ results. Commercially available systems of this type with specially designed wash bottles are called minigenerators. The three most commonly used minigenerator systems provide $_{56}Ba^{137}$—a gamma source with a half life of 2.6 min; $_{49}In^{113}$—a gamma source with a half life of 100 min; $_{39}Y^{90}$—a beta source with a half life of 64 hr.

PROCEDURE

Before beginning any part of this experiment read all of Appendix D on instructions for working with radioactive materials. Also familiarize yourself with the operation of the counting system supplied for this experiment, by reading either the instructions in Experiment 48 or the manual furnished with the equipment.

1. Check with your instructor regarding what short-half-life radioisotope you will be using and how it is to be obtained. These radioisotopes will trigger detectors in the same manner as any other source but will have a short half life in comparison to other sources you have used. Hence, you will need to make some preliminary plans relative to the best length of counting intervals and the appropriate spacing of the intervals.

2. You will need one timepiece, such as a watch or a continuously running electric timer, to keep track of the total elapsed time. Also, the scaler must be turned on and off for known time intervals to give the count rate at previously chosen intervals during the decay process. This can be done either by watching the elapsed timer (or watch) or by using an auxiliary automatic preset timer which starts and stops with the **On** and **Off** count switch.

3. During the first part of the decay period, the following suggestions should work satisfactorily. If your source is indium, you may count 2 min, rest 8 min, count 2 min, rest 8 min, and so on. In the case of I^{128}, count 1 min, rest 4 min, count 1 min; for silver, count 15 sec, rest 15 sec, count 15 sec. If your source is different from these ask your instructor for appropriate counting and rest intervals. The rest periods will provide time to reset the scaler to zero and record the data.

4. Another factor which must be considered is the time relation of the count rate obtained to the total elapsed time scale. If you count for 1 min and get 4000 counts, you have a count rate of 4000 cpm, but at what time did the activity rate have this value? It will be approximately correct to assume that the count rate was 4000 cpm at the middle of the 1-min period. Hence, you should start your counting interval 30 sec before the time, $t = 0$, on your time scale so that the middle of the counting period will be the activity rate at time $t = 0$. In like manner, all later counting intervals must be assigned the time which has elapsed at the middle of the interval. This requires considerable coordination of the two lab partners.

5. After you have obtained five trials of the background count rate and all adjustments are ready, ask your instructor for the radioactive sample and be sure to handle it with forceps or some other appropriate holder. Place it in the sample tray and set the tray on the upper shelf nearest the Geiger tube. Select a beginning time that will simplify the timekeeping problems and start the counting one-half interval before time $t = 0$. Continue counting and recording until the activity is reduced to at least one–forth of A_0.

6. When you have finished the run, remove the source from your area of work and make a second 5-min trial measurement of the background count. Use the average of the two background runs to correct all count rates obtained from the sample after the latter have been converted to counts per minute. Record the value of ln A.

7. Plot two graphs: (a) A versus t, (b) ln A versus t. From the graph of ln A versus t, which should approximate a straight line, determine the slope (or λ) and compute the half life T from Equation [3]. From the graph of A versus t, find the value of A_0 when $t = 0$, and then determine the time at which $A = 1/2A_0$. This should also be the half life. Compare the average of your two values with the accepted value listed in the radioisotope tables (Appendix D). Record the percent discrepancy and discuss some possible ways to improve your experiment (Question 8).

8. When finished, first turn the tube voltage down; then turn off all switches and return all radioactive material to the instructor.

1. For a measurement of counting rate, do you think a scaler is more accurate than a count-rate meter? Explain (Consider the statistical nature of radioactive emission in both cases.)

2. State some advantages that each of the above-named instruments has over the other.

3. In almost all cases, the count rate is higher as the sample is placed closer to the end of the Geiger tube. Why is it true? (The absorption in the air is negligible at these distances.)

4. If you were going to determine the half life of an isotope for which the half life was of the order of 2 weeks, what changes in procedure would you suggest?

5. Some radioisotopes are shipped in solution in small bottles. (Ask your instructor to show you one.) Suggest a procedure for preparing a sample from such a liquid for a half life determination.

6. Explain why a nucleus changes its atomic number by emitting a beta particle but does not change in atomic number when it absorbs a neutron.

7. Explain why a nucleus does not change in mass number when a beta particle is emitted.

8. Describe how you might proceed to obtain a more reliable value for the half life of the radioisotope you used.

9. Would your results have been better if you had counted for 10-sec periods, 5 sec on each side of the 10-min intervals? Explain (Examine your graph of A versus t.)

10. Set $N = 0$ in Equation [2] and determine the time required for all of the original atoms to decay. Does your curve of A versus t support this answer? Explain.

11. Derive Equation [2] from Equation [1] by integrating and evaluating the constant of integration.

12. Suggest a method of determining the count rate of the sample the instant it was removed from the neutron generator.

NAME *(Observer)* _____ Date _____

(Partner) _____ Course _____

RECORD OF DATA AND RESULTS

Experiment 50—Measurement of Radioactive Half Life

THE BACKGROUND COUNT RATE

Time of counting	Number of counts	Count rate (cpm)	Time of counting	Number of counts	Count rate (cpm)
Sum			Sum		
Average of all trials for background count					

DATA FOR HALF LIFE — SOURCE USED _____

Reckoning time for count rate	Time of counting period	Number of counts	Count rate A (cpm)	A (corrected)	log A or ln A
T (measured) =		T (tables) =		Percent Discrepancy =	

APPENDIXES

APPENDIX A

TRIGONOMETRIC RELATIONS

Certain trigonometric functions of an angle may be defined as the ratios of pairs of sides of a right triangle containing the angle in question. The three most commonly used trigonometric functions are the *sine*, *cosine*, and *tangent*. In the accompanying right triangle.

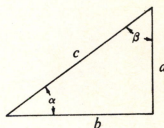

$$\text{sine of } \alpha = \sin \alpha = \frac{\text{side opposite angle}}{\text{hypotenuse}} = \frac{a}{c}$$

$$\text{cosine of } \alpha = \cos \alpha = \frac{\text{side adjacent to angle}}{\text{hypotenuse}} = \frac{b}{c}$$

$$\text{tangent of } \alpha = \tan \alpha = \frac{\text{side opposite angle}}{\text{side adjacent to angle}} = \frac{a}{b}$$

Two other useful relations for a triangle of any shape are the sine law

$$a/\sin \alpha = b/\sin \beta = c/\sin \gamma$$

and the cosine law

$$c^2 = a^2 + b^2 - 2ab \cos \gamma$$

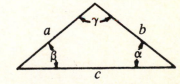

A great variety of relations exist between trigonometric functions. Some knowledge of these is essential for certain computations in elementary physics. Those which are most commonly encountered are

$$\sin \alpha / \cos \alpha = \tan \alpha$$
$$\sin^2 \alpha + \cos^2 \alpha = 1$$
$$\sin(\alpha \pm \beta) = \sin \alpha \cos \beta \pm \cos \alpha \sin \beta$$
$$\cos(\alpha \pm \beta) = \cos \alpha \cos \beta \mp \sin \alpha \sin \beta$$

APPENDIX B

TABLES OF PHYSICAL CONSTANTS

1. DENSITIES (gm/cm³)

SOLIDS

Aluminum	2.70
Brass	(about) 8.5
Copper	8.92
Cork	0.2–0.4
Glass (crown)	2.4–2.6
Iron (gray cast)	(about) 7.10
Iron (wrought)	7.86
Lead	11.34
Nickel	8.90
Oak	(about) 0.8
Silver	10.57
Steel	(about) 7.8
Zinc	7.15

LIQUIDS

Alcohol (ethyl)	0.79
Carbon bisulfide	1.27
Carbon tetrachloride	1.60
Gasoline	0.70–0.75
Glycerine	1.26
Kerosene	(about) 0.80
Mercury	13.60

GASES (at STP)

Air (dry)	0.001293
Carbon dioxide	0.001974
Hydrogen	0.000089
Oxygen	0.001430

2. VELOCITY OF SOUND (m/sec)

GASES AT 0°C

Air	331.5
Natural gas	432
Hydrogen	1270
Oxygen	317

LIQUIDS at 15°C

Alcohol	1240
Water	1440
NaCl, 10% sol	1470

SOLIDS

Aluminum	5104
Brass	3500
Copper	3560
Iron	5130
Steel	5000

3. COEFFICIENTS OF ELASTICITY—YOUNG'S MODULES (dynes/cm²)

Aluminum	6.5×10^{11}	Brass	10×10^{11}
Copper	12.0×10^{11}	Iron	19×10^{11}
Steel	22×10^{11}		

4. HEAT CONSTANTS FOR MATERIALS

Substance	Coefficient of linear expansion	Specific heat (°C)	Melting point (°C)
Aluminum	24.0×10^{-6}/°C	0.214	660
Brass	18.9×10^{-6}/°C	0.0917	900
Copper	16.8×10^{-6}/°C	0.092	1083
Glass (common)	8.5×10^{-6}/°C	0.161	1100 (?)
Iron	11.4×10^{-6}/°C	0.115	1535
Lead	29.4×10^{-6}/°C	0.0306	327
Nickel	12.8×10^{-6}/°C	0.105	1455
Silver	18.8×10^{-6}/°C	0.056	961
Steel	13.2×10^{-6}/°C	0.107	1483
Tin	26.9×10^{-6}/°C	0.054	232
Zinc	26.3×10^{-6}/°C	0.093	419

5. THERMAL COEFFICIENT OF CUBICAL EXPANSION OF LIQUIDS (20°C)

Alcohol	1.12×10^{-3}	Mercury	0.181×10^{-3}
Carbon Tetrachloride	1.236×10^{-3}	Petroleum	0.955×10^{-3}
Glycerine	0.505×10^{-3}	Water	0.207×10^{-3}

6. DENSITIES AND PRESSURES OF SATURATED WATER VAPOR

Temperature C	Temperature F	Grams per cubic meter	Pressure (centimeters of mercury)	Pressure (inches of mercury)
0°	32.0°	4.84	0.46	0.18
2°	35.6°	5.54°	0.53	0.21
4°	39.2°	6.33°	0.61	0.24
6°	42.8°	7.22°	0.70	0.28
8°	46.4°	8.21°	0.80	0.32
10°	50.0°	9.33°	0.92	0.36
12°	53.6°	10.57°	1.05	0.41
14°	57.2°	11.96°	1.20	0.47
16°	60.8°	13.50°	1.36	0.54
18°	64.4°	15.22°	1.55	0.61
20°	68.0°	17.12°	1.75	0.69
22°	71.6°	19.22°	1.98	0.78
24°	75.2°	21.54°	2.23	0.88
26°	78.8°	24.11°	2.51	0.99
28°	82.4°	26.93°	2.82	1.11
30°	86.0°	30.04°	3.17	1.25
32°	89.6°	33.45°	3.55	1.40
34°	93.2°	37.18°	3.98	1.57
36°	96.8°	41.28°	4.44	1.75

7. HYGROMETRY

Dry-bulb thermometer (°F)	Difference between Dry-Bulb and Wet-Bulb Thermometers														
	1°	2°	3°	4°	5°	6°	7°	8°	9°	10°	11°	12°	13°	14°	15°
50	93	87	81	74	68	62	56	50	44	39	33	28	22	17	12
52	94	88	81	75	69	63	58	52	46	41	36	30	25	20	15
54	94	88	82	76	70	65	59	54	48	43	38	33	28	23	18
56	94	88	82	77	71	66	61	55	50	45	40	35	31	26	21
58	94	89	83	77	72	67	62	57	52	47	42	38	33	28	24
60	94	89	84	78	73	68	63	58	53	49	44	40	35	31	27
62	94	89	84	79	74	69	64	60	55	50	46	41	37	33	29
64	95	90	85	79	75	70	65	61	56	52	48	43	39	35	31
66	95	90	85	80	76	71	66	62	58	53	49	45	41	37	33
68	95	90	85	81	76	72	67	63	59	55	51	47	43	39	35
70	95	90	86	81	77	72	68	64	60	56	52	48	44	40	37
72	95	91	86	82	78	73	69	65	61	57	53	49	46	42	39
74	95	91	86	82	78	74	70	66	62	58	54	51	47	44	40
76	96	91	87	83	78	74	70	67	63	59	55	52	48	45	42
78	96	91	87	83	79	75	71	67	64	60	56	53	50	46	43
80	96	91	87	83	79	76	72	68	65	61	57	54	51	47	44
84	96	92	88	84	80	77	73	70	66	63	59	56	53	50	47
88	96	92	88	85	81	78	74	71	67	64	61	58	55	52	49
90	96	92	89	85	82	78	75	72	68	64	62	58	56	53	50

8. SIZES AND PROPERTIES OF COPPER WIRE

B & S Gauge	Diameter (mm)	Diameter (mils)	Area (cir mils)	Ohms/ 1000 ft at 20°C	Ohms/km at 20°C	Current capacity (A)
5	4.62	181.9	33,100	.313	1.03	77
6	4.12	162.0	26,250	.395	1.30	65
7	3.67	144.3	20,820	.498	1.63	55
8	3.26	128.5	16,510	.628	2.06	46
9	2.91	114.4	13,090	.792	2.60	39
10	2.59	101.9	10,380	1.00	3.27	33
11	2.31	90.7	8,234	1.26	4.13	27
12	2.05	80.8	6,530	1.59	5.21	23
13	1.83	72.0	5,178	2.00	6.57	19
14	1.63	64.1	4,107	2.53	8.28	16
15	1.45	57.1	3,257	3.18	10.45	14
16	1.29	50.8	2,583	4.02	13.2	11
17	1.15	45.3	2,048	5.06	16.6	8.1
18	1.02	40.3	1,624	6.38	20.9	6.7
19	0.91	35.9	1,288	8.05	26.4	5.7
20	.81	32.0	1,022	10.15	33.3	4.8
21	.72	28.5	810	12.8	42.0	4.0
22	.64	25.4	642	16.1	53.0	3.4
23	.57	22.6	509	20.4	66.8	2.8
24	.51	20.1	404	25.7	84.2	2.4
25	.45	17.90	320	32.4	106.2	2.0
26	.40	15.94	254	40.8	134	1.7
27	.36	14.20	202	51.5	169	1.4
28	.32	12.64	160	64.9	213	1.2
29	.29	11.26	127	81.8	269	1.0
30	.25	10.03	100.5	103	339	
31	.23	8.93	79.7	130	427	
32	.20	7.95	63.2	164	538	
33	.18	7.08	50.1	207	679	
34	.16	6.31	39.8	261	856	
35	.14	5.62	31.5	329	1079	
36	.13	5.00	25.0	415	1361	

9. RESISTIVITIES AND TEMPERATURE COEFFICIENTS

Substance	Resistivity		Temperature coefficient (per °C)
	Ohm-cm	Ohm cir mils per ft	
Aluminum	2.63×10^{-6}	16.0	.00423
Copper (Annealed)	1.72×10^{-6}	10.5	.00393
Copper (hard drawn)	1.60×10^{-6}	9.75	.00408
Iron (annealed)	9.70×10^{-6}	59.0	.00625
Iron (hard drawn)	15.0×10^{-6}	91.5	
Mercury	95.8×10^{-6}	585	.00089
Nickel	6.93×10^{-6}	42.2	.00622
Silver	1.63×10^{-6}	9.9	.00380
Tin	11.5×10^{-6}	70.0	.00440
Zinc	5.75×10^{-6}	35.0	.00406
German silver	33×10^{-6}	201	.00036
Manganin	44×10^{-6}	268	.00000

10. ELECTROCHEMICAL DATA

Element	Atomic Weight	Valence	Electrochemical equivalent (gm/coulomb)
Aluminum	27.0	3	.0000932
Chromium	52.0	3	.0001796
Copper (cupric)	63.6	2	.0003294
Copper (cuprous)	63.6	1	.0006588
Gold	197.2	3	.0006812
Hydrogen	1.008	1	.0000104
Iron (ferric)	55.8	3	.0001929
Iron (ferrous)	55.8	2	.0002894
Lead	207.2	2	.0001074
Oxygen	16.00	2	.0000829
Nickel	58.7	2	.0003041
Platinum	195.2	4	.0005058
Silver	107.9	1	.0001118

11. INDEX OF REFRACTION FOR SODIUM D LINE

Alcohol (ethyl)	1.36	Canada Balsam	1.53
Carbon bisulfide	1.63	Crown glass	1.52
Glycerin	1.47	Heavy flint glass	1.65
Water	1.33	Heaviest flint glass	1.90
Salt water	1.34–1.38	Quartz	1.46

12. WAVELENGTHS OF SOME PROMINENT SPECTRAL LINES (ANGSTROM)*

Element	Wavelength	Element	Wavelength	Element	Wavelength
Argon	4159 v	Mercury	4047 v	Lithium	4132 v
	4164 v		4078 v		4603 b
	4182 v		4358 b		6104 o
	4191 v		4916 g		6708 r
	4198 v		5461 g	Potassium	4046 v
	4201 v		5770 y		6911 r
	4259 b		5791 y		6939 r
	4272 b		6907 r		
	4300 b	Neon	5401 g	Sodium	5890 y
	4334 b		5852 y		5896 y
	4345 b		5882 y		
	4511 b		6030 o	Strontium	4077 v
	4596 b		6074 o		4607 b
	4628 bg		6164 o		6878 r
	4702 bg		6217 ro		7070 r
	5496 g		6266 ro		
	5651 y		6334 r		
Helium	4388 b		6383 r		
	4471 b		6402 r		
	4713 b		6506 r		
	5016 g		6599 r		
	5876 y		6929 r		
	6678 r		7032 r		
Hydrogen	4102 v				
	4340 b				
	4861 bg				
	6563 r				

*The letters to the right of the wavelength numbers denote the color of the particular line. r = red, o = orange, y = yellow, v = violet, etc.

THE USE AND CARE OF ELECTRICAL INSTRUMENTS

NOTE: The instructions and precautions given here are to be used and adhered to for all experiments which require electrical equipment.

1. *General Precautions.* Electrical apparatus is usually delicate and expensive and requires careful and intelligent handling. Many electrical instruments require the same care in handling as would an expensive watch. *You are responsible for all damage to apparatus while in your care*.

The three most frequent causes of injury to electrical instruments are

a. Rough Handling.
b. Improper connections.
c. Overheating resulting from excessive currents.

When connecting a circuit, *always* insert a switch. Leave one wire disconnected from the battery and have the switch open until your circuit has been checked by the instructor.

When a circuit is dismantled, the leads should be removed first from the battery.

2. *The Use of a Reversing Switch.* A reversing switch is designed to send a direct current in either direction through a portion of the circuit. Figure C-1*a* shows such a switch connected in the circuit and in the open position. The current cannot flow from one side of the switch to the other except through the diagonal connections. The symbol ✕ means that the wires cross but are not connected. Note the battery symbol and the + and − markings on the terminals. Figure C-1*b* shows that the reversing switch is closed to the right and the direction of the current through the resistance *R* (note the symbol) is down, whereas in Figure C-1*c* the switch is closed to the left and the direction of the current through *R* is up. Note that current through the battery has the same direction in both cases.

3. *Double-Pole Double-Throw Switch.* This type of switch has six binding posts on it like the one above but

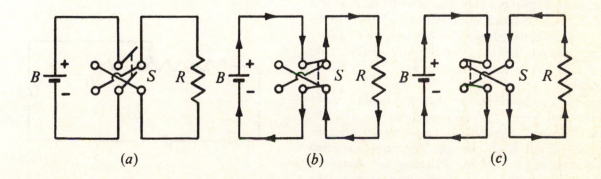

(*a*)   (*b*)   (*c*)

Figure C–1 Connections for a Reversing Switch.

does not have the diagonal cross-wires. It is used to connect either of two circuits to a source (Figure C-2). When the switch is closed to the right, lamp L_1 is connected to the battery, and when the switch is closed to the left, lamp L_2 is connected to the battery.

4. *Devices for Control and Detection of Current.* A *rheostat* is a variable resistor, usually used in a circuit to control the current or to act as a potential divider. In Figure C-3*a*, *B* and *D* are the end terminals of the rheostat coil, and *C* is the sliding-contact point. As shown, the current enters at *B* and leaves by way of the contact point *C*. Sometimes it may be more convenient to use *D* instead of *B*. By moving sliding contact *C*, the resistance may be more convenient to use *D* instead of *B*. By moving sliding contact *C*, the resistance may be adjusted to any desired value. Figure C-3*b* shows a wiring diagram of a circuit containing a rheostat. On the head of slide *C* the manufacturers usually indicate the current-carrying capacity and the approximate resistance of the rheostat. Never send a larger current through a rheostat than it is designed to carry; otherwise it may be burned out. Suppose a rheostat is marked 25 ohms and 4.8-A capacity. If connected across a 120-V source, the current would be 4.8 A if all its resistance were in the circuit, but it would be overloaded if any amount less than all its resistance were used. Always determine what the maximum current is likely to be for the voltage of the source you are to use.

Resistance boxes may be of either the dial or plug type. Some *dial types* have four dials whose unit values in ohms are 100, 10, 1, and 0.1, respectively. If these four dials were set on the numbers 2, 7, 3, and 5, respectively, the total resistance would be $2(100) + 7(10) + 3(1) + 5(0.1) = 273.5$ ohms. In the plug-type resistance box, the plugs act as shunts across the resistance coils and the desired resistance coils and the desired resistance is put in the circuit by removing the plug from the box. Do not handle with your hands the part of the plug that makes the electrical contact.

Most resistance boxes are designed for a maximum of *one watt* of power and will be *damaged or burned out* if more than one watt is expended in any one coil of the box. The coils unplugged in a plug box, or set on the dials of a dial box, are all connected in series. If the current in the circuit is known, the relation $P = I^2R$ will give the power being expended in the coils. A small total resistance setting usually means a large current with large power expenditure.

Portable *galvanometers* are used for many types of experimental work. Our common form has a small, button-like clamp which supports the suspension when the instrument is not in use. Be sure the clamp is released before attempting to read any deflection, but always see that the suspension is clamped when carrying the instrument. A *galvanometer* is a very delicate instrument and *must never be connected directly to a battery*.

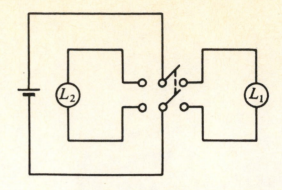

Figure C–2 Double-Pole, Double-Throw Switch.

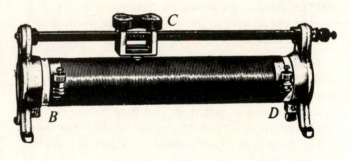

Figure C–3*a* Rheostat (Courtesy of Central Scientific Co.).

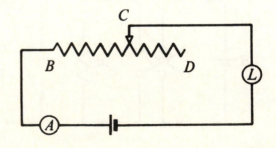

Figure C–3*b* Connections for Rheostat.

5. *Connections for Voltmeters and Ammeters*. Direct current ammeters and voltmeters may be injured by sending a current through them in the wrong direction. The needle can move backward very little, and the amount of current being sent through the coil cannot be detected. Before connecting a direct current instrument in a circuit, find the battery is connected to the + terminal of your instrument.

Sometimes a direct current ammeter or voltmeter has the zero in the center of the scale. The pointer may then move in either direction. Instruments with zero center do not have their teminals marked.

Loose connections are a very common cause of trouble in electrical experiments. They often introduce extraneous resistance and thereby cause calculations to be in error in many cases. Never twist two wires together; always use a continuous piece which is long enough to serve your purpose. For best contact when fastening a wire to a screw-type binding post, always loop the wires in the direction in which the binding posts are to be turned. This procedure will pull the wire closer to the screw as the cap is tightened.

6. *Damage Resulting from Overheating*. Only experience and extreme care will enable you to avoid instrument damage from overheating. Until you acquire experience and knowledge of electrical principles, depend on the instructor for guidance and then follow instructions.

The instruments most likely to be burned out are ammeters and galvanometers. Unless you are positive that other conditions in the circuit will keep the current below the rated capacity of the ammeter, always connect a rheostat in the circuit in series with the ammeter. If a meter has more than one range, always connect to the highest range unless you are sure that a lower range will carry the current. You can always change to the lower range if the reading on the high range does not exceed the maximum of the lower range. If the lower range is used first, the meter may be burned out.

Never short circuit a battery to test its condition for use. Such a procedure heats the battery tremendously and shortens its life. If you are in doubt about the current capacity of your battery, ask your instructor to check it for you.

7. *How to Determine the Polarity of an Unknown Source of Current*. You should have some idea of the magnitude of the potential difference of the terminals of the source to be used before attempting to make any kind of measurements. However, if the current is furnished to the laboratory tables from a central switchboard, you may not always know the polarity of the terminals at the table. The polarity may be determined by connecting a safe range of a voltmeter to the terminals, but this procedure often puts a strain on the meter if it is connected backward for a large deflection. A safer method is as follows. Select a slide-wire rheostat with a current capacity and resistance that indicate the current capacity will not be exceeded when connected to the source

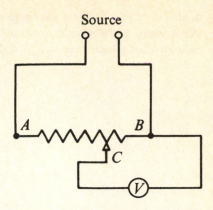

Figure C–4 A Rheostat Used as a Potential Divider.

at the end terminals A and B (Figure C-4). Then slide the contact C completely to one end of the rheostat and connect the voltmeter V, as shown in the figure. When C is in the position shown, the potential difference between C and B will be practically zero, and no current will flow through the voltmeter. Now as C is moved slowly from B toward A, the needle of the voltmeter will begin to move and thereby tell you which terminal of the source is positive. The wire leading to the positive terminal of the meter must come from the positive terminal of the source if the meter reads properly as the slide C is moved.

8. *The Use of Cables and Jacks*. In some laboratories the power is furnished by a bank of batteries, and permanent connections are made to a pair of jacks installed in the individual tables. The *red jack* is connected to the positive terminal of the battery bank and the *black jack to the negative terminal*. When connected to the pair of red and black lead wires, the red lead should be plugged into the red jack and the black lead into the black jack. The other end of the red cable should lead to the + terminal of the meters in the circuit.

9. *Power Supplies*. Many laboratories are now equipped with small low-voltage power supplies to furnish the needed direct current for student use. These units operate as rectifiers to convert alternating current to direct current and, by means of a transformer, step the ac line voltage down to come convenient range for the usual laboratory experiments. These units have a control knob which allows the output voltage to be varied from, say 0–20 V, and furnish a current of 5–10 A. This arrangement may make it unnecessary to use a rheostat for some types of experiments. Some power supplies have meters on the face of the instrument to indicate the voltage and current, while others do not.

The primary purpose of these meters is to indicate the approximate value of the voltage and/or current supplied to the entire circuit. They also indicate the range of values likely to be encountered in the measurements to be made, but meters calibrated for greater precision should be used for actual measurements in the various parts of the circuit.

10. *Nuclear Counting Devices.* See Experiment 47.

APPENDIX **D**

INFORMATION RELATIVE TO RADIOISOTOPES*

LABORATORY PROCEDURES FOR HANDLING RADIOISOTOPES

1. *General Warning.* Radioactivity is not to be feared but to be respected. In today's scientific laboratories many people constantly work with and handle radioactive materials. These people are required to abide by a set of safety rules and regulations for protection against the hazards of radiation. Through either the ignorance or the carelessness of the experimenter, it is possible for radioactive material to gain entrance into the body via the hands to the eyes, nose, or mouth and then continue its dangerous activity from within the human body.

Materials usually made available to students taking elementary courses are encased in a plastic capsule of some type. However, it is still possible for a capsule to leak or be partially opened by some means. Students should exercise caution and handle the radioactive sources carefully and only as necessary. The radiation received from any source in this laboratory during the course of an experiment will be small and does not represent any danger.

2. Eating, drinking, and the use of cosmetics in the laboratory are not permitted while working with radioactive materials.

3. Check with your laboratory instructor on the procedure for checking out and returning your radioactive sources.

All sources must be fully accounted for before leaving the laboratory.

4. When handling unsealed radioactive materials, wear rubber or plastic gloves and a laboratory coat or apron if possible.

5. Never pipette radioactive solutions by mouth suction. Special suction devices are available for this purpose.

6. All wounds, spills, and other emergencies should be reported to the instructor immediately.

7. If, in the course of work, any personal contamination is suspected, ask your instructor for a survey with a suitable meter. If contamination is found, follow the cleansing instructions given by your instructor.

8. In general, all radioactive sources and contaminated materials are to be retained within the laboratory.

9. Before leaving the laboratory, be sure all written records, if required, have been completed.

10. Personal warnings about working with radioactive materials.

    a. Do Not rub your eyes, nose, ears, or mouth.

    b. Do Not put your pencil in your mouth.

    c. Be Sure to wash your hands after checking in the sources used and before leaving the laboratory.

*See the following table for half life and energy relations of some common radioisotopes.

SOME COMMON RADIOISOTOPES

Isotope	Half Life	Principal Radiation (MeV)			Maximum Activity (µCi)*	
		Alpha	Beta	Gamma	Sealed	Unsealed
Americium-241	458 yr	5.48, 5.44		0.060	10	1
Barium-133	10.7 yr			0.356, 0.081	10	1
Calcium-45	163 days		0.257		10	10
Carbon-14	5730 yr		0.156		50	50
Cesium-137 Barium-137m	30.2 yr 2.55 min		0.512, 1.18	0.662	10	1
Chlorine-36	3.01×10^5 yr		0.709		10	1
Cobalt-57	271 days			0.122, 0.136	10	1
Cobalt-60	5.27 yr		0.318	1.17, 1.33	10	1
Indium-116m	54.2 min		1.00, 0.87 0.60	1.29, 1.10 0.417, 2.11	10	1
Iodine-131	8.04 days		0.606, 0.333 0.257	0.365, 0.637 0.284, 0.080 0.723	10	10
Lead-210 Bismuth-210	22.3 yr 5.01 days	4.69, 4.65	0.017, 0.061 1.16	0.047 0.266, 0.305	10	1
Manganese-54	313 days			0.835	10	1
Phosphorus-32	14.3 days		1.71		10	10
Polonium-210	138 days	5.31		0.802	1	0.1
Potassium-42	12.4 hr		3.52, 1.97	1.53	10	10
Radium-226	1600 yr	4.78, 4.60		0.186	1	0.1
Silver-108	2.41 min		1.65	0.633	10	1
Sodium-22	2.60 yr		0.545	1.28	10	10
Strontium-90 Yttrium-90	29 yr 64.0 hr		0.546 2.28		1	0.1
Sulfur-35	87.2 days		0.168		50	50
Thallium-204	3.77 yr		0.764		50	50
Tin-113 Indium-113m	115 days 99.5 min			0.255 0.392	10	10
Uranium-238	4.51×10^9 yr	4.20, 4.15		0.050	10	1

*Figures in the "Maximum Activity" column represent the most radioactivity obtainable in a sealed or unsealed source without a specific license.

LASER SAFETY

It is recommended that all experiments in this manual use He–Ne lasers with a power rating of 1.0 mW or less. Studies have shown that such lasers are safe; however, the following precautions and safety procedures are recommended:

1. Treat all laser beams with respect.
2. Never look directly at a laser beam with the naked eye or any optical device. Do not rely on protective eyeware.
3. Never allow the laser beam to shine in the area of anyone's eyes.
4. Never place highly reflective objects (such as rings, watches, glassware) in the laser beam path.
5. Turn the laser on only when actual measurements are taking place and be aware of where the beam is shining.
6. Cover outside windows to protect passers-by.
7. Generaly it is wise to keep the laser work area highly illuminated to keep the eye pupil small.
8. Treat the laser as any other electrical device. It should never be tampered with while the power cord is connected.

7234